Aquatic Macrophytes: Ecology, Functions and Services

Sanjeev Kumar • Kuldeep Bauddh •
Ritu Singh • Narendra Kumar •
Rajesh Kumar
Editors

Aquatic Macrophytes: Ecology, Functions and Services

Springer

Editors
Sanjeev Kumar
Department of Geology, School of Earth and Environmental Sciences
Babasaheb Bhimrao Ambedkar University
Lucknow, Uttar Pradesh, India

Kuldeep Bauddh
Department of Environmental Sciences
Central University of Jharkhand
Ranchi, India

Ritu Singh
Department of Environmental Science
Central University of Rajasthan
Ajmer, India

Narendra Kumar
Department of Environmental Sciences
Babasaheb Bhimrao Ambedkar University
Lucknow, Uttar Pradesh, India

Rajesh Kumar
Department of Environmental Science
Central University of Rajasthan
Ajmer, India

ISBN 978-981-99-3821-6 ISBN 978-981-99-3822-3 (eBook)
https://doi.org/10.1007/978-981-99-3822-3

This Springer imprint is published by the registered company Springer Nature Singapore Pte Ltd.
The registered company address is: 152 Beach Road, #21-01/04 Gateway East, Singapore 189721, Singapore

Preface

Aquatic macrophyte basically refers to those plants that grow in or around water and are visible to the naked eye. It comprises vascular (emergent, submerged, or floating) plants, bryophytes, and macro-algae flourishing in waterbodies. Aquatic macrophytes play a crucial role in assessing the ecological status of waterbodies. As primary producers, they form the basis of food webs, share a major part of highly productive aquatic ecosystems, and have a significant impression on ecosystem functions and services. They play a significant role in maintaining water quality, enhancing biodiversity, supporting aquatic food webs, etc. These plants constitute the elementary biotic component of the ecosystem and have an important role in structuring communities of the aquatic environment and nutrient cycling. Macrophytes are also known to strongly influence the micro-climate and biogeochemical processes occurring in littoral zones of marine ecosystems and sediment dynamics of freshwater systems. Aquatic macrophytes are also important for mitigating the impacts of climate change. They serve as the most effective carbon sinks and play an important role in carbon sequestration, which helps in reducing greenhouse gas emissions and mitigating the effects of climate change. In addition, they provide important ecosystem services, such as water purification and flood control, which can help to reduce the negative impacts of climate change on human societies. Apart from these, aquatic macrophytes play an important role in the area of environmental clean-up. Remediation through these plants is considered as an easy, cost-effective, eco-friendly, and energy-efficient method of decontamination. These macrophytes help to remove organic and inorganic impurities as well as engineered nanoparticles from contaminated water. Their self-purification utilities confirm the maintenance of water quality. Aquatic macrophytes is a key component of wetland systems also, contributing much of the total ecosystem biomass. Besides their presence in urban areas improves the aesthetic value of landscapes, contributing to human well-being and quality of life. An understanding of the functions of aquatic macrophytes in wetlands systems is critical for understanding the elementary processes of the ecosystem and associated issues, such as restoration of ecological integrity in the wetland ecosystem, wastewater treatment, and management of hostile invasive/alien species.

Despite their critical importance, aquatic macrophytes face numerous threats, including habitat loss, pollution, and invasive species. In the past three decades, due to climate-induced changes and altered land use patterns, significant changes in the status, growth profile, general distribution, and abundance of these communities have been recorded, which are likely to have a huge ecological impact on the structural and functional part of the aquatic ecosystems on a global scale. Since aquatic macrophytes are considered as the keystone species of aquatic ecosystems, they are much prone to climate-induced changes; therefore, the knowledge of the impacts of changing climate on the growth patterns, distribution, abundance, and productivity of these plants with its probable implications are very much required in present time. The study of their ecological functions and services is crucial for understanding the impacts of these threats on aquatic ecosystems and developing effective management strategies to protect them. For instance, understanding how aquatic macrophytes respond to pollution can help to identify pollution hotspots and develop strategies to reduce pollution levels. Similarly, understanding how invasive species impact aquatic macrophytes can help to develop strategies to control invasive species and prevent their spread.

This book is an attempt to document the information available on aquatic macrophytes' ecological functions and services, which are essential for the sustainable management of aquatic ecosystems and the benefits of human society. The book highlights several aspects of aquatic macrophytes, such as their role in nutrient recycling, biogeochemical processes in the water column and sediments, biomass production, wetland ecosystems, water resource management, carbon sequestration, environmental clean-up, and bioenergy production. The book presents the current status of aquatic macrophytes and highlights the major challenges towards exploiting the benefits served by aquatic macrophytes as ecosystem services. This book will be beneficial as a source of educational material for graduates and post-graduate students, faculties, researchers, policymakers, and industrial personnel who are engaged in assessing the functions of aquatic macrophytes in the natural environment. It could also serve as a reference book for research scholars, scientists, academicians, and readers from diverse backgrounds across various fields such as ecology, environmental science, biology, biogeochemistry, wetland conservation, phytoremediation, biomonitoring, wastewater management, and bioenergy production. We hope that the chapters of this book will provide readers with valuable insights into the varied aspects of the ecological functions and services offered by aquatic macrophytes.

Lucknow, India — Sanjeev Kumar
Ranchi, India — Kuldeep Bauddh
Ajmer, India — Ritu Singh
Lucknow, India — Narendra Kumar
Ajmer, India — Rajesh Kumar

Contents

About the Authors

Sanjeev Kumar has obtained his M.Sc. and Ph.D. in Environmental Science from Babasaheb Bhimrao Ambedkar University, Lucknow, India. Currently, he is working as an Assistant Professor in the Department of Geology at the same university. Dr. Kumar has made significant contributions in the field of sustainable agriculture, focusing on various aspects such as soil quality management, agriculture gaseous emissions, slow-release fertilizers, and phytoremediation. His research endeavors aim to develop innovative approaches for enhancing agricultural practices while minimizing environmental impacts.

Kuldeep Bauddh is currently an Assistant Professor in the Department of Environmental Sciences at the Central University of Jharkhand, Ranchi, India. He received his M.Sc. and Ph.D. in Environmental Science from Babasaheb Bhimrao Ambedkar University, Lucknow, India. Dr. Bauddh is currently teaching thrust areas of environmental sciences like environmental pollution and management, environmental impact assessment, environmental toxicology, etc. His major research areas include environmental pollution and management, phytoremediation, and restoration of abandoned mining sites.

Ritu Singh is an Assistant Professor in the Department of Environmental Science, School of Earth Sciences at the Central University of Rajasthan, Ajmer, India. She received her Ph.D. in Environmental Science from the Indian Institute of Toxicology Research (CSIR-IITR), Lucknow, and Babasaheb Bhimrao Ambedkar University, Lucknow, Uttar Pradesh, India. She is actively involved in teaching and research in the field of environmental chemistry, environmental toxicology, and wastewater treatment. Her research primarily focuses on the potential of metal nanoparticles and nanocomposites for various environmental applications, including remediation of emerging contaminants and wastewater treatment. She has a keen interest in the study of nano-agro particles, exploring their applications in agriculture and environmental remediation.

Narendra Kumar is a Professor in the Department of Environmental Science and is currently the Head of the Department of Geology, Babasaheb Bhimrao Ambedkar University, Lucknow, India. Dr. Kumar has made significant contributions to various areas of environmental science, including phytoremediation, air and water pollutants, and wastewater treatment. His work involves exploring the capabilities of different plant species in removing or degrading pollutants from contaminated soil, water, and air. In addition to his expertise in phytoremediation, he has also contributed to the field of river ecology. He studies the ecological health and dynamics of rivers, including their biodiversity, water quality, and the impact of anthropogenic activities on river ecosystems.

Rajesh Kumar is a Professor in the Department of Environmental Science and presently serves as the Dean of the School of Earth Sciences at the Central University of Rajasthan, Ajmer, India. He earned his M.Sc. (Tech.) and Ph.D. in Geophysics specialization in Meteorology, from Banaras Hindu University (BHU), Varanasi, India. With 23 years of experience in teaching and research, Dr. Kumar has made significant contributions to various fields within environmental science. His expertise lies in the study of air pollution and its impact on human health, climate change, arid and desert environments, as well as the application of remote sensing and Geographic Information System (GIS). His current research focuses on the fields of air pollution, climate change, and the Himalayan cryosphere.

An Introduction to the Functions and Ecosystem Services Associated with Aquatic Macrophytes

1

Sanjeev Kumar, Ritu Singh, Dhananjay Kumar, Kuldeep Bauddh, Narendra Kumar, and Rajesh Kumar

Abstract

In recent decades, aquatic macrophytes have been recognized as a significant component of aquatic ecosystems, which play an important role in providing environment and ecosystem services. Aquatic macrophytes also play a key role in the production of bioenergy, biochar formation, climate change, photosynthesis, etc. The presence and activity of macrophytes enhance sediment formation, carbon sequestration, natural water purification, remediation of organic and inorganic pollutants, etc. in the aquatic ecosystem. They also provide natural medicines, fibers, biochemicals, and other resources to the human society. There are several evidences that macrophyte diversity enhances the functioning, services, and stability of the ecosystem and provides multiple benefits to human well-being. This chapter delves into the functions and ecosystem services associated with aquatic macrophytes, which play a crucial role in maintaining the health and sustainability of aquatic ecosystems and significantly impact the well-being of human societies.

S. Kumar (✉) · D. Kumar
Department of Geology, School of Earth and Environmental Sciences, Babasaheb Bhimrao Ambedkar University, Lucknow, Uttar Pradesh, India

R. Singh · R. Kumar
Department of Environmental Science, School of Earth Sciences, Central University of Rajasthan, Ajmer, Rajasthan, India
e-mail: ritu_ens@curaj.ac.in; rajesh.kumar@curaj.ac.in

K. Bauddh
Department of Environmental Science, School of Earth Sciences, Central University of Jharkhand, Ranchi, Jharkhand, India
e-mail: kuldeep.bauddh@cuj.ac.in

N. Kumar
Department of Environmental Science, Babasaheb Bhimrao Ambedkar University, Lucknow, Uttar Pradesh, India

S. Kumar et al. (eds.), *Aquatic Macrophytes: Ecology, Functions and Services*,
https://doi.org/10.1007/978-981-99-3822-3_1

Keywords

Ecosystem services · Environment · Macrophyte · Remediation · Pollutant

1.1 Introduction

Aquatic macrophytes constitute the elementary biotic component of the ecosystem and have an important role in structuring communities of the aquatic environment and nutrient cycling (Thomaz 2021; Lind et al. 2022). Basically, aquatic macrophytes are large aquatic plants that grow in water bodies such as lakes, ponds, rivers, and wetlands. There are many types of aquatic macrophytes, including submerged plants, floating-leaved plants, and emergent plants (Chambers et al. 2008), as shown in Fig. 1.1. Submerged plants grow entirely underwater and can be found at various depths in the water column, while floating-leaved plants have leaves that float on the water surface and roots that are anchored in the substrate (Chambers et al. 2008). Emergent plants grow partially or fully above the water surface, with their roots anchored in the substrate and their leaves and stems rising above the water (Chambers et al. 2008).

Aquatic macrophytes play an important role in the food chain and food web of aquatic ecosystems (Bornette and Puijalon 2011). They provide food, shelter, and habitat for a wide range of organisms, from tiny microorganisms to large fish and other aquatic animals (Bornette and Puijalon 2011; Engelhardt and Ritchie 2001). In the food chain, aquatic macrophytes are the primary producers or the first organisms to convert sunlight into organic matter through the process of photosynthesis (Bornette and Puijalon 2011). They produce a range of organic compounds, such as carbohydrates, proteins, and lipids, that are used as food by herbivorous aquatic animals, such as snails, insects, and some fish. Aquatic macrophytes also provide habitat for a variety of organisms, including fish, amphibians, and aquatic invertebrates. These animals use the plants as shelter and breeding grounds and also as a source of food. For example, many fish species feed on insects that live on or near the plants or on smaller fish that use the plants as cover. The presence of aquatic macrophytes can also affect the structure and function of the food web in an aquatic ecosystem. For example, in systems with dense macrophyte growth, predators may be able to hunt and capture prey more easily, while herbivores may consume a greater proportion of the available primary production. This signifies that aquatic macrophytes have a significant role in supporting the diversity and productivity of aquatic ecosystems, and their presence is an important indicator of the overall health of these systems (Ferna'ndez-Ala'ez et al. 1999; Pı́palová 2002; Akasaka et al. 2010; Mikulyuk et al. 2011; Fernandez-Alaez et al. 2018).

Aquatic macrophytes are a major component of aquatic ecosystems and provide many services to the environment, ecosystem, and humans (Thomaz 2021). Aquatic macrophytes play a crucial role in assessing the ecological status of water bodies. They are known to strongly influence the micro-climate and biogeochemical processes occurring in littoral zones of marine ecosystems and sediment dynamics of

Fig. 1.1 Various types of aquatic macrophytes

freshwater systems. They serve as the most effective carbon sinks and play an important role in carbon sequestration (Ankit et al. 2020; Lolu et al. 2019). Apart from these, aquatic macrophytes play an important role in the area of environmental clean-up (Ankit et al. 2020; Kumar et al. 2019). Remediation through these plants is considered as an easy, cost-effective, eco-friendly, and energy-efficient method of decontamination (Ankit et al. 2020; Kumar et al. 2019). These macrophytes help to remove organic and inorganic impurities as well as engineered nanoparticles from contaminated water (Ankit et al. 2020; Kumar et al. 2019; Anand et al. 2018; Mishra et al. 2013). Their self-purification utilities confirm the maintenance of water quality (Fawzy et al. 2012; Dhote and Dixit 2009). Aquatic macrophytes are the key component of wetland systems, contributing much of the total ecosystem biomass

(Adhikari et al. 2009; Ankit et al. 2020). Despite their important ecological roles, aquatic macrophytes can also cause problems in some situations. For example, excessive growth of macrophytes can lead to oxygen depletion in the water column and cause the death of fish and other aquatic animals. They can also impede water flow and navigation in rivers and canals and interfere with recreational activities such as swimming and boating. Overall, aquatic macrophytes are important and complex components of aquatic ecosystems, and their management requires careful consideration of both their ecological benefits and potential drawbacks. The present chapter highlights the various functions and services of aquatic macrophytes to better understand the elementary processes of the ecosystem and associated areas such as restoration of ecological integrity in wetland ecosystem, wastewater treatment, management of hostile invasive/alien species, etc.

1.2 Ecosystem Functions and Services Associated with Aquatic Macrophytes

Ecosystem services are the ecological characteristics, functions, or processes that directly or indirectly contribute to human well-being, the benefits people derive from functioning ecosystems. The Millennium Ecosystem Assessment (2005) categorized ecosystem services into four different types, i.e., provisioning services, regulating services, cultural services, and supporting services (Fig. 1.2). Provisioning services are the benefits that humans derive from ecosystems that involve the production of natural resources. These services include food, water, timber, fibers, medicines, and other products that are directly extracted from ecosystems. For example, forests provide timber for construction, fuel, and paper production. Agriculture provides food crops, livestock, and fish. These services are critical for human survival and form the basis of many economies worldwide. Regulating services refer to the benefits that ecosystems provide by regulating the physical and chemical conditions of the environment. These services include climate regulation, flood control, water purification, and pollination. For example, wetlands and forests absorb and store carbon dioxide, which helps in regulating the Earth's climate. Wetlands also filter and purify water, reducing the risk of floods and erosion. Pollination services

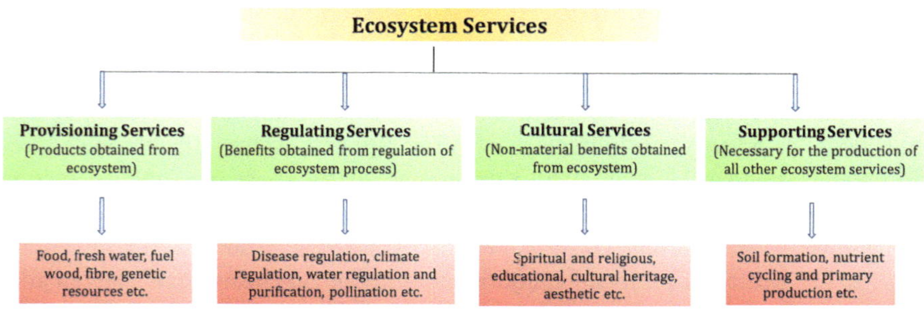

Fig. 1.2 Categories of ecosystem services

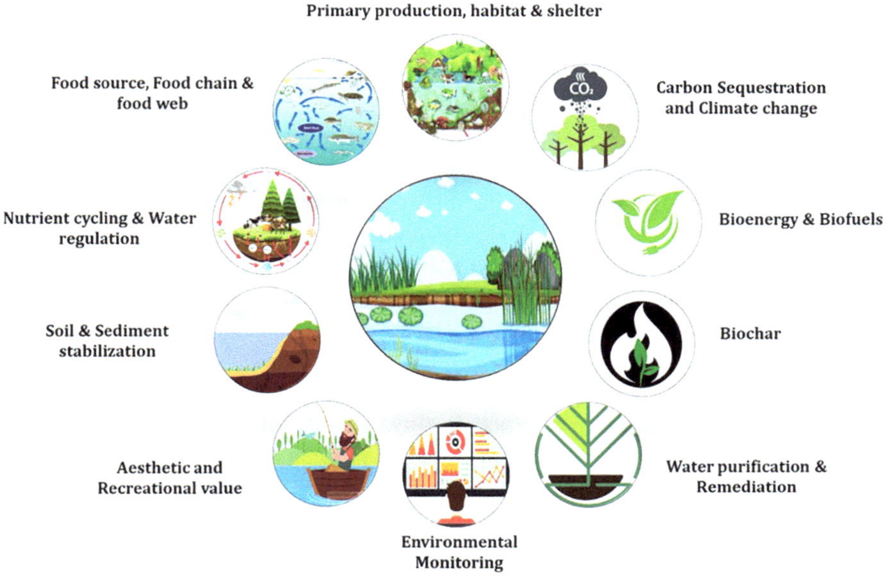

Fig. 1.3 Various ecosystem functions and services provided by aquatic macrophytes

provided by bees and other insects are critical for crop production and food security. Cultural services are the non-material benefits that humans derive from ecosystems. These services include recreation, tourism, aesthetic enjoyment, and cultural and spiritual values. For example, natural parks, beaches, and other scenic landscapes provide opportunities for outdoor recreation and tourism. Ecosystems also have cultural and spiritual significance for many societies and communities, as they provide a sense of connection to the natural world and a source of inspiration and identity. Supporting services are the underlying services that are necessary for the production of other ecosystem services. These services include soil formation, nutrient cycling, and primary productivity. For example, soil formation is essential for the growth of crops, while nutrient cycling and primary productivity are necessary for the production of food, fiber, and other ecosystem services. These services are critical for maintaining the structure and function of ecosystems and for sustaining human well-being (Daily et al. 2009; de Groot et al. 2010; Mace et al. 2012). Each of these categories provides a range of benefits that are critical for maintaining healthy and functioning ecosystems and for sustaining human societies and economies. Aquatic macrophytes are the key component of freshwater and marine ecosystems, and they offer various ecological services, which are critical for the health and sustainability of the ecosystems and for the well-being of human societies, as represented in Fig. 1.3. All the ecosystem services are interrelated with each other and cannot be considered in isolation. The following sections presents an overview of the function and services of the aquatic macrophytes.

1.2.1 Primary Production

Primary production refers to the conversion of light energy into organic matter by photosynthetic organisms. Aquatic macrophytes have a significant influence on primary production in aquatic ecosystems (Kazanjian et al. 2018; Nõges et al. 2010). They contribute to primary production directly through photosynthesis and indirectly by altering nutrient availability and water chemistry (Thomaz 2021; Reitsema et al. 2020; Amir et al. 2019). Aquatic macrophytes have high photosynthetic rates and contribute a significant amount of organic matter to the ecosystem. They provide a substrate for the attachment of algae and other small organisms, which can grow and multiply on their surfaces. This can increase the overall photosynthetic activity in the ecosystem, leading to higher primary production. Further, aquatic macrophytes provide shelter and protection for small organisms from predation, which can lead to higher growth rates and higher biomass of these organisms. This, in turn, could lead to increased nutrient recycling, thus stimulating primary production (Thomaz 2021). Besides, aquatic macrophytes could alter the physical and chemical characteristics of the water column, creating microhabitats that are more conducive to the growth and survival of photosynthetic organisms. For example, they can reduce water flow and increase sediment deposition, which can provide a nutrient-rich environment for plant growth. Aquatic macrophytes can also influence primary production by altering the light regime in the water column. They can absorb and scatter light, which can reduce the amount of light available to other photosynthetic organisms, and can also create shaded areas that are more suitable for the growth of certain types of algae. Overall, the influence of aquatic macrophytes on primary production is significant, and their presence or absence could have important implications for the structure and function of aquatic ecosystems (Reitsema et al. 2020; Amir et al. 2019).

1.2.2 Food Source

Aquatic macrophytes provide food for a variety of aquatic animals and humans (He et al. 2021; Yasuno et al. 2021). Many herbivorous fishes, waterfowl, crustaceans, and other aquatic organisms feed on aquatic plants as a significant part of their diet. Additionally, some aquatic plants, such as water chestnuts and lotus roots, are consumed by humans as food. They are commonly used in Asian cuisine. Some other common examples of aquatic macrophytes that provide food include water lilies, duckweed, and water hyacinth. These plants contain essential nutrients like carbohydrates, proteins, vitamins, minerals, fats, and fibers that can be used by aquatic organisms as a source of energy and building blocks for growth (He et al. 2021; Yasuno et al. 2021). For example, watercress is a popular leafy green that is often used in salads and sandwiches. It is high in vitamin C, iron, and calcium. Other aquatic plants that are commonly used as food include seaweed, which is a popular ingredient in sushi and other Japanese dishes, and kelp, which is used in many different types of food products. In addition to providing direct food for aquatic organisms, aquatic macrophytes also play an important role in the food chain by supporting the growth of phytoplankton and other small organisms that are the foundation of the aquatic food web (He et al. 2021; Yasuno et al. 2021). Therefore, the presence of aquatic macrophytes can have cascading effects on the overall health and productivity of aquatic ecosystems.

1.2.3 Habitat and Shelter

Aquatic macrophytes provide important habitat and shelter for a wide variety of aquatic organisms (Haroon 2022; Dorn et al. 2001). These plants create a complex three-dimensional structure in the water column, providing hiding places, resting areas, and breeding grounds for many aquatic species. The leaves, stems, and roots of aquatic macrophytes provide shelter for small aquatic invertebrates, such as snails, insects, and crustaceans. These invertebrates, in turn, provide food for larger aquatic animals, such as fish and amphibians, which also use the macrophytes as habitat (Haroon 2022; Dorn et al. 2001). In addition to providing shelter, aquatic macrophytes also help to create diverse microhabitats within aquatic ecosystems. Different species of macrophytes have different structures and chemical compositions, creating a variety of niches for different species of aquatic organisms (Haroon 2022; Quirino et al. 2021; Dorn et al. 2001). The roots and rhizomes of aquatic macrophytes also provide a substrate for periphyton, which are communities of algae, bacteria, and other microorganisms that grow on surfaces in aquatic environments. These periphyton communities provide additional food and habitat for aquatic organisms (Quirino et al. 2021; Dorn et al. 2001). The presence of aquatic macrophytes is critical for maintaining the diversity and abundance of aquatic organisms in freshwater and saltwater ecosystems, making their conservation and management important for the health and sustainability of these systems.

1.2.4 Nutrient Recycling

Aquatic macrophytes play a significant role in nutrient cycling (Kazanjian et al. 2018; O'Brien et al. 2014). These plants absorb nutrients such as nitrogen and phosphorus from the water column and sediments, which are further used for growth and reproduction. Their roots and rhizomes provide a substrate for the growth of microorganisms, which break down the organic matter and release nutrients (Kazanjian et al. 2018; O'Brien et al. 2014; Lawniczak et al. 2010). They can take up nutrients from the sediment also and prevent their release into the water column, reducing the potential for nutrient pollution. Additionally, they release oxygen during photosynthesis, which can enhance microbial activity and promote nutrient cycling. Furthermore, when macrophytes die or shed their leaves, the organic matter they release serves as a source of nutrients for other organisms (Kazanjian et al. 2018). Microbes in the sediment break down this organic matter, releasing nutrients back into the water column, where they can be taken up by other plants or used by other organisms. Macrophytes also influence the water flow and turbulence, thereby affecting sediment deposition and nutrient availability. Altogether, aquatic macrophytes help to maintain nutrient cycling and improve water quality in aquatic ecosystems (O'Brien et al. 2014; Lawniczak et al. 2010). However, their excessive growth could lead to eutrophication and other ecological problems, so it is important to manage these systems carefully to maintain a healthy and balanced ecosystem.

1.2.5 Soil/Sediment Stabilization

Aquatic macrophytes play a crucial role in influencing soil/sediment in aquatic environments (Ali et al. 2021; Roelofs et al. 2002). They can affect sediment composition, structure, and stability, as well as the biogeochemical processes that occur within sediments. Aquatic macrophytes can help to stabilize sediment by reducing water velocity, which prevents erosion and sediment transport (Ali et al. 2021; Roelofs et al. 2002). The plant roots also provide physical binding of the sediment, preventing it from being resuspended. Macrophytes also trap sediment and organic matter within their root systems, leading to the accumulation of sediment and creation of new habitats. This accumulation of sediment can also help to increase water clarity by removing suspended particles. Besides, macrophytes alter sediment biogeochemistry (Thomaz 2021). They take up nutrients such as nitrogen and phosphorus from the sediment, thereby reducing nutrient concentrations in the water column. This can reduce the likelihood of harmful algal blooms and promote the growth of desirable macrophyte species. Additionally, the macrophytes increase the oxygen concentration in sediment via photosynthesis, creating a more hospitable environment for the microbial communities that carry out the decomposition of organic matter and recycle nutrients (Amir et al. 2019; Roelofs et al. 2002). Overall, aquatic macrophytes can play an important role in shaping sediment dynamics in aquatic ecosystems, with implications for water quality, habitat creation, and nutrient cycling (Thomaz 2021; Reitsema et al. 2020; Amir et al. 2019).

1.2.6 Hydrologic Cycle and Regulation

Aquatic macrophytes play an important role in the water cycle and regulation (Keitel et al. 2016; Sharip et al. 2012). Aquatic macrophytes absorb and store large amounts of water from the surrounding environment, which helps in regulating the water level in water bodies and reduces the volume of water available for runoff or evapotranspiration (Keitel et al. 2016; Sharip et al. 2012). This could further help in preventing flooding during periods of heavy rainfall and maintain water availability during dry spells. Aquatic macrophytes release water vapor into the atmosphere through transpiration, which contributes to the formation of clouds and precipitation (Madsen et al. 2001; Kurilenko and Osmolovskaya 2006). This process also helps to regulate the temperature and humidity of the surrounding environment. In addition, they help in filtering and purifying water by absorbing nutrients and pollutants, which in turn improves the water quality and reduces the risk of harmful algal blooms. Aquatic macrophytes also help in stabilizing the sediment in water bodies, reducing erosion and sedimentation, which further assist in maintaining the water flow and preventing the sediments from accumulating in the water body (Keitel et al. 2016; Sharip et al. 2012; Kurilenko and Osmolovskaya 2006). Thus, the presence of aquatic macrophytes has important effects on the hydrologic cycle of aquatic ecosystems, contributing to water storage, nutrient cycling, and sedimentation processes.

1.2.7 Aesthetic and Recreational Value

Aquatic macrophytes provide aesthetic and recreational value in aquatic environments (Hunt et al. 2019; Tallar and Suen 2017). Their presence can enhance the beauty of water bodies and provide opportunities for recreational activities such as fishing, boating, and bird-watching. The diverse forms and colors of aquatic macrophytes can add visual interest to water bodies, creating a natural landscape that attracts people to visit and enjoy them (Smith et al. 2015; Hunt et al. 2019). Floating-leaved plants such as water lilies and lotus can be particularly visually striking, with their large, showy flowers and broad leaves. They also provide habitat for fish and other aquatic animals, making them popular fishing spots. Fishermen often target areas with dense macrophyte growth because these areas are likely to be home to a variety of fish species. In addition to fishing, aquatic macrophytes provide opportunities for other recreational activities such as kayaking, canoeing, and paddle boarding. These activities are often enjoyed in areas with calm, shallow water where aquatic macrophytes are abundant. The aesthetic and recreational value provided by aquatic macrophytes can contribute to the economic and social well-being of local communities, making their conservation and management important for the sustainability of these systems (Pflüger et al. 2010; Smith et al. 2015; Hunt et al. 2019; Tallar and Suen 2017).

1.2.8 Water Purification

Aquatic macrophytes help in removing various nutrients, pollutants, and organic matter from water bodies through various mechanisms (Ankit et al. 2020; Kumar et al. 2019; Anand et al. 2018; Mishra et al. 2013). One of the primary ways that aquatic macrophytes help to purify water is through their ability to take up and use nutrients such as nitrogen and phosphorus (Anand et al. 2018). These nutrients are often present in excessive amounts in water bodies due to agricultural runoff, sewage discharge, and other human activities. By absorbing these nutrients, the macrophytes help in reducing the excess concentration of nutrients in the water, which would prevent algal blooms and other forms of water pollution (Töre and Özkoç 2022; Ankit et al. 2020). Moreover, aquatic macrophytes provide habitat for microorganisms that break down the contaminants present in the water bodies. The roots of the plants provide a surface for these microorganisms to attach to, and the plants themselves can take up some of the pollutants and convert them into non-toxic forms (Fawzy et al. 2012; Anand et al. 2018). Additionally, the physical presence of aquatic macrophytes improves the water quality by reducing sedimentation and erosion. The plants help to stabilize sediments on the bottom of the water body, which prevents the further release of nutrients and other pollutants into the water column (Srivastava et al. 2008; Fawzy et al. 2012; Anand et al. 2018). In summary, aquatic macrophytes play a vital role in maintaining the health of aquatic ecosystems and in ensuring the availability of clean water for societal use (Ankit et al. 2020; Kumar et al. 2019; Anand et al. 2018).

1.2.9 Remediation

Aquatic macrophytes act as remediation agents by removing or reducing pollutants from aquatic ecosystems (Anand et al. 2018; Bhaskaran et al. 2013; Yadav et al. 2023). This process is known as phytoremediation, and it involves the use of plants to clean up contaminated soil or water. Aquatic macrophytes can remediate pollutants through a variety of mechanisms. For example, they can absorb nutrients such as nitrogen and phosphorus from the water column, which can reduce the growth of harmful algae and improve water quality. Additionally, some macrophytes can take up heavy metals such as lead, cadmium, and mercury from the water and store them in their tissues, effectively removing these pollutants from the ecosystem. This process is known as phytoextraction. In addition to nutrient and metal removal, aquatic macrophytes can also help to break down organic pollutants such as pesticides and herbicides through a process called phytodegradation (Mishra et al. 2013; Anand et al. 2018). In this process, the plant roots release enzymes and other chemicals that break down the pollutants, making them less toxic or even non-toxic. Overall, the use of aquatic macrophytes as remediation agents has several potential benefits. First, it can provide a natural, low-cost method for removing pollutants from aquatic ecosystems. Second, it can promote the restoration and conservation of aquatic habitats by reducing the impact of pollution on aquatic animals and plants. Finally, it can help to improve water quality and reduce the risk of harmful algal blooms and other negative impacts of pollution on human health and the environment (Anand et al. 2018; Bhaskaran et al. 2013; Yadav et al. 2023).

However, it is important to note that the effectiveness of phytoremediation using aquatic macrophytes can vary depending on a variety of factors, including the type and concentration of pollutants present in the ecosystem, the type of macrophyte species used, and the environmental conditions of the ecosystem. As such, careful monitoring and management are necessary to ensure the effectiveness and safety of using aquatic macrophytes for remediation purposes.

1.2.10 Carbon Sequestration

Carbon sequestration is the process of removing carbon dioxide from the atmosphere and storing it in carbon sinks, such as plants, soils, and oceans (Nag et al. 2023; Ankit et al. 2020; Lolu et al. 2019; Pal et al. 2017). Aquatic macrophytes are able to sequester carbon from the atmosphere through the process of photosynthesis. During photosynthesis, carbon dioxide is taken up from the surrounding water and used to build organic molecules, such as carbohydrates, which are stored in the plant tissues (Bloom et al. 2010; Kayranli et al. 2010). The carbon in these organic molecules can then be stored for long periods of time, potentially for centuries or even millennia, in the sediment at the bottom of the water body. The amount of carbon sequestered by aquatic macrophytes can vary depending on a variety of factors, such as the species of macrophyte, the environmental conditions in the water body, and the level of disturbance or management of the ecosystem (Lolu et al. 2019; Pal et al. 2017). However, studies have shown that aquatic macrophytes can be an important carbon sink in aquatic ecosystems, especially in wetland habitats such as marshes and swamps (Nag et al. 2023; Ankit et al. 2020). Carbon sequestration by aquatic macrophytes can have important implications for mitigating climate change. By removing carbon dioxide from the atmosphere and storing it in aquatic ecosystems, macrophytes can help to reduce the concentration of greenhouse gases in the atmosphere, thereby reducing the rate of global warming (Bloom et al. 2010; Kayranli et al. 2010; Lolu et al. 2019; Pal et al. 2017).

1.2.11 Climate Change

Aquatic macrophytes play several roles in climate change, both as a result of their ability to sequester carbon and their responses to changing environmental conditions (Zhou et al. 2021; Pereira et al. 2021; Hossain et al. 2017). As mentioned in the above section, aquatic macrophytes sequester significant amounts of carbon from the atmosphere through the process of photosynthesis. This carbon is then stored in the plant tissue and in the sediment at the bottom of the water body, making it unavailable for release into the atmosphere as carbon dioxide (Zhou et al. 2021; Pereira et al. 2021). This carbon sequestration by aquatic macrophytes helps to reduce the concentration of greenhouse gases in the atmosphere, thereby mitigating climate change. The conversion rate of atmospheric carbon varies along the longitudinal gradient as the plant's growth is slower at high latitudes due to less nutrient, insolation, and cold temperature (Adhikari et al. 2009). Aquatic macrophytes also affect the emission of methane (CH_4), a potent greenhouse gas, from aquatic ecosystems. Some studies have shown that dense macrophyte growth can reduce CH_4 emission by creating anoxic zones in the sediment, which limit the activity of methane-producing microorganisms. In contrast, other studies have suggested that macrophytes can also enhance CH_4 emission by promoting the growth of methane-producing microorganisms. The quantity of generation and absorption of atmospheric CH_4 depends on the water table of wetlands (Moore and Dalva 1993). It has been reported that 20–25% of existing global CH_4 emissions are emitted by wetlands, which is around 15–227 Tg CH_4 per year, and in this, the contribution of rice paddies alone is around 60–80 STg per year (Bloom et al. 2010).

Aquatic macrophytes also play a role in regulating the temperature of aquatic ecosystems. The dense growth of macrophytes can provide shade and cooling effects in the water column, which can help to mitigate the effects of rising water temperatures due to climate change. Additionally, macrophytes help to regulate the oxygen content of aquatic ecosystems, which could also be impacted by changes in temperature and other environmental factors. Coastal wetlands, which are often dominated by aquatic macrophytes, could provide a buffer against storm surges and sea level rise. These wetlands could help to protect coastal communities and infrastructure from the impacts of climate change-induced extreme weather events (Zhou et al. 2021; Pereira et al. 2021; Hossain et al. 2017).

However, the responses of aquatic macrophytes to climate change are complex and not always predictable. Changes in water temperature, precipitation patterns, and other environmental factors could impact the growth and distribution of macrophytes, potentially altering the balance of aquatic ecosystems. For instance, in some cases, warming temperatures and altered precipitation patterns may lead to increased growth of invasive macrophyte species, which could outcompete native species and reduce the biodiversity of aquatic ecosystems. The roles of aquatic macrophytes in climate change are complex and multifaceted, highlighting the importance of understanding the interactions between these plants and their environments in order to develop effective strategies for mitigating and adapting to the impacts of climate change (Xia et al. 2022; Zhou et al. 2021).

1.2.12 Biochar

Biochar is a type of charcoal obtained from the thermochemical conversion of biomass in an oxygen-limited environment. Aquatic macrophytes are one such organic material that can be used to synthesize biochar (Kumari et al. 2021). Aquatic macrophytes are often harvested from lakes, rivers, and other bodies of water to manage their growth and prevent them from becoming invasive. These harvested plants can then be dried and converted into biochar through pyrolysis. Besides pyrolysis, gasification and carbonization method has also been used to convert aquatic plants into biochar (Yaashikaa et al. 2020; Cui et al. 2022).

Biochar synthesized from aquatic macrophytes has several potential benefits. It can be used as a soil amendment, helping to improve soil fertility, crop productivity, and water retention (Bird et al. 2012). It can also be used to sequester carbon, helping to mitigate climate change (Cui et al. 2016, 2022). In addition, biochar has been shown to improve water quality when applied to lakes and other bodies of water, as it can help to remove excess nutrients and pollutants (Gong et al. 2017; Li et al. 2018; Michalak et al. 2019; Mokrzycki et al. 2021). However, the quality of biochar synthesized from aquatic macrophytes can vary depending on the species of plant used and the conditions under which it was harvested and pyrolyzed. It is, therefore, important to carefully control these factors in order to produce high-quality biochar that can be effectively used for various purposes. Parallelly, harvesting large amounts of macrophytes could disrupt aquatic habitats and impact biodiversity; thus, it is important to judiciously assess the environmental impacts and ensure sustainable practices are followed in the production of biochar from aquatic macrophytes.

1.2.13 Bioenergy and Biofuels

Aquatic macrophytes have the potential to produce biofuels, which are renewable fuels made from organic matter, such as plants or algae (Ankit et al. 2020; Stabenau et al. 2018; Anand et al. 2017). Aquatic macrophytes can be used as a feedstock for biofuel production, either alone or in combination with other biomass sources (Röhl et al. 2019; Anyanwu et al. 2022; Pinho and Mateus 2023). One of the most promising types of biofuel that can be produced from aquatic macrophytes is bioethanol. Bioethanol is an alcohol that is produced by fermenting sugars and other carbohydrates found in the plant material (Ankit et al. 2020; Stabenau et al. 2018; Anand et al. 2017). Aquatic macrophytes such as water hyacinth and duckweed are particularly rich in sugars and starches, making them potential sources of bioethanol. In addition to bioethanol, aquatic macrophytes can also be used to produce other types of biofuels, such as biodiesel and biogas. Biodiesel is a fuel made from vegetable oils or animal fats, which can be produced by extracting the oil from the plant material and then processing it into a fuel. Biogas, on the other hand, is a mixture of gases such as methane and carbon dioxide, which are produced by the anaerobic digestion of organic matter (Anyanwu et al. 2022; Pinho and Mateus 2023). Aquatic macrophytes can be used as a feedstock for biogas production, either alone or in combination with other organic materials (Anyanwu et al. 2022; Sricoth et al. 2018).

In comparison to the biofuel produced from agro-waste material such as rice husk and biochar, freshwater macrophytes (*S. porticalis* and *Nymphaea L.*) have been reported to produce more bioenergy (Anyanwu et al. 2022). *Typha angustifolia* and *Eichhornia crassipes* have also been reported to have higher potential for biofuel generation owing to their faster growth rate, high dry mass production, and high content of cellulose, hemicellulose, etc. (Sricoth et al. 2018). The production of biofuels from aquatic macrophytes has several potential benefits. For instance, it provides a renewable source of energy that is less carbon-intensive than fossil fuels, helping to mitigate climate change. It could provide economic opportunities for local communities, particularly in areas where aquatic macrophytes are considered to be a nuisance or an invasive species. Furthermore, it could help to reduce the reliance on imported fossil fuels, promoting energy independence and security.

However, there are also challenges associated with the production of biofuels from aquatic macrophytes. These challenges include the high costs associated with harvesting and processing the plant material, as well as the potential environmental impacts of large-scale cultivation and harvesting. Additionally, the use of aquatic macrophytes for biofuel production may compete with other important ecosystem services provided by these plants, such as their role in providing habitat and food for aquatic animals. As such, the development of sustainable and environmentally responsible methods for the production of biofuels from aquatic macrophytes will be an important area of research in the coming years.

1.2.14 Environmental Monitoring

Aquatic macrophytes are commonly used as bioindicators of the health of aquatic ecosystems because they are sensitive to changes in water quality and habitat conditions (Töre and Özkoç 2022; Galal and Farahat 2015; Kolada 2010). They can indicate changes in water quality by displaying characteristic symptoms of nutrient deficiencies or toxicity. For instance, if a water body has high nutrient levels, macrophytes can become overgrown, leading to an imbalance in the ecosystem. Similarly, if the water is contaminated with pollutants, macrophytes can display signs of stress, such as leaf discoloration or stunted growth. The changes in macrophyte communities could indicate the introduction of invasive species that may outcompete native plants, alter habitat availability, and disrupt ecosystem processes. Besides, aquatic macrophytes provide important habitats for aquatic organisms, and changes in their distribution could indicate changes in habitat quality (Töre and Özkoç 2022; Jaramillo et al. 2019; Kolada 2010). For example, the disappearance of certain macrophyte species can indicate changes in water depth, sediment composition, or water flow. In aquatic ecosystems, macrophytes support a diverse range of aquatic organisms, and their presence or absence can indicate significant changes in biodiversity. The monitoring of the macrophyte communities could yield useful insights into the health of the entire aquatic ecosystem. Additionally, aquatic macrophytes are sensitive to changes in climate, including changes in temperature, precipitation, and water availability. Monitoring the changes in macrophyte distribution and abundance could provide insights into the impacts of climate change on aquatic ecosystems. This indicates that aquatic macrophytes are valuable tools for environmental monitoring and can provide valuable information on the health and functioning of aquatic ecosystems (Töre and Özkoç 2022; Galal and Farahat 2015; Kolada 2010).

1.3 Impact of Climate Change on Aquatic Macrophytes

Climate change refers to the long-term changes in temperature, precipitation, and other climatic variables that have been observed over the past century and are projected to continue in the future. Over the past century, the Earth's average surface temperature has increased by about 1.1 °C, with the majority of this warming occurring in the past few decades (Xia et al. 2022; Zhou et al. 2021; Pereira et al. 2021; Hossain et al. 2017). This warming trend is largely driven by the increase in atmospheric concentrations of greenhouse gases, such as carbon dioxide, methane, nitrous oxide, etc. While some natural factors, such as volcanic eruptions and solar activity, can influence the Earth's climate, the current temperature trend cannot be explained by these factors alone (Xia et al. 2022; Zhou et al. 2021). The evidence suggests that human activities, such as burning fossil fuels, deforestation, and land-use changes, are the primary drivers of climate change in recent times. In addition to rising temperatures, other observed changes include changes in precipitation patterns, melting of glaciers and sea ice, and rising sea levels. These changes have significant impacts on natural systems, including changes in ecosystems and biodiversity, changes in agricultural productivity, and increases in extreme weather events such as droughts, floods, and heat waves (Xia et al. 2022; Zhou et al. 2021; Pereira et al. 2021).

Climate change is a threat to aquatic macrophytes also. It could have substantial impacts on the growth, distribution, and composition of aquatic macrophyte communities, with potential consequences on the functioning of aquatic ecosystems (Xia et al. 2022; Zhou et al. 2021). One of the most significant impacts of climate change on aquatic macrophytes is the alteration of water temperature regimes. As temperatures rise, many macrophyte species may experience reduced growth rates, altered phenology (timing of life cycle events), and changes in reproductive success. Additionally, warmer water temperatures could promote the growth of invasive plant species, which may outcompete native macrophytes for resource utilization (Pereira et al. 2021; Hossain et al. 2017).

Changes in precipitation patterns and water availability could also have a significant impact on macrophyte communities. Drought conditions could lead to reduced water levels and increased water temperatures, which in turn could put stress on macrophyte populations and lead to declines in growth and reproductive success. Conversely, increased precipitation could lead to flooding, resulting in damage or uprooting of macrophyte populations (Xia et al. 2022; Zhou et al. 2021). Changes in nutrient availability could also impact aquatic macrophyte communities. Increased temperatures may accelerate nutrient cycling and increase nutrient availability, which would promote the growth of certain macrophyte species. However, excessive nutrient inputs from agricultural and urban runoff could lead to eutrophication, leading to harmful algal blooms and promoting the growth of invasive macrophytes (Hossain et al. 2017; Reitsema et al. 2020).

The interactions between macrophytes and other organisms in the ecosystem could also get influenced by climate change. For instance, the changes in macrophyte growth rates and distribution alter the availability of food and habitat for fish and invertebrates, which could have cascading effects on the entire ecosystem (Xia et al. 2022; Pereira et al. 2021; Hossain et al. 2017; Reitsema et al. 2020). Overall, the impacts of climate change on aquatic macrophytes are complex and varied and depend on a range of factors such as species composition, ecosystem context, and the specific nature of the climate change drivers. However, these impacts have important implications for the functioning and resilience of aquatic ecosystems and highlight the need for effective management strategies that can help mitigate or adapt to the effects of climate change.

1.4 Conclusion

Aquatic macrophytes are an essential component of aquatic ecosystems. This chapter discusses the various functions and ecological services offered by the aquatic macrophytes, which are critical for the health and sustainability of freshwater and marine ecosystems and for the well-being of human societies. Understanding and managing these services is essential for the effective conservation and management of aquatic environments; however, this has become a challenge in recent decades. Climate change and anthropogenic activities, directly or indirectly, have disturbed the growth, distribution, and composition of the aquatic macrophyte communities, which will have potential consequences on the functioning of aquatic ecosystems. The complex and interactive nature of these threats underscores the need for effective management strategies that can help mitigate or adapt to the effects of climate change on aquatic macrophytes and the ecosystems they support.

References

Adhikari S, Bajracharaya RM, Sitaula BK (2009) A review of carbon dynamics and sequestration in wetlands. J Wetl Ecol 2:42–46

Akasaka M, Takamura N, Mitsuhashi H, Kadono Y (2010) Effects of land use on aquatic macrophyte diversity and water quality of ponds. Freshw Biol 55:902–922

Ali F, Jilani G, Bai L, Wang C, Tian L, Jiang H (2021) Effects of previous drying of sediment on root functional traits and rhizoperformance of emerged macrophytes. Front Environ Sci Eng 15(6):1–18

Amir M, Paul D, Samal RN (2019) Sources of organic matter in Chilika lagoon, India inferred from stable C and N isotopic compositions of particulates and sediments. J Mar Syst 194:81–90

Anand S, Bharti SK, Dviwedi N, Barman SC, Kumar N (2017) Macrophytes for the reclamation of degraded waterbodies with potential for bioenergy production. In: Bauddh K, Singh B, Korstad J (eds) Phytoremediation potential of bioenergy plants. Springer, Singapore

Anand S, Kumar S, Kumar N (2018) Remediation of inorganic substances and heavy metals through aquatic macrophytes from flashlight manufacturing industry effluent. Geophytology 47(2):29–40

Ankit BN, Tiwari J, Kumar S, Korstad J, Bauddh N (2020) Removal of organic pollutants from contaminated water bodies by using aquatic Macrophytes coupled with bioenergy production and carbon sequestration. In: Bharagava R (ed) Emerging eco-friendly green Technologies for Wastewater Treatment. Microorganisms for sustainability, vol 18. Springer, Singapore

Anyanwu IN, Okeke CS, Nwankwo SC, Nwachukwu MO, Michael MO, Opara VC, Chamba EB (2022) Aquatic macrophytes (*Spirogyra porticalis* and *Nymphaea L.*) as substrates for biofuel production: potentials and challenges. Scientific African 18:e01412

Bhaskaran K, Vijaya Nadaraja A, Tumbath S, Babu Shah L, Veetil PGP (2013) Phytoremediation of perchlorate by free floating macrophytes. J Hazard Mater 260:901–906

Bird MI, Wurster CM, de Paula Silva PH, Paul NA, De Nys R (2012) Algal biochar: effects and applications. GCB Bioenergy 4(1):61–69

Bloom AA, Palmer PI, Fraser A, Reay DS, Frankenberg C (2010) Large-scale controls of methanogenesis inferred from methane and gravity space borne data. Science 327:322–325

Bornette G, Puijalon S (2011) Response of aquatic plants to abiotic factors: a review. Aquat Sci 73(1):1–14

Chambers PA, Lacoul P, Murphy KJ, Thomaz SM (2008) Global diversity of aquatic macrophytes in freshwater. Hydrobiologia 595:9–26

Cui X, Hao H, He Z, Stoffella PJ, Yang X (2016) Pyrolysis of wetland biomass waste: potential for carbon sequestration and water remediation. J Environ Manag 15(173):95–104
Cui X, Wang J, Wang X, Khan MB, Lu M, Khan KY, Song Y, He Z, Yang X, Yan B, Chen G (2022) Biochar from constructed wetland biomass waste: a review of its potential and challenges. Chemosphere 287:132259
Daily GC, Polasky S, Goldstein J, Kareiva PM, Mooney HA, Pejchar L, Ricketts TH, Salzman J, Shallenberger R (2009) Ecosystem Services in Decision Making: time to deliver. Front Ecol Environ 7(1):21–28
De Groot R, Alkemade R, Braat L, Hein L, Willemen L (2010) Challenges in integrating the concept of ecosystem services and values in landscape planning, management and decision making. Ecol Complex 7(3):260–272
Dhote S, Dixit S (2009) Water quality improvement through macrophytes—a review. Environ Monit Assess 152:149–153
Dorn NJ, Cronin G, Lodge DM (2001) Feeding preferences and performance of an aquatic lepidopteran on macrophytes: plant hosts as food and habitat. Oecologia 128:406–415
Engelhardt KAM, Ritchie ME (2001) Effects of macrophyte species richness on wetland ecosystem functioning and services. Nature 441:687–689
Fawzy MA, Badr NES, El-Khatib A, Abo-El-Kassem A (2012) Heavy metal biomonitoring and phytoremediation potentialities of aquatic macrophytes in River Nile. Environ Monit Assess 184(3):1753–1771
Ferna'ndez-Ala'ez C, Ferna'ndez-Ala'ez M, Be'cares E (1999) Influence of water level fluctuation on the structure and composition of the macrophyte vegetation in two small temporary lakes in the northwest of Spain. Hydrobiologia 415:155–162
Fernandez-Alaez C, Fernandez-Alaez M, Garcia-Criado F, Garcio-Giron J (2018) Environmental drivers of aquatic macrophyte assemblages in ponds along an altitudinal gradient. Hydrobiologia 812:79–98
Galal TM, Farahat EA (2015) The invasive macrophyte *Pistia stratiotes* L. as a bioindicator for water pollution in Lake Mariut, Egypt. Environ Monit Assess 187:701
Gong YP, Ni ZY, Xiong ZZ, Cheng LH, Xu XH (2017) Phosphate and ammonium adsorption of the modified biochar based on Phragmites australis after phytoremediation. Environ Sci Pollut Res 24(9):8326–8335
Haroon AM (2022) Review on aquatic macrophytes in Lake Manzala, Egypt. The Egypt J Aquat Res 48:1
He Y, Zhao K, Zhang H, He L, Niu Y, Zhang M, Xu J (2021) Linking macrophyte community structure with food chain length: a case study in the largest freshwater lake in China and ecological restoration implications. Ecol Indic 123:107363
Hossain K, Yadav S, Quaik S, Pant G, Maruthi AY, Ismail N (2017) Vulnerabilities of macrophytes distribution due to climate change. Theor Appl Climatol 129(3):1123–1132
Hunt LM, Morris DM, Drake DAR, Buckley JD, Johnson TB (2019) Predicting spatial patterns of recreational boating to understand potential impacts to fisheries and aquatic ecosystems. Fish Res 211:111–120
Jaramillo E, Duarte C, Labra FA, Lagos NA, Peruzzo B, Silva R et al (2019) Resilience of an aquatic macrophyte to an anthropogenically induced environmental stressor in a Ramsar wetland of southern Chile. Ambio 48:304–312
Kayranli B, Scholz M, Mustafa A, Hedmark A (2010) Carbon storage and fluxes within freshwater wetlands: a critical review. Wetlands 30:111–124
Kazanjian G, Flury S, Attermeyer K et al (2018) Primary production in nutrient-rich kettle holes and consequences for nutrient and carbon cycling. Hydrobiologia 806:77–93
Keitel J, Zak D, Hupfer M (2016) Water level fluctuations in a tropical reservoir: the impact of sediment drying, aquatic macrophyte dieback, and oxygen availability on phosphorus mobilization. Environ Sci Pollut Res 23:6883–6894
Kolada A (2010) The use of aquatic vegetation in lake assessment: testing the sensitivity of macrophyte metrics to anthropogenic pressures and water quality. Hydrobiologia 656:133–147

Kumar D, Anand S, Poonam, Tiwari J, Kisku GC, Kumar N (2019) Removal of inorganic and organic contaminants from terrestrial and aquatic ecosystems through phytoremediation and biosorption. In: Sobti R, Arora N, Kothari R (eds) Environmental biotechnology: for sustainable future. Springer, Singapore

Kumari K, Swain AA, Kumar M et al (2021) Utilization of *Eichhornia crassipes* biomass for production of biochar and its feasibility in agroecosystems: a review. Environ Sustain 4:285–297

Kurilenko VV, Osmolovskaya NG (2006) Ecological-biogeochemical role of macrophytes in aquatic ecosystems of urbanized territories (an example of small water bodies of St. Petersburg). Russ J Ecol 37:147–151

Lawniczak AE, Zbierska J, Choiński A, Szczepaniak W (2010) Response of emergent macrophytes to hydrological changes in a shallow lake, with special reference to nutrient cycling. Hydrobiologia 656:243–254

Li J, Fan J, Zhang J (2018) Preparation and evaluation of wetland plant-based biochar for nitrogen removal enhancement in surface flow constructed wetlands. Environ Sci Pollut Res 25:13929–13937

Lind L, Eckstein RL, Relyea RA (2022) Direct and indirect effects of climate change on distribution and community composition of macrophytes in lentic systems. Biol Rev 97:1677–1690

Lolu AJ, Ahluwalia AS, Sidhu MC, Reshi ZA (2019) Carbon sequestration potential of Macrophytes and seasonal carbon input assessment into the Hokersar wetland, Kashmir. Wetlands 39:453–472

Mace GM, Norris K, Fitter AH (2012) Biodiversity and ecosystem services: a multilayered relationship. Trends Ecol Evol 27(1):19–26

Madsen JD, Chambers PA, James WF, Koch EW, Westlake DF (2001) The interaction between water movement, sediment dynamics and submersed macrophytes. Hydrobiologia 444:71–84

Michalak I, Baśladyńska S, Mokrzycki J, Rutkowski P (2019) Biochar from a freshwater macroalga as a potential biosorbent for wastewater treatment. Water 11(7):1390

Mikulyuk A, Sharma S, Van Egeren S, Erdmann E, Nault ME, Hauxwell J (2011) The relative role of environmental spatial and land-use patterns in explaining aquatic macrophyte community composition. Can J Fish Aquat Sci 68:1778–1789

Millennium Ecosystem Assessment (2005) Ecosystems and human well-being: synthesis. Island Press, Washington DC

Mishra S, Mohanty M, Pradhan C (2013) Physico-chemical assessment of paper mill effluent and its heavy metal remediation using aquatic macrophytes—a case study at JK paper mill, Rayagada, India. Environ Monit Assess 185:4347–4359

Mokrzycki J, Michalak I, Rutkowski P (2021) Biochars obtained from freshwater biomass—green macroalga and hornwort as Cr (III) ions sorbents. Biomass Conversion and Biorefinery 11:301–313

Moore TR, Dalva M (1993) The influence of temperature and water table position on carbon dioxide and methane emissions from laboratory columns of peatland soil. J Sci 44:651–664

Nag SK, Das Ghosh B, Nandy S, Aftabuddin M, Sarkar UK, Das BK (2023) Comparative assessment of carbon sequestration potential of different types of wetlands in lower Gangetic basin of West Bengal, India. Environ Monit Assess 195(1):154

Nõges T, Luup H, Feldmann T (2010) Primary production of aquatic macrophytes and their epiphytes in two shallow lakes (Peipsi and Võrtsjärv) in Estonia. Aquat Ecol 44:83–92

O'Brien JM, Lessard JL, Plew D et al (2014) Aquatic Macrophytes Alter metabolism and nutrient cycling in lowland streams. Ecosystems 17:405–417

Pal S, Chattopadhyay B, Datta S, Mukhopadhyay SK (2017) Potential of wetland macrophytes to sequester carbon and assessment of seasonal carbon input into the East Kolkata wetland ecosystem. Wetlands 37(3):497–512

Pereira KM, Hefler SM, Trentin G, Rolon AS (2021) Influences of landscape and climatic factors on aquatic macrophyte richness and composition in ponds. Flora 279:151811

Pflüger Y, Rackham A, Larned S (2010) The aesthetic value of river flows: an assessment of flow preferences for large and small rivers. Landsc Urban Plan 95:68–78
Pinho HJ, Mateus DM (2023) Bioenergy routes for valorizing constructed wetland vegetation: an overview. Ecol Eng 187:106867
Pípalová I (2002) Initial impact of low stocking density of grass carp on aquatic macrophytes. Aquat Bot 73:9–18
Quirino BA, Lansac-Tôha FM, Thomaz SM, Heino J, Fugi R (2021) Macrophyte stand complexity explains the functional α and β diversity of fish in a tropical river-floodplain. Aquat Sci 83(1):12
Reitsema RE, Preiner S, Meire P, Hein T, Dai Y, Schoelynck J (2020) Environmental control of macrophyte traits and interactions with metabolism and hydromorphology in a groundwater-fed river. River Res Appl 37:294. https://doi.org/10.1002/rra.3708
Roelofs J, Brouwer E, Bobbink R (2002) Restoration of aquatic macrophyte vegetation in acidified and eutrophicated shallow soft water wetlands in The Netherlands. Hydrobiologia 478:171–180
Röhl M, Roth S, Schütz W, Zehnsdorf A, Herbes C (2019) Biogas production from submerged macrophytes—a case study of regional biomass potentials in Germany. Energy Sustain Soc 9(1):1–12
Sharip Z, Schooler SS, Hipsey MR et al (2012) Eutrophication, agriculture and water level control shift aquatic plant communities from floating-leaved to submerged macrophytes in Lake Chini, Malaysia. Biol Invasions 14:1029–1044
Smith AJ, Duffy BT, Novak MA (2015) Observer rating of recreational use in wadeable streams of New York state, USA: implications for nutrient criteria development. Water Res 69:195–209
Sricoth T, Meeinkuirt W, Pichtel J, Taeprayoon P, Saengwilai P (2018) Synergistic phytoremediation of wastewater by two aquatic plants (Typha angustifolia and Eichhornia crassipes) and potential as biomass fuel. Environ Sci Pollut Res 25:5344–5358
Srivastava J, Gupta A, Chandra H (2008) Managing water quality with aquatic macrophytes. Rev Environ Sci Biotechnol 7:255–266
Stabenau N, Zehnsdorf A, Rönicke H, Wedwitschka H, Moeller L, Ibrahim B, Stinner W (2018) A potential phosphorous fertilizer for organic farming: recovery of phosphorous resources in the course of bioenergy production through anaerobic digestion of aquatic macrophytes. Energy Sustain Soc 8(1):1–10
Tallar RY, Suen JP (2017) Measuring the aesthetic value of multifunctional lakes using an enhanced visual quality method. Water 9(4):233
Thomaz SM (2021) Ecosystem services provided by freshwater macrophytes. Hydrobiologia 850: 1–21
Töre GY, Özkoç ÖB (2022) Recent developments in aquatic macrophytes for environmental pollution control: a case study on heavy metal removal from lake water and agricultural return wastewater with the use of duckweed (Lemnaceae). In: Phytoremediation Technology for the Removal of heavy metals and other contaminants from soil and water. Elsevier, Amsterdam, pp 75–127
Xia W, Zhu B, Zhang S, Liu H, Qu X, Liu Y, Rudstam L, Anderson JT, Ni L, Chen Y (2022) Climate, hydrology, and human disturbance drive long-term (1988–2018) macrophyte patterns in water diversion lakes. J Environ Manag 319:115726
Yaashikaa PR, Kumar PS, Varjani S, Saravanan A (2020) A critical review on the biochar production techniques, characterization, stability and applications for circular bioeconomy. Biotechnology Reports 1(28):e00570
Yadav N, Ahn HJ, Rane NR, Kurade MB, Li X, Park YK, Khan MA, Chung WJ, Chang SW, Jeon BH (2023) Comprehensive study on removal of bisphenol-S and its metabolic fate using aquatic macrophytes. Chem Eng J 455:140967

Yasuno N, Shimada T, Fujimoto Y (2021) Semiaquatic spiders *Alopecosa cinnameopilosa* rely on prey derived from macrophyte-based food web: evidence from Lake Izunuma, Japan. Wetl Ecol Manag 29:507–517

Zhou L, Yan W, Sun X, Shao J, Zhang P, Zhou G, Zhou X (2021) Regulation of climate, soil and hydrological factors on macrophyte biomass allocation for coastal and inland wetlands in China. Sci Total Environ 774:145317

Factors Structuring Aquatic Macrophytes

2

Takudzwa C. Madzivanzira, Julie A. Coetzee, and Tatenda Dalu

Abstract

Aquatic macrophytes are a key component of freshwater ecosystems, providing habitats for aquatic organisms, and play an integral role in food webs and nutrient cycles. Understanding the factors that influence macrophyte growth, distribution, structure and community composition is indispensable for their integrated management, which are explored in this chapter. Among these are biotic (herbivory, macrophyte properties and competition and pathogens and diseases) and abiotic (water chemistry including temperature, substrate composition/embeddedness and hydrological conditions) factors. Anthropogenic stressors further drive these biotic and abiotic factors individually or in combination, causing either the extinction of important native macrophytes or the uncontrolled proliferation of macrophytes, usually invasive alien species, which has been recognised as an important issue of aquatic ecosystem management in freshwater systems globally. Among the notorious aquatic macrophytes of global concern are the invasive water hyacinth (*Pontederia crassipes*) and giant salvinia (*Salvinia molesta*), which have detrimental impacts in invaded freshwater systems. The global problem of nuisance macrophytes needs to be holistically handled at all levels to prevent ecological and socioeconomic impacts associated with their proliferation. Solutions to curb the nuisance growth of aquatic macrophytes include mechanical removal, biological control and chemical treatment although

T. C. Madzivanzira (✉) · T. Dalu (✉)
Aquatic Systems Research Group, School of Biology and Environmental Sciences, University of Mpumalanga, Nelspruit, South Africa

South African Institute for Aquatic Biodiversity, Makhanda, South Africa

J. A. Coetzee
South African Institute for Aquatic Biodiversity, Makhanda, South Africa

Centre for Biological Control, Department of Botany, Rhodes University, Makhanda, South Africa

S. Kumar et al. (eds.), *Aquatic Macrophytes: Ecology, Functions and Services*,
https://doi.org/10.1007/978-981-99-3822-3_2

integrated control is the most cost-effective control option. The control efforts need to be integrated at catchment and regional scales, facilitating the integration and partnerships of institutions to ensure functional aquatic systems and conservation of global biodiversity.

Keywords

Eutrophication · Freshwater · Climate change · Nutrients · Invasive · Plant · Water hyacinth

2.1 Introduction

Aquatic macrophytes are macroscopic forms of aquatic plants (including mosses, macroalgae, ferns and angiosperms), with vegetative parts that grow actively either intermittently or permanently floating on, submerged below and growing up through the water surface (Pieterse 1990; Piedade et al. 2022). This includes macrophytes that live in temporary and permanent freshwater systems (Chambers et al. 2008; Murphy et al. 2019). Macrophytes include (1) free-floating macrophytes that are typically floating on or under the water surface; (2) emergent macrophytes rooted in soil or soils periodically or intermittently inundated, with vegetation extending above the water surface; (3) floating-leaved macrophytes that are rooted to the sediment with leaves floating on the surface of the water; and (4) submerged macrophytes growing completely submersed under water, with roots attached to or closely associated with the substrate (Chambers et al. 2008). The global diversity of macrophytes is well reported in Chambers et al. (2008), and there are 3457 macrophyte species within 456 genera and 93 families. Out of these, 49.8% are monocotyledons, 44.0% dicotyledons, 6.1% ferns and fern allies and 0.2% clubmosses and horsetails (Murphy et al. 2019).

Macrophytes play crucial roles in the structure and functioning of freshwater systems (Fig. 2.1) (Chambers et al. 2008; Swe et al. 2021; Haroon 2022; Piedade et al. 2022). For example, they are important in nutrient cycling and stabilising and maintaining clear waters in turbid and hypertrophic systems (Scheffer et al. 1993). This helps in suppressing algal growth through several mechanisms that include nutrient competition (Mjelde and Faafeng 1997) and/or allelopathic chemical release, which are toxic and harmful to algae (Gross et al. 2007). Macrophytes also remove heavy metals in freshwater systems that potentially affect the health of various faunal groups and humans (Ladislas et al. 2011; Prajapati et al. 2017; Netshiongolwe et al. 2020). Healthy macrophytes provide crucial habitat structure for macroinvertebrates and fish (Chambers et al. 2008; Choi and Kim 2020; Piedade et al. 2022) and provide a food source for primary consumers and detritivores (when they decompose), thereby providing an important base of aquatic food webs (Newman 1991).

Macrophytes are, however, recognised as a major issue in aquatic ecosystem management across many ecosystem types of the world if they are allowed to grow

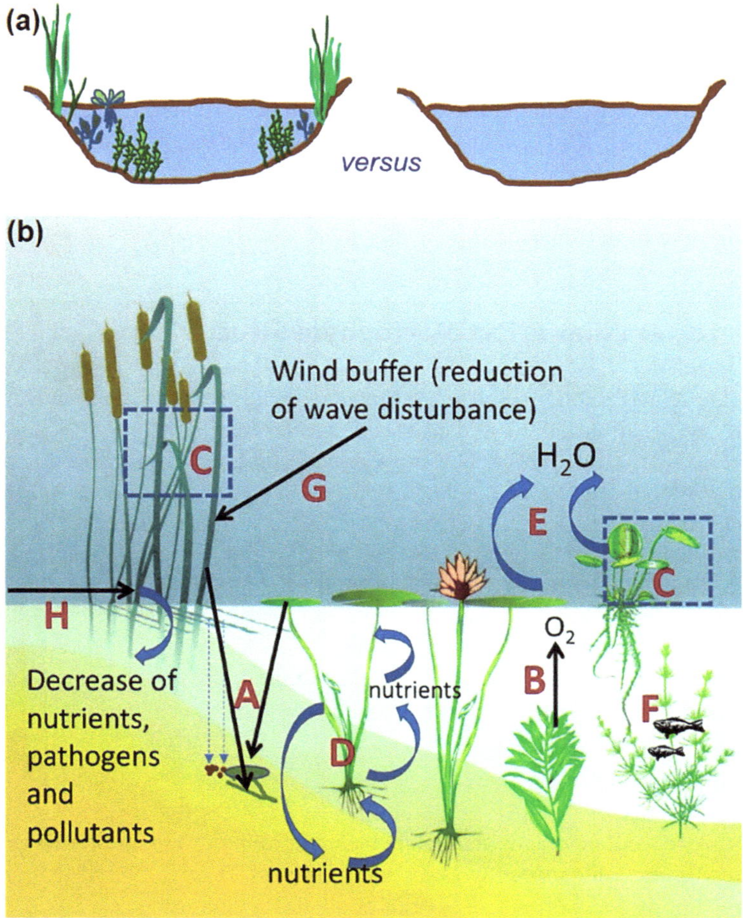

Fig. 2.1 (**a**) An image trying to depict what one could ask if macrophytes were to be removed from an aquatic ecosystem. (**b**) Ecosystem services provided by aquatic macrophytes, ranging from *supporting services* [e.g. **A**, formation of sediment from siltation and detritus accumulation; **B**, photosynthesis resulting in oxygen production; **C**, primary production; **D**, nutrient cycling (i.e. through movement across sediment, water and plant compartments); **E**, water cycling; and **F**, habitat provisioning]; *regulating services* [e.g. **G**, erosion regulation through wave disturbance reduction; **E**, water regulation; **D** and **H**, water purification through nutrient and pollutant retention; **H**, disease regulation via pathogen reduction; **C**, biotic resistance]; to *provisioning services* [**C**, fibre, ornamental resources, food, genetic resources]. Adopted from Thomaz (2021)

and proliferate uncontrollably (Coetzee et al. 2014; Kagami et al. 2019; Haroon 2022). Recently, aquatic macrophytes have been receiving great attention because of the devastating effects they cause when they grow and proliferate uncontrollably (Hussner et al. 2021; Haroon 2022). Thus, the diversity and abundance of macrophytes are extensively affected by the water physicochemical environment, sediment texture or embeddedness and biotic interactions among different organisms

(Haroon and Abd Ellah 2021). The challenge faced by environmental managers and various authorities is to determine and properly manage the degree and magnitude to which disturbance, climatic, hydrogeomorphic and physicochemical factors in combination, individually or in tandem affect macrophyte communities (Lougheed et al. 2001; Coetzee et al. 2021; Datta et al. 2021). In view of the significant roles played by macrophytes, understanding the factors that influence their distribution patterns is indispensable for their integrated management (Dar et al. 2014), and this forms the main objective of this chapter.

2.2 Factors that Affect Macrophyte Structuring

Several biotic and abiotic factors have been shown to interact, thereby affecting the productivity, distribution, abundance and species composition of macrophytes (Fig. 2.2) (Germ et al. 2021). The most notable biotic factors include herbivory, macrophyte functional traits, competition among the macrophyte species and pathogens and diseases (Germ et al. 2021). Abiotic factors include light, water chemistry including temperature, substrate composition/embeddedness and hydrological conditions (Harvey et al. 1987; Piedade et al. 2022). These are discussed in detail in the sections below.

2.3 Biotic Factors

2.3.1 Herbivory

All floating-leaved, emergent and submersed macrophytes are subject to extensive herbivory losses, which directly alters their biomass, relative abundance and productivity (Lodge and Lorman 1987; Liu et al. 2021; Ghirardi et al. 2022). Several studies (e.g. Liu et al. 2021; Filho et al. 2021; Ghirardi et al. 2022; Masese et al. 2022) suggest that many small (i.e. invertebrates) and large grazers (i.e. fish, waterfowls, hippos) may affect macrophytes through consumption and non-consumptive destruction. Reductions in macrophyte root biomass or shoots by grazers can result in 100% macrophytes loss, with reductions that exceed 50% reported for invasive crayfish, insect larvae, fish, snails and waterfowl (Liu et al. 2021; Filho et al. 2021; Ghirardi et al. 2022; Masese et al. 2022). For example, redbreast tilapia *Coptodon rendalli* causes significant structuring of macrophytes through consumption and tends to prefer certain species over others. In Lake Kariba, *C. rendalli* preferred *Vallisneria aethiopica* over *Ceratophyllum demersum* (Chifamba 1990). In Spain, extensive removal of macrophytes by crayfish is posited to have led to the local extinction of three macrophytes, *C. demersum*, *Myriophyllum alterniflorum* and *Utricularia australis* (Montes et al. 1993). Non-consumptive destruction can also affect macrophytes, for example, shoots of submersed macrophyte species may be clipped near the sediment by rusty crayfish, *Faxonius rusticus* (Lodge et al. 1994). An undetermined macrophyte percentage destruction by coypu,

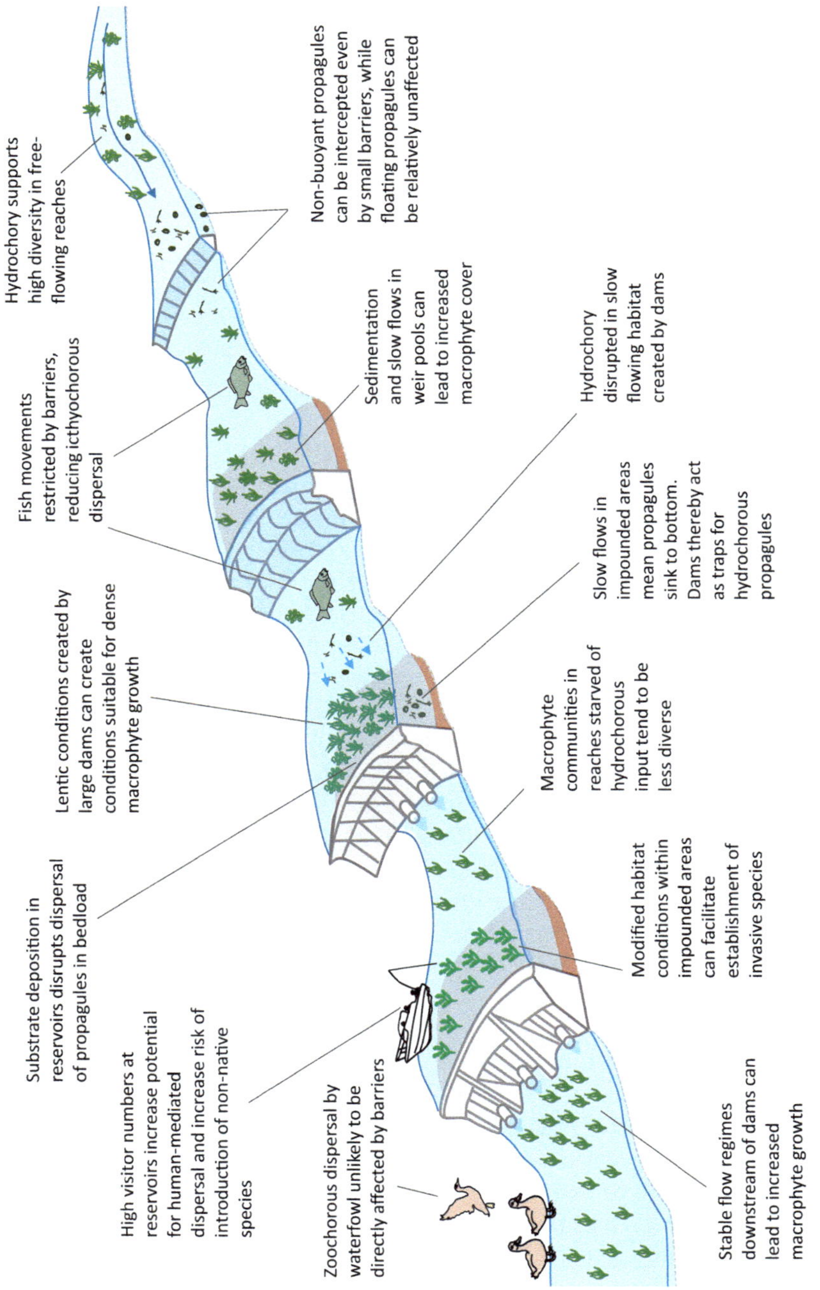

Fig. 2.2 Some examples of the potential barrier impacts to macrophyte species dispersal and population structure within aquatic ecosystems. Adopted from Jones et al. (2020)

muskrats and waterfowl goes into building nests and waste (Anderson and Low 1976; Howell and Richardson 2019). All macrophytes sustain substantial damage, except for emergent macrophytes, which appear to be less susceptible to invertebrate damage or consumption than submersed and floating-leaved forms due to their physical structure and lack of leaves (Lodge et al. 1989; Piedade et al. 2022). Waterfowls influence emergent and submersed macrophytes across spatiotemporal scales, with snow geese reducing macrophytes substantially on their summer breeding grounds at migration stopover sites and wintering grounds (Lodge et al. 1989). The importance of herbivory in structuring macrophyte communities is perhaps most evident in biological control programmes implemented against invasive macrophytes where host-specific herbivores, mostly insect's species, significantly reduce macrophyte biomass, allowing recovery of native macrophyte communities (Coetzee et al. 2021).

2.3.2 Macrophyte Properties

Since herbivore and macrophyte growths are highly dependent on the nutrients, nitrogen (N) and phosphorus (P) (Rao et al. 2020), macrophytes that grow rapidly may be good food and energy source for herbivores, since N-P contents of those macrophytes are relatively high (Grutters et al. 2016). For example, grass carp and rudd prefer macrophytes characterised by low C:N ratios (Dorenbosch and Bakker 2011). Apart from the elemental stoichiometry, macrophytes differ in their chemical feeding deterrent properties and in their respective ratios of feeding deterrents to attractants (Cronin et al. 2002; Choi et al. 2002). The caterpillar *Acentria ephemerella* is able to develop on chemically defended *M. spicatum* but performs better on *Potamogeton perfoliatus*, which is less-defended and highly nutritious (Choi et al. 2002). In a similar situation, fish eat more of the well-defended *M. spicatum* when it contains high N (Dorenbosch and Bakker 2011). Foliar traits interestingly differ between the native and introduced macrophytes, with herbivores restricting the introduced macrophyte growth, thereby allowing the growth of native macrophytes by selective consumption, i.e. if the exotic species are the favoured macrophytes and vice versa (Penuelas et al. 2010; Grutters et al. 2016).

2.3.3 Macrophyte Interactions

The community structure of macrophytes is merely not an assemblage of species in each area; however, its dynamics, structure and functioning are governed by positive, negative or indifferent interactions between and/or among macrophyte species, besides environmental interactions. Species can interact through exploitative, interference and allelopathy competition (Gopal and Goel 1993). Exploitative competition is the most common where two macrophytes of the same or different species compete whenever a specific valuable resource such as light, nutrients and space that the species share is limited (Harper 1977). In contrast to this, interference competition involves actively denying resource access by an individual competitor to the other (Begon et al. 1986). Allelopathy is where a competing organism produces chemicals, which when released into the environment retard or impair the growth of the other species, providing the organism with a competitive advantage against the latter (Gopal and Goel 1993). The ability of macrophytes to produce allelochemicals may have evolved to suppress or even kill their neighbours, thus eliminating the competition for limited resources. While most studies have looked at allelopathy of macrophytes on other biotic components, few studies have shown how macrophytes affect other macrophytes, but nevertheless, some interactions between macrophytes could involve the release of allelochemicals (Thiébaut et al. 2018; Puche et al. 2020). This was demonstrated in the laboratory whereby invasive water primrose *Ludwigia hexapetala* released allelochemicals (i.e. myricitrin, pruning and quercitrin), which affected two exotic species, the emergent growth form of *Myriophyllum aquaticum* and submerged *Egeria densa* and *C. demersum* (Thiébaut et al. 2018). Several macrophytes such as *Elodea nuttallii* and *M. spicatum* have been shown to affect phytoplankton (Gross 2003; Maredová et al. 2021; Wijewardene et al. 2022) and inhibit the germination seeds and/or the growth of seedlings (Gopal and Goel 1993) via the release of allelochemicals. Many macrophytes, however, release allelopathic chemicals into the freshwater system with very little impacts on performance of the other macrophytes due to long co-evolutionary history (Thorpe et al. 2009).

2.3.4 Pathogens and Diseases

Literature on the role of pathogens and diseases in determining the distribution and abundance of macrophytes is scant (Shearer 1994). However, there are examples where fungal pathogens are used in biological control of invasive macrophytes. Of the 60 fungal taxa reported around the world to be pathogenic to the floating *P. crassipes*, 10 are known to generate diseases as they have been found to be highly virulent (Charudattan 1996; Shabana 2005), and some of these have been developed into mycoherbicides for *P. crassipes* control in Africa (Bateman 2001). In the USA, the feasibility of using some native pathogens in the control of invasive *Hydrilla verticillata* has also been investigated, with the identification of *Mycoleptodiscus terrestris* from hydrilla growing in Lake Houston in Texas in the late 1980s (Joye 1990). Both field and lab experimental studies have demonstrated that *M. terrestris* can significantly reduce the biomass and abundance of hydrilla after inoculation compared with plants that are untreated (Shearer 2010). Pathogens have also been highlighted as a cause of *M. spicatum* declines in Madison lakes, USA (Shearer 1994).

2.3.5 Abiotic Factors

2.3.5.1 Light

As macrophytes photosynthesise, the availability of light is considered as a major limiting factor for primary production and growth (Harvey et al. 1987; Chambers et al. 2008). The quality, intensity and duration of the light matters the most as this affects macrophytes, with the availability of light being affected by water depth, turbidity and shading (Harvey et al. 1987). Riparian vegetation shades aquatic ecosystems, thereby reducing the duration of light availability to macrophytes. Turbidity is increased by algal blooms in nutrient-rich freshwater systems and clay and silt (mostly from run-off), which can result in decreases in water transparency and light availability, thus inhibiting submerged macrophyte growth (Harvey et al. 1987).

Aquatic ecosystems with low light penetration and intensity are dominated by free-floating and rooted floating-leaved macrophytes adapted to grow their leaves above the water surface, where there is plenty of light (Lacoul and Freedman 2006). Submerged macrophytes are conversely abundant in ecosystems where water column light is abundantly available (Lacoul and Freedman 2006). The light regime is the main primary driver for macrophyte niche within lakes and reservoirs' littoral zones (Harvey et al. 1987), which typically consists of angiosperms at shallow depths, with bryophytes and charophytes occupying the deeper depths of the littoral zone (Chambers et al. 2008). This characteristic zonation is driven primarily by a combination of light availability and macrophytes' adaptation to differing light conditions.

2.3.5.2 Water Temperature

Water temperature within the thermal tolerance range of macrophytes promotes their growth and reproduction. The distribution of certain macrophytes is mainly temperature-driven, indicating that temperature is equally as important as light in structuring macrophytes (Kõrs et al. 2012). Macrophytes exhibit different temperature tolerances, and many of them have an optimal photosynthetic rate of between 20 and 35 °C, which also explains their global spatial distributions (Bornette and Puijalon 2011).

2.3.5.3 Nutrients

The distribution and growth of macrophytes also require nutrients, particularly N and P, which are assimilated in the form of nitrate (NO_3^-), ammonium (NH_4^+) and phosphate (PO_4^{3-}), respectively (Dalu et al. 2012; Rao et al. 2020). These two macronutrients are important components of all the biotic components and are closely linked to the aquatic carbon, N and P cycles, determining both primary production and microbial mineralisation of the organic matter in aquatic ecosystems. Nitrogen contributes to structural component and metabolic and generic compounds in macrophyte cells. Nitrogen is an essential part of chlorophyll and also the building block of most proteins and amino acids and qualitative defence compounds. Phosphorus contributes mainly to the complexity of the nucleic acid structure (Rao et al.

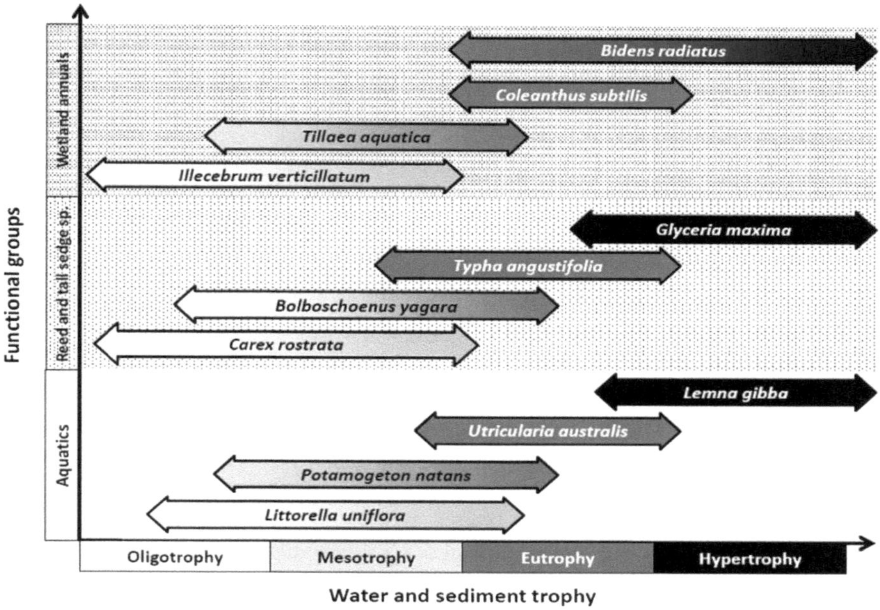

Fig. 2.3 Relationship between aquatic macrophyte species, sediment and water trophic level in common carp (*Cyprinus* sp.) ponds. Position and arrow length indicate species' range width in relation to conditions where the species is still able to grow (i.e. trophy), whereas the area shaded represents the species' trophic optimum. Note that macrophyte species with a specific relationship to the trophy were the only ones that were selected, and species with broad trophic ranges were not included. Adopted from Francová et al. (2019)

2020). The nucleic acid is a requirement for protein synthesis regulation, and therefore, P is important in cell division as well as the development of new tissues within macrophytes (Rao et al. 2020).

An example of nutrient effects on macrophytes is highlighted by Francová et al. (2019) who observed that the nutrient loading level resulted in oligotrophic or mesotrophic status of carp ponds, which caused a substantial spectrum of species reflected in an increased diversity of macrophytes but low overall biomass (Fig. 2.1). This allowed growth in oligo-mesotrophic (e.g. *Carex rostrata*, *Littorella uniflora*) to hypertrophic (e.g. *Potamogeton crispus*, *Typha latifolia*) waters. The study observed that in ponds with low trophic status, macrophyte species (e.g. *Glyceria maxima*) that require high nutrients were most likely rare (Fig. 2.3).

Both N and P are generated within aquatic ecosystems from the decomposition of organic matter and nutrient cycles, as well as from external sources such as water flowing into the aquatic ecosystem containing fertilisers washed away from agricultural fields, and municipal (i.e. sewage, industrial) deposition from towns, which are discussed in the later sections of the chapter. Submerged and floating macrophytes are able to filter nutrients directly from water columns, while rooted-emergent macrophytes take up nutrients from sediment (Preiner et al. 2020). Not much literature is available on the secondary and micronutrient requirements of macrophytes despite their importance. Secondary nutrients are required in moderate amounts (e.g. potassium, calcium, magnesium, sulphur), and trace or micro-nutrients are required in tiny amounts (boron, chlorine, copper, iron, manganese, molybdenum, zinc).

2.3.5.4 Hydrological Variations

Water-level fluctuation triggers shoreline erosion and, depending on fluctuation range, also affects macrophyte species composition and/or disappearance of sensitive macrophytes (Dalu et al. 2012). The degree to which macrophytes tolerate hydrological fluctuations (i.e. flooding events and drawdown) determines macrophyte community structure (Špoljar et al. 2021; Zhao et al. 2021). Rapid water-level rises induced by summer floods after exceptionally aberrant strong rains generally cause large reductions in the shoot spatial extent and biomass and disappearance of submerged macrophytes in aquatic systems due to low-light availability (Zhu et al. 2012). Flooding increases the water depth, compromising light penetration, and the most affected are emergent macrophytes as they fail to reach the water surface (Casanova and Brock 2000). Deep waters can also inhibit macrophytes' growth and decrease their mechanical resistance (Zhu et al. 2012).

Drawdown also has catastrophic effects on macrophytes, primarily through exposure of both above-ground vegetation parts and beneath-ground rhizome and root systems to desiccation under either hot or freezing conditions (Wagner and Falter 2002). Freshwater systems repeatedly disturbed by drawdown events are then likely to be dominated by desiccation-tolerant species (Maltchik et al. 2007). Drawdown can also indirectly affect macrophytes through its effect on the habitat by forming frost heaves on dewatered sediments and subsequent mechanical damage to rhizomes and roots of macrophytes (Wagner and Falter 2002). For example, a study by Kang et al. (2022) assessed how *Vallisneria spinulosa* responds (morphologically and reproductively) to the water depth gradients of 0.6–1.5 m and to water drawdown of up to 0.3 m. It was observed that *V. spinulosa* grew best at water depths of 1 m and manifested high biomass (Kang et al. 2022). At the low water depth, *V. spinulosa* resulted in it growing to produce small plants with less vegetative biomass, while drawdown produced *V. spinulosa* with more tuber biomass. *Vallisneria spinulosa* growing in deeper waters (1.5 m) were able to adapt water depth decline as expressed by increased stolon weight, ramet number and below-ground biomass. The fluctuations of water levels can also affect substrate particle size distribution and sediment particle size, which is an important factor structuring macrophytes as it controls root attachment surfaces, nutrient dynamics and intra-sediment chemistry (Wagner and Falter 2002).

Mechanical resistance of macrophytes is significantly affected by hydraulic forces from water wave action and wind, which results in mechanical damage (uprooting, leaf damage, stem breakage), which has adverse effects on the growth of a plant (Zhu et al. 2012). Root anchorage strength and the stem mechanical properties are essential macrophyte traits for mechanical resistance to reduce this type of damage in aquatic habitats, and hence, macrophytes that possess these traits will dominate in such systems (Zhu et al. 2012). Macrophytes may also respond to high wave action and winds over time by changing and altering their morphology (Bornette and Puijalon 2011). Wave action also disturbs the fine textured sediment, thereby increasing turbidity, which decreases light available for submersed macrophytes (Madsen and Cedergreen 2002). Silt presence due to wave action was negatively correlated with species richness and diversity in Lake Schilling, USA, which was previously dominated by curly leaf pondweed, *Potamogeton crispus* (Schmid et al. 2021). The fragrant water lily, *Nymphaea odorata*, later dominated Lake Silver as it produces thicker rhizomes, which are able to support large floating leaves (Schmid et al. 2021). Once floating leaves reach the water surface, *N. odorata* no longer experience detrimental turbidity effects, which is the reason why water lilies and other macrophyte species (morphologically similar to water lilies) are often dominant in shallow, turbid freshwater systems (Lacoul and Freedman 2006). Therefore, as observed in Lake Silver, *N. odorata* is not only inhibited by suspended silt due to high wave action but benefits from high sediment nutrient availability as a result of the high cation exchange capacity of silt (Gerbersdorf et al. 2007).

2.3.5.5 Substrate Composition

Free-floating macrophyte dominance in an aquatic ecosystem creates an anoxic and dark condition that affects aquatic life beneath (Abdel-Tawwab et al. 2006; Coetzee et al. 2014). Free-floating macrophytes create environmental alternate stable states when compared to submerged plant growth, and the two can co-exist. Since floating macrophytes directly acquire water column nutrients, the submerged rooted macrophytes take their required nutrients from both the water column (Madsen and Cedergreen 2002) and substrate (Chambers et al. 1989). A deficit in nutrients in the water column can result in the dominance of the submerged rooted macrophytes (Lu et al. 2013). In contrast, substrates that are nutrient-poor will result in submerged macrophytes strongly competing with floating macrophytes for nutrients from the water column, but considering the cover and shading effect of floating macrophytes, the latter are likely to dominate (Lu et al. 2013). Thus, depending on water quality, submerged and floating-leaved macrophytes can be excellent competitors for particular resources mainly due to an asymmetry in their competitive abilities. As submerged macrophytes access nutrients from the sediment and water column, they can limit the floating macrophyte growth, whereas the floating-leaved macrophytes have more light access and can cause reductions in light for the submerged macrophytes (Kosten et al. 2009).

Rooted submerged macrophytes are often influenced by the characteristics of the substrate that surrounds them through organic content and redox potentials, as well as through impacts on their rates of nutrient uptake (Madsen and Adams 1989; Engloner et al. 2013; Shields and Moore 2016). Sandy substrates are least preferred by macrophytes as sand is too unstable for proper rooting, and areas without sand provide macrophytes with a better grip (Madsen and Adams 1989). Macrophytes can become depressed in sections that contain purely sand sediments, and furthermore, in lakes and reservoirs, wave action washes away silt leaving behind rough and less fertile substrates like gravel and sand. Once the fine substrates are gone, continued wave action threatens macrophytes growing in the sand with abrasion or uprooting (Madsen and Adams 1989). Silt/muck-dominated habitats appear to be the most preferred environments by a diverse range of macrophytes as recorded in south-eastern South Dakota lakes (Kading and Xu 2021).

2.3.6 Anthropogenic Activities

2.3.6.1 Climate Change

While global biodiversity is alarmingly declining, the decline is more pronounced in aquatic ecosystems than other ecosystems (Dudgeon et al. 2006; Reid et al. 2019). Macrophytes' loss is accelerating, especially the submerged macrophytes (Zhang et al. 2017). Human-induced stressors such as climate change, eutrophication, habitat loss, pollution and invasive species are causing macrophytes to be vulnerable (Kundzewicz et al. 2008). Climate change is a major stressor driven by the release of greenhouse gases (GHG) (e.g. carbon dioxide (CO_2), fluorinated gases, methane, nitrous oxide (NO_2)) with drastic implications for freshwater ecosystems (Sala et al. 2000; Hartmann et al. 2013; Biesbroek et al. 2022). Changes in climate affect water quality and quantity, either directly by the observed and projected rising concentrations of CO_2 that affects the water salinity levels and associated temperature changes, which can impact on growth and reproduction, or indirectly through changes in precipitation levels and regimes, water-level variations and melting ice cover and glaciers (Rosenzweig et al. 2007; Meehl et al. 2007; Biesbroek et al. 2022). The global atmospheric GHG concentration is presently at the highest levels, and average temperatures have increased by over 1 °C since the 1900s and are predicted to rise by 1.5 °C between 2030 and 2052. Given the scenario that air temperatures continue to rise at the current rate, aquatic ecosystems will get warmer too (IPCC 2018; Saintilan et al. 2018).

Rising atmospheric CO_2 concentrations may increase dissolved inorganic carbon (DIC) levels and stimulate photosynthetic rates and, hence, macrophytes' productivity (Mormul et al. 2020). Macrophyte growth is also impaired under high dissolved organic carbon (DOC) concentrations, particularly by the humic substances, which turn the water brown (i.e. brownification) and attenuate light. Increased CO_2 concentrations benefit free-floating macrophytes more and limit submerged macrophytes, which leads to lower macrophyte abundances and diversities. Surprisingly, information on the potential increased atmospheric carbon dioxide level effects on submerged aquatic biotic components is scant, even though species composition and diversity are highly correlated with pCO_2 (Mormul et al. 2020). As a result of species-specific responses to elevated HCO_3^- and carbon dioxide availability, effects of increasing the atmospheric carbon dioxide would likely to vary with water body and the macrophyte species. Soft water lakes are characterised by poor DIC concentrations and often occur in temperate and boreal regions and at high altitudes (Lind et al. 2022). Soft water lakes are limited in bicarbonate and characterised by low acidity neutralisation capacity (Arts 2002; Lind et al. 2022).

Native isoëtid macrophyte species do well in these low bicarbonate conditions (Smolders et al. 2002), where macrophytes that utilise CO_2 from the sediment of the aquatic system (e.g. *Lobelia dortmanna*) dominate. An increase in CO_2 concentration and acidification in these soft water lakes results in the large-scale invasion of the water column with fast-growing rooted submerged macrophytes' carbon dioxide users (e.g. elodeids), thereby replacing the native rooted submerged slow-growing isoëtids (Arts 2002). Dissolved inorganic carbon-rich waters, which have strong alkaline pH (greater than 8.2), where carbon dioxide availability is the limiting factor for the growth of carbon dioxide users (e.g. *Myriophyllum triphyllum*), are often dominated by hydrogen carbonate users, while CO_2 increases would support the growth as well as the competitive strength of carbon dioxide users in soft water lakes (Hussner et al. 2015).

Several studies (e.g. Rosenzweig et al. 2007; Hintz et al. 2020; Magadza et al. 2020) exemplify the warming trend for aquatic ecosystems. The effects of climate change on macrophyte communities are well summarised in Lind et al. (2022) and include drastic changes in their physiology, phenology, morphology, biomass, productivity, distribution, species composition and population dynamics, although such effects are species specific. A temperature rise enhances macrophyte growth of certain species to a greater extent. *Elodea canadensis* and *L. major* that were exposed to different temperature treatments, which represents the natural seasons, showed a difference in growth responses (Silveira and Thiébaut 2017). Macrophytes also exhibit variations in their warming response, depending on geographic location and biological properties, with floating macrophytes together with those inhabiting shallow zones being exposed to higher temperatures than other macrophyte types (Santamaría 2002; McKee et al. 2002). In a laboratory experiment on climate change, McKee et al. (2002) found that the total abundance of three macrophyte species (i.e. *E. nuttallii*, *L. major*, *Potamogeton natans*) was not significantly affected by warming; however, the macrophyte community structure and composition changed, i.e. the relative *L. major* percentage increased with associated increases in growth rate under a continuous warming treatment, while *P. natans* increased its floating leaf surface (Lind et al. 2022). In most high-altitude aquatic systems, reduced ice cover because of warming has allowed prolonged growing seasons, which ultimately results in increased algal abundance and productivity (Karst-Riddoch et al. 2005; Dalu et al. 2022). Algal abundance increases reduced the amount of light available and ultimately decreased macrophyte cover (Karst-Riddoch et al. 2005). The rate of photosynthesis increases with warming up to an optimum point, beyond which it decreases; therefore, high temperatures of aquatic systems cause a resultant negative effects/impacts on the net primary production (Tait and Schiel 2013; Lind et al. 2022). Since this response will depend on individual macrophyte species to some extent and the growth conditions, an increase in the water temperature may cause major changes in the macrophyte species community and distribution, although precisely the response is dependent on species and region (Kirschbaum 2004; Lind et al. 2022).

Climate change is causing increased weather conditions' variability, with intense precipitation periods and increased drought frequencies (Meehl et al. 2007;

Rosenzweig et al. 2007; Kundzewicz et al. 2008; IPCC 2018). Changes in water levels associated with drought will impact macrophytes particularly those that are dependent on specific hydrological conditions, for example, in a large subtropical Florida lake following a natural drought, *Chara* sp. rapidly expanded and dominated the littoral habitats for 1 year, and thereafter, *H. verticillata* and *P. illinoiensis*, the vascular taxa, became dominant (Havens et al. 2005). Severe drought in Argentina's floodplain lake region resulted in a decreased free-floating macrophyte dominance (O'Farrell et al. 2011).

Increases in water levels due to increased precipitation in Turkish shallow lakes negatively affected submerged macrophytes (Tan and Beklioglu 2005). Heavy rainfall washes away nutrients from the terrestrial environment, resulting in aquatic ecosystem nutrient enrichment, thereby causing significant shift in species composition (Vaithiyanathan and Richardson 1999). This was described in a study from the Everglades, USA, where the macrophytes' community changed with P enrichment (Vaithiyanathan and Richardson 1999). Evidence from these studies implies that both heavy rainfall and droughts can cause either increases or decreases in macrophyte species diversity, biomass and cover (O'Farrell et al. 2011; Lind et al. 2022).

2.3.6.2 Invasive Species

The spread of invasive alien species (IAS) is facilitated increasingly by anthropogenic activities, which break geographical barriers for species distribution, often through intentional movements of species by humans for food, trade, recreation and other economic interests across the world, although unintentional introductions also occur through stowaway species of contamination (Nuñez et al. 2012; Blackburn et al. 2014). Most invasive macrophyte species have been introduced by humans for ornamental purposes (e.g. water hyacinth *P. crassipes*) (Hill et al. 2020). Once IAS are released into the aquatic systems, they spread via flow and floods that enable the connectivity of water bodies. Propagules of invasive macrophytes can also be dispersed by birds and boats. Among the notorious macrophytes are water primroses (*Ludwigia hexapetala*, *Ludwigia grandiflora*, *Ludwigia peploides* subsp. *montevidensis*), dense waterweed (*Egeria densa*), water hyacinth (*Pontederia crassipes*), giant salvinia (*Salvinia molesta*) and parrot feather (*Myriophyllum aquaticum*), which are all natives to South America. These notorious macrophytes have been introduced to almost all the continents where they have established and extending their invasive ranges (Thouvenot et al. 2013). These taxa are highly invasive because of their documented impacts on freshwater systems where they form dense masses, which cover aquatic system surfaces, thereby obstructing waterways making it difficult for navigation, decrease dissolved oxygen, reduce native macrophyte diversity and even threaten human health by providing refugia for vectors such as mosquitoes (Hill et al. 2020; Piedade et al. 2022). With their allelopathy effects, they can affect other native macrophytes (Gopal and Goel 1993).

Several factors contribute to invasive macrophytes' dominance over native macrophytes (Lind et al. 2022). These include the absence of natural enemies or competitors and disturbances, e.g. eutrophication, climate change and altered hydrology, which selectively affects the native macrophytes over the invasive ones

(Lind et al. 2022). Climate change may cause invasive macrophytes to potentially establish in previously unsuitable habitats through thermal acclimation, thereby increasing their invasion range. Several studies (e.g. Lacoul and Freedman 2006; Tattersdill 2017) reported a shift in the geographical range of invasive macrophytes, e.g. the establishment of the threadleaf crowfoot *Ranunculus trichophyllus* into high-elevation Himalayan lakes, which previously were non-vegetated (e.g. Lacoul and Freedman 2006), and the spread of the pondweed *Elodea canadensis* northward in Europe (e.g. Tattersdill 2017). Increased temperatures because of climate change have been shown to promote *E. canadensis* establishment and spread (Silveira and Thiébaut 2017). The invasive free-floating *Salvinia natans* was found to benefit from warming, whereas *E. nuttallii* decreased in biomass (Netten et al. 2011). Further projections have predicted that the Iceland climate will be suitable for *M. aquaticum* and *Egeria densa* by 2070 (Gillard et al. 2017), while *P. crassipes* is predicted to invade beyond 35 °S and N (Kriticos and Brunel 2016).

Invasive macrophytes also facilitate the establishment and spread of other IAS generally termed as an invasional meltdown (Simberloff 2006). In Lake Kariba, Zimbabwe, *P. crassipes* facilitates the spread of invasive Australian redclaw crayfish *Cherax quadricarinatus*, as juveniles were found on floating macrophytes (Marufu et al. 2018), with impacts cascading to fisheries (Madzivanzira et al. 2022). Invasive crayfish species have been shown to affect macrophyte species through direct herbivory (Madzivanzira et al. 2020, 2022). In Lake Kariba (Zambia and Zimbabwe), aquatic macrophytes dominated crayfish diet across different size classes as revealed by gut content analysis and stable isotopes (Marufu et al. 2018), whereas, in Lake Naivasha (Kenya), the red swamp crayfish *Procambarus clarkii* introductions resulted in notable declines in densities of the water lily *Nymphaea nouchali*, highlighting the direct consumptive impacts by crayfish (Lowery and Mendes 1977). *Cherax quadricarinatus* invasion in the Pilbara region of Australia resulted in 100% macrophyte cover loss and subsequent reorganisation of the community (Pinder et al. 2019).

2.3.6.3 Eutrophication

Anthropogenic nutrient enrichment above natural acceptable levels of aquatic systems is one of the pervasive forms of aquatic ecosystem change, which strongly influences both abiotic and biotic factors (Alexander et al. 2017). Eutrophication has affected most aquatic ecosystems globally from freshwater to saline waters and is mostly caused by industrial and urban sewage, erosion run-off and leached nutrients from farming lands (Smith et al. 2009). The amount of nutrients (mainly N and P) entering the aquatic systems has been increasing gradually since human beings first began clearing land for settlements and agriculture, with nutrients' supply dramatically increasing in many aquatic ecosystems mid-1900s (Smith et al. 2009). This increase in nutrient release was driven by human population growth increases and the improvements of households' sewage networks, which discharged into local water bodies with minimal treatment (Alexander et al. 2017).

Several studies (e.g. Toivonen and Huttunen 1995; Portielje and Roijackers 1995; Houlahan and Findlay 2004; Thiébaut and Muller 1998; Lougheed et al. 2001; Wu et al. 2021) have established that nutrient enrichment in aquatic ecosystems causes changes in macrophyte species composition, richness and density. Nutrient enrichment drives competition for light between macrophyte taxa and between macrophytes and their benthic, attached and water column phytoplankton (Hilton et al. 2006), which is a well-described process for aquatic systems whereby nutrients promote phytoplankton proliferation, which outcompetes littoral macrophytes for light (Moss 1998). Macrophytes are initially lost from pelagic waters where light penetration is less, and as the process becomes worse, the submerged macrophytes will eventually disappear from algae-dominated aquatic systems (Moss 1998). When established, the algae-dominated state of the aquatic ecosystem is stable, and it is challenging to change an aquatic ecosystem between two stable states that are either algae-dominated or macrophyte-dominated (Scheffer and van Nes 2007). O'Hare et al. (2018) concluded that an individual macrophyte morphotype replacement can proceed only where physical conditions are suitable (Fig. 2.4). Thus, the reality is

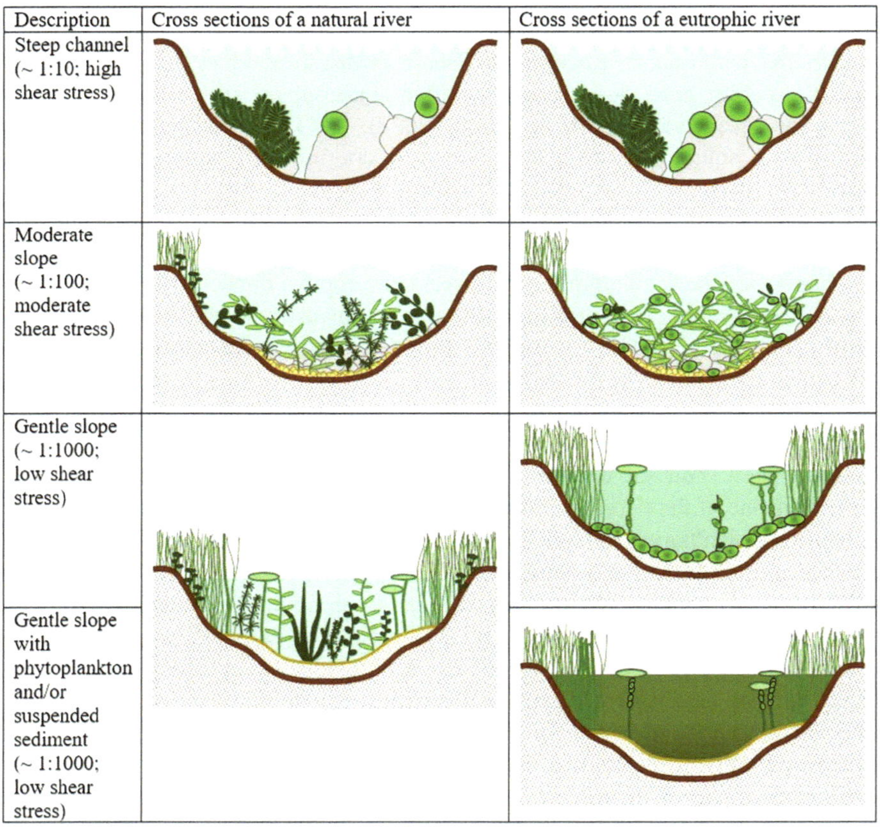

Fig. 2.4 A representation of river cross sections for macrophytes indicating natural and eutrophic conditions of different bed slopes. Adapted from O'Hare et al. (2018)

different with the physical habitat characteristics of an aquatic system, which determine the potential for different macrophyte morphotypes in the first instance.

Eutrophication has caused declines in submerged macrophytes due to the excessive growth of filamentous and periphytic algae (Phillips et al. 1978). In Northern Vosges Mountain streams in France, disturbed sites with high nutrient loading had low macrophytes' species richness due to high filamentous algae biomass (Thiébaut and Muller 1998). Highly sensitive macrophytes, for example, stoneworts, are mostly affected by eutrophication, and the presence of stoneworts is generally indicative of a pristine or slightly disturbed habitat (Van den Berg et al. 1999). Eutrophication seriously threatens stoneworts, which is reflected by their establishment in the shallow regions of the littoral zones migrating from the deeper regions, followed by rapid declines and extinction (Van den Berg et al. 1999; Kolada 2021) and hence the widespread use of stoneworts as a bioindicator in ecological assessments in many countries (Poikane et al. 2018; Kolada 2021). A survey of lakes and reservoirs in Poland revealed that stoneworts were present in systems that had significantly low N, P and chlorophyll-*a* concentrations (Kolada 2021). Under eutrophic conditions, other macrophytes proliferate, for example, the enormous mats of *P. crassipes* in Lake Chivero (Zimbabwe), Hartbeespoort reservoir (South Africa) and Lake Victoria (Uganda), which further strains the native macrophyte community.

2.4 Management

This section mainly focuses on the management of notorious invaders, which have established and perform well in disturbed aquatic ecosystems. The most effective strategy in IAS management is preventing their initial arrival. It is, however, very difficult to prevent the introduction and spread of invasive macrophytes mainly due to the connectivity of aquatic ecosystems. In the event of failures to prevent the invasive macrophytes from arriving, early detection programmes and rapid response should be implemented to contain the invasion (Robertson et al. 2020). For the issue of invasive macrophytes, their complete eradication is virtually impossible because of their ecology. Macrophytes are negatively perceived as aquatic weeds especially during the periods of nuisance growth when they form dense masses, which interfere with various human activities (Thiemer et al. 2021). Macrophytes removal is desired to prevent flooding of adjacent land, prevent clogging of hydropower electric plants, facilitation of irrigation, trade and commerce, disease control, recreational activities such as swimming, angling, boating and water skiing. Management options to control nuisance macrophytes include mechanical (cutting, dredging), biological (biocontrol agents such as herbivorous insects and fish or shading) and chemical (herbicide, salt) control (Madsen 1997; Thiemer et al. 2021). When evaluating the available management techniques, an assumption is erroneously made that not doing anything is environmentally neutral (Madsen 1997). The environmental repercussions of not taking action against invasive macrophyte species such as Eurasian watermilfoil, giant *Salvinia* and *Hydrilla*, may be very high. Unmanaged, these invasive macrophyte species can have deleterious effects on native macrophytes, water quality and the abundance and diversity of fish and invertebrates (Madsen 1997).

2.4.1 Mechanical Control

Mechanical control methods involve manually removing macrophytes by hand pulling, raking with weed rakes or motor-driven harvesters with underwater cutting blades; deep sediment dredging; nutrient removal; and water-level drawdown. Mechanical control methods have been implemented in many countries in the world. In Collins Lake, USA, biomass of *P. crispus* remained significantly lower 10 years after dredging than pre-dredging levels (Tobiessen et al. 1992). Manual control successfully controlled approximately 1500 ha of *Salvinia molesta* that had invaded an Indian hydroelectric reservoir (Cook 1976). On Lake Chivero, Zimbabwe, ~500 t of *P. crassipes* was removed per day in the early 1980s, which slowed the regeneration of *P. crassipes* (Chikwenhere and Phiri 1999). This management strategy was, however, discontinued due to lack of resources, and the lake is presently infested by the invasive macrophyte.

Despite the cost as well as the limited applicability of mechanically harvesting *P. crassipes*, it is at most times the only available solution for their control in European freshwater systems. For example, approximately 8 km (560,000 m^2) of the Mare 'e Foghe River (central eastern Sardinia, Italy) was covered by a dense mat of *P. crassipes*, where mechanical control was implemented. By December 2010, ~6700 t of plant biomass had been removed, costing €175,000. A 25,000 m^2 stretch of the Mare 'e Foghe River (about 400–500 m long) invaded by *P. crassipes* remained, and an additional €500,000 was set aside for operations to continue up to 2013 (Brundu et al. 2012).

To successfully manage macrophytes and keep their densities manageable, weed harvesters must be operated a number of times during the season of nuisance growth to effectively cut the vegetation. When mechanically controlling macrophytes, the cut vegetation should be dumped where it cannot re-enter the water as fragments left to float can produce new macrophytes. Since most macrophytes are perennial, they have underground rhizomes and portions that can re-sprout new shoots; therefore, it is essential to harvest the below-ground growth (rhizomes and roots) for an effective mechanical control. With larger plants such as *Typha* species, this can be difficult.

Mechanical control methods are labour intensive and costly and have short-term effects due to the regrowth of macrophytes. A manual removal effort to control *S. molesta* in lower Colorado River, USA, failed because of dense masses of the macrophyte, which could not be entirely removed in combination with its extremely rapid regrowth. Some macrophytes, for example, *Elodea* sp., appear to resist cutting, and macrophyte survival is guaranteed in the long term (Baattrup-Pedersen et al. 2002). Cutting *Elodea* sp. produces and spreads its fragments with a high potential for regeneration, and its residual fragments tend to form several lateral branches and sprouts after being cut (Mielecki and Pieczynska 2005). In the cut regions, light penetration into the water column promotes fast regrowth (Baattrup-Pedersen et al. 2002). Mechanical harvesting is also non-selective and affects desirable native macrophytes in the treatment area, as well as other organisms such as amphibians, insects, macroinvertebrates and reptiles that can also become trapped by harvesters (Haller et al. 1980; Engel 1990; Booms 1999). Mechanical removal requires periodic repetitions to control macrophytes to maintain them at acceptable levels (Murphy 1988).

2.4.2 Chemical Control

Herbicidal chemicals are used to manage both terrestrial and aquatic plants, except in Europe where chemical control in aquatic ecosystems is banned. Herbicide application is relatively an easy method and, in some situations, may be the only practical control method. Common herbicides include glyphosate (Rodeo, Pondmaster), fluridone (Sonar), endothall (Aquathol, Hydrothol), 2,4-D, chelated copper compounds and diquat, which are registered with the US Environmental Protection Agency (EPA) and, when used as recommended and directed, pose no significant threat to both public health and the environment. Herbicides vary in their effectiveness, toxicity and aquatic use restrictions (Ortiz et al. 2020). Herbicide selection depends largely on the macrophyte to be treated, for example, submersed and floating macrophytes are often treated with fluridone and diquat; emergent macrophytes are effectively treated with glyphosate. Controlling nuisance macrophytes using herbicides can range in scale from backpack sprayers to large-scale treatments, which target the entire aquatic ecosystems using aircraft and boats. To successfully control submersed macrophytes and some emergent and floating macrophytes not normally treated via the foliar application, herbicides are applied directly to the water around target plants (Ortiz et al. 2020). The chemical must also remain at prescribed concentrations in the water column, and macrophytes must be exposed to the chemical for a certain period of time (ranging from hours to months) (Gettys et al. 2014).

Herbicides have been used to control *P. crassipes* in various countries in the world. The Zambezi River Authority (ZRA) aerial sprayed 2,4-D at the rate of 6 L/ha in Lake Kariba in 1998 and successfully controlled and reduced *P. crassipes* mats (ZRA 1999). Water and fish samples were taken before, during and after spraying to determine any detrimental effects, and none were detected (ZRA 1999). In the Burdekin floodplain complex, Australia, four aerial spray applications of glyphosate (mixture ratio of 10 mg/L) were completed to target *P. crassipes* ($<$10 m above the water surface) between 2013 and 2015 (Waltham and Fixler 2017). In Canada, diquat is the only herbicide registered for aquatic use under a permit to directly control macrophytes although other herbicides are occasionally permitted under an emergency registration from the Pest Management Regulatory Agency (PMRA) of Canada (Breckels and Kilgour 2018). In 2016, 500 ha of *Phragmites* sp. in Lake Erie was treated with Aquasurf and glyphosate. Imazapyr and the surfactant Ag-Surf II have been used in 2013 to control *Spartina* sp. in tidal mudflats in the British

Columbia coast and were approved in 2016 in Lake Isle, Canada, to control *Butomus umbellatus* (Breckels and Kilgour 2018). Spraying of diquat herbicide on weed mechanical harvesters before operation was done to reduce the risk of transferring macrophyte fragments from one aquatic ecosystem to the other as reported in macrophyte control protocols in New Zealand (Howard-Williams et al. 1996).

Fish kills may occur after herbicides have been applied, even when the chemical used is not prescribed to be directly harmful and toxic to fish. Suffocation can cause fish to die, rather than from direct chemical poisoning as masses of decomposing macrophytes killed by the chemical create anoxic conditions. When chemicals are used to control macrophytes, one half (or even less) of the aquatic system should be treated at a time to give allowance for the fish to freely move to oxygen-rich areas without chemical treatment. Herbicides should also be applied during the season when the water temperatures are cooler.

Herbicide control of invasive macrophytes is often a limitation in developing countries, as it is costly and requires highly skilled personnel, and often the herbicides are perceived as poisons. For example, spraying *P. crassipes* with 2,4-D in Lake Chivero, Zimbabwe, in the early 1970s generated an outcry and claims of increased abortions in women and deliveries of deformed babies (Phiri et al. 2000). This shows the extreme social problems associated with *P. crassipes*. Furthermore, in developing countries, the water in most invaded sites is used for drinking, washing and fishing, and therefore, the use of herbicides to control nuisance macrophytes contaminates these sites, thereby threatening human health (Phiri et al. 2000).

2.4.3 Biological Control

Introducing organisms that compete with or eat nuisance macrophytes represents another control method. Herbivorous animals (examples discussed in the herbivory chapter) can be stocked into an aquatic ecosystem to consume the nuisance macrophytes. Among fishes, grass carp *Ctenopharyngodon idella* is the most researched and used organism in biological control of macrophytes (Silva et al. 2014). This fish was introduced in Europe, America and Africa to control macrophytes and for fish production through polyculture (Silva et al. 2014). High numbers of *Carassius* sp. were introduced to control Nuttall's waterweed *Elodea nuttallii*, which had been dominating in Lake Zwemlust, Netherlands (Van Donk and Otte 1996). The weevils *Neochetina eichhorniae* and *N. bruchi* and the water hyacinth borer *Niphograpta albiguttalis* are all effective biological control agents on *P. crassipes*, particularly in tropical regions of the world (Hill and Julien 2004). These insects feed on *P. crassipes* reducing their size, its vegetative propagation and seed production (Akers et al. 2017). In Lake Guiers, Senegal, green mats of the water lettuce *Pistia stratiotes*, which persisted prior to the release of the control agent, *Neohydronomus affinis*, turned dark brown, died and started to sink (Diop 2006). *Salvinia molesta* has been effectively controlled by the salvinia weevil, *Cyrtobagous salviniae*, in many tropical and subtropical countries, including Congo,

South Africa, Senegal, Sri Lanka, the Philippines, Australia, Papua New Guinea and Namibia (Coetzee and Hill 2020).

Biological control of tropical macrophytes in more temperate regions is often limited due to climate incompatibility between the control agents and the country of introduction. In South Africa, seven arthropods and one pathogen species have been released against *P. crassipes*, and while good control was achieved in some subtropical areas, success was limited in areas with cold winter temperatures (Hill and Coetzee 2017). Recently, the planthopper *Megamelus scutellaris* was released in 2013 to control *P. crassipes* (Hill and Coetzee 2017). Through an inundative approach to biological control, which relies on frequent releases of mass-reared planthoppers, *M. scutellaris* has established and is affecting *P. crassipes* even in the cooler regions of South Africa where the other biological agents have traditionally struggled to establish and have noticeable effects (Coetzee et al. 2022). Globally, other organisms used to control nuisance macrophytes include *Eccritotarsus catarinesis* for *P. crassipes*; *Piaractus mesopotamicus* for *C. demersum*, *Egeria najas* and *E. densa*; *Mycoleptodiscus terrestris* for *Hydrilla verticillata*; and *Agasicles hygrophila* for *Alternanthera sessilis* and *A. philoxeroides* (Silva et al. 2014).

Although the introduction of a biological organism can efficiently diminish nuisance macrophytes, a multitude of negative ecological impacts can manifest if the science behind releasing host-specific control agents is not followed. There are reports of the polyphagous *Ctenopharyngodon idella* destroying local macrophytes and causing irreversible changes to the ecosystem (Zhao et al. 2020). Macrophytes covering more than 80% of Lake Donghu area in China (~30 km^2) in the 1960s experienced a dramatic decline in the 1970s after *C. idella* introduction and then almost disappeared in 1979. During the same period, algal abundance in the lake increased substantially, and from the 1980s, algal blooms occurred annually (Zhao et al. 2020). The release of biocontrol agents therefore requires intensive research work and a rigorous risk analysis procedure prior to the release of an organism (Weyl et al. 2017).

Biological control is slow initially, which often leads to impatience and increased pressure to use herbicides to achieve quick control (van Wyk and van Wilgen 2002). Like other control methods, biological control will not completely eradicate nuisance macrophytes, and therefore, low infestation levels, with occasional outbreaks from time to time, will remain a feature of the aquatic ecosystems under biological control (Diop 2006). However, its relatively low cost offsets these disadvantages (Diop 2006).

After management programmes have been implemented, native macrophyte community restoration should be the end goal (Nichols 1991). Native macrophyte restoration is a biological control approach that aims to restore native macrophyte communities following the invasion in a disturbed aquatic ecosystem (Madsen 1997). In communities recently invaded by IAS, a propagule bank will restore the native macrophytes after management of the invasive macrophytes (Getsinger et al. 1997). For communities that have had monospecific invasive macrophyte dominance for quite a long period of time, native macrophytes may have to be reintroduced after successful management programmes (Madsen 1997). A healthy native macrophyte community is resilient and might slow invasion (or reinvasion) by invasive macrophytes and will provide the environmental as well as the habitat needs of an aquatic ecosystem. However, even the healthy and well-developed native macrophyte communities may eventually succumb to invasion by invasive macrophytes (Madsen 1997).

2.5 Conclusions

This chapter provided a discussion on the factors that structure macrophyte communities in freshwater ecosystems. Macrophyte communities are affected by biotic (herbivory, macrophyte properties, competition among the macrophyte species and pathogens and diseases) and abiotic (water chemistry including temperature, substrate composition/embeddedness and hydrological conditions) factors. The abiotic and biotic factors should be within ranges that do not alarmingly decimate or cause the undesirable proliferation of macrophytes. The global problem of invasive macrophyte species needs to be holistically handled at all levels to prevent impacts discussed in this chapter and associated costs on the ecosystems and to control them. Successful control of nuisance macrophytes will depend on implementation of a sound management plan (van Wyk and van Wilgen 2002). These plans need to be implemented using ecological principles focusing on the different approaches (or a combination) suitable for different regions (van Wyk and van Wilgen 2002). Environmental managers should ensure that approaches that will be used in combination are compatible (Ueckermann and Hill 2001).

Nuisance macrophytes do not recognise national boundaries, and control efforts mounted in one country or region may be thwarted when there is a steady influx of invasive macrophytes and propagules from neighbouring regions. Although most countries have regulations and laws to prevent unauthorised introductions of IAS, new macrophyte invasions, however, do occur regularly. Preventing the establishment and reestablishment of invasive macrophytes before they reach a problematic level should therefore be a standard operating procedure and an early and timely intervention, which serves as a cheaper option than to act when the macrophytes have formed dense masses. Management will need to integrate control efforts at the catchment scale, facilitating the integration of institutions and partnerships nationally, regionally and internationally to ensure the continuity in ownership of initiatives for the control of invasive macrophytes (van Wyk and van Wilgen 2002). Cooperation and coordination of efforts by various stakeholders including the public are required to ensure success.

References

Abdel-Tawwab M, El-Marakby HI, Ahmad MH (2006) Cannibalism in Oreochromis niloticus (Linnaeus): effect of stocking density, feed quality and submerged macrophytes. Indian J Fish 53(3):245–251

Akers RP, Bergmann RW, Pitcairn MJ (2017) Biological control of water hyacinth in California's Sacramento–San Joaquin River Delta: observations on establishment and spread. Biocontrol Sci Tech 27(6):755–768. https://doi.org/10.1080/09583157.2017.1342220

Alexander TJ, Vonlanthen P, Seehausen O (2017) Does eutrophication–driven evolution change aquatic ecosystems? Philos Trans R Soc B 72(1712):20160041

Anderson MG, Low JB (1976) Use of sago pondweed by waterfowl on the Delta Marsh, Manitoba. J Wildl Manag 40(2):233–242

Arts GH (2002) Deterioration of Atlantic soft water macrophyte communities by acidification, eutrophication and alkalinisation. Aquat Bot 73(4):373–393

Baattrup-Pedersen A, Larsen SE, Riis T (2002) Long-term effects of stream management on plant communities in two Danish lowland streams. Hydrobiologia 481:33–45

Bateman R (2001) IMPECCA: An International, Collaborative Program to Investigate the Development of a mycoherbicide for use Against Water hyacinth in Africa. In: Julien MH, Hill MP, Center TD, Jianqing D (eds) Biological and Integrated Control of Water Hyacinth, *Eichhornia crassipes.* Proceedings of the Second Meeting of the Global Working Group for the Biological and Integrated Control of Water Hyacinth, *Eichhornia crassipes*. Australian Centre for International Agricultural Research (ACIAR), Canberra, pp 57–61

Begon M, Harper JL, Townsend CR (1986) Ecology: individuals, populations and communities. Blackwells, Oxford

Biesbroek R, Wright SJ, Eguren SK, Bonotto A, Athanasiadis IN (2022) Policy attention to climate change impacts, adaptation and vulnerability: a global assessment of National Communications (1994–2019). Clim Pol 22(1):97–111

Blackburn TM, Essl F, Evans T, Hulme PE, Jeschke JM, Kühn I, Kumschick S, Marková Z, Mrugała A, Nentwig W, Pergl J, Pyšek P, Rabitsch W, Ricciardi A, Richardson DM, Sendek A, Vilá M, Wilson JRU, Winter M, Genovesi P, Bacher S (2014) A unified classification of alien species based on the magnitude of their environmental impacts. PLoS Biol 12(5):e1001850. https://doi.org/10.1371/journal.pbio.1001850

Booms TL (1999) Vertebrates removed by mechanical weed harvesting in Lake Keesus, Wisconsin. J Aquat Plant Manag 37:34–36

Bornette G, Puijalon S (2011) Response of aquatic plants to abiotic factors: a review. Aquat Sci 73: 1–14. https://doi.org/10.1007/s00027-010-0162-7

Breckels RD, Kilgour BW (2018) Aquatic herbicide applications for the control of aquatic plants in Canada: effects to nontarget aquatic organisms. Environ Rev 26(3):333–338. https://doi.org/10.1139/er-2018-0002

Brundu G, Stinca A, Angius L, Bonanomi G, Celesti-Grapow L, D'Auria G et al (2012) *Pistia stratiotes* L. and *Eichhornia crassipes* (M art.) Solms.: emerging invasive alien hydrophytes in Campania and Sardinia (Italy). EPPO Bulletin 42(3):568–579

Casanova MT, Brock MA (2000) How do depth, duration and frequency of flooding influence the establishment of wetland plant communities? Plant Ecol 147:237–250

Chambers PA, Prepas EE, Bothwell ML, Hamilton HR (1989) Roots versus shoots in nutrient uptake by aquatic macrophytes in flowing waters. Can J Fish Aquat Sci 46(3):435–439

Chambers PA, Lacoul P, Murphy KJ, Thomaz SM (2008) Global diversity of aquatic macrophytes in freshwater. In: Balian EV, Lévêque C, Segers H, Martens K (eds) Freshwater animal diversity assessment. Springer, Cham, pp 9–26. https://doi.org/10.1007/978-1-4020-8259-7_2

Charudattan R (1996) Pathogens for biological control of water hyacinth. In: Charudattan R, Labrada R, Center TD, Kelly-Begazo C (eds) Strategies for water hyacinth control. Food and Agriculture Organization, Rome, pp 189–196

Chifamba PC (1990) Preference of *Tilapia rendalli* (Boulenger) for some species of aquatic plants. J Fish Biol 36(5):701–705. https://doi.org/10.1111/j.1095-8649.1990.tb04324.x
Chikwenhere GP, Phiri G (1999, November) History of water hyacinth and its control efforts on Lake Chivero in Zimbabwe. In: Proceedings of the First IOBC Global Working Group Meeting for the Biological and Integrated Control of Water Hyacinth. Weeds Research Division, Agricultural Research Council, South Africa. ARC, Pretoria
Choi J-Y, Kim S-K (2020) Effects of aquatic macrophytes on spatial distribution and feeding habits of exotic fish species *Lepomis macrochirus* and *Micropterus salmoides* in shallow reservoirs in South Korea. Sustainability 12(4):1–14. https://doi.org/10.3390/su12041447
Choi C, Bareiss C, Walenciak O, Gross EM (2002) Impact of polyphenols on growth of the aquatic herbivore *Acentria ephemerella*. J Chem Ecol 28(11):2245–2256. https://doi.org/10.1023/a:1021049332410
Coetzee JA, Hill MP (2020) *Salvinia molesta* D. Mitch. (Salviniaceae): impact and control. CAB Reviews 15(033):1–11
Coetzee JA, Jones RW, Hill MP (2014) Water hyacinth, *Eichhornia crassipes* (Pontederiaceae), reduces benthic macroinvertebrate diversity in a protected subtropical lake in South Africa. Biodivers Conserv 23(5):1319–1330
Coetzee JA, Bownes A, Martin GD, Miller BE, Smith R, Weyl PSR, Hill MP (2021) A review of the biocontrol programmes against aquatic weeds in South Africa. African Entomol 29(3): 935–964
Coetzee JA, Miller BM, Kinsler D, Sebola K, Hill MP (2022) It's a numbers game: Inundative biological control of water hyacinth (*Pontederia crassipes*) yields success at a high elevation, hypertrophic impoundment in South Africa. Biocontrol Sci Tech 32:1302
Cook CDK (1976) 28. Salvinia in Kerala, S. India and its control. In: Proceedings of a Regional Seminar on Noxious Aquatic Weeds in South East Asia, New Dehli. 12–17 December 1973, p 241
Cronin G, Lodge DM, Hay ME, Miller M, Hill AM, Horvath T, Bolser RC, Lindquist N, Wahl M (2002) Crayfish feeding preferences for freshwater macrophytes: the influence of plant structure and chemistry. J Crustac Biol 22(4):708–718. https://doi.org/10.1163/20021975-99990285
Dalu T, Nhiwatiwa T, Clegg B (2012) Aquatic macrophytes in a tropical African reservoir: diversity, communities and the impact of reservoir–level fluctuations. Transactions of the Royal Society of South Africa 67(3):117–125
Dalu T, Mwedzi T, Wasserman RJ, Madzivanzira TC, Nhiwatiwa T, Cuthbert RN (2022) Land use effects on water quality, habitat, and macroinvertebrate and diatom communities in African highland streams. Sci Total Environ 846:157346
Dar NA, Pandit AK, Ganai BA (2014) Factors affecting the distribution patterns of aquatic macrophytes. Limnol Rev 14(2):75–81
Datta A, Maharaj S, Prabhu GN, Bhowmik D, Marino A, Akbari V, Rupavatharam S, Sujeetha JAR, Anantrao GG, Poduvattil VK, Kumar S (2021) Monitoring the spread of water hyacinth (*Pontederia crassipes*): challenges and future developments. Front Ecol Evol 9:631338
Diop I (2006) Management of invasive aquatic weeds with emphasis on biological control in Senegal. PhD Thesis. Rhodes University, Grahamston, South Africa
Dorenbosch M, Bakker ES (2011) Herbivory in omnivorous fishes: effect of plant secondary metabolites and prey stoichiometry. Freshw Biol 56(9):1783–1797. https://doi.org/10.1111/j.1365-2427.2011.02618.x
Dudgeon D, Arthington AH, Gessner MO, Kawabata ZI, Knowler DJ, Lévêque C, Naiman RJ, Prieur-Richard AH, Soto D, Stiassny ML, Sullivan CA (2006) Freshwater biodiversity: importance, threats, status and conservation challenges. Biol Rev 81(2):163–182
Engel S (1990) Ecological impacts of harvesting macrophytes in Halverson Lake, Wisconsin. J Aquat Plant Manag 28:41–45
Engloner A, Szalma E, Sipos K, Dinka M (2013) Occurrence and habitat preference of aquatic macrophytes in a large river channel. Community Ecol 14(2):243–248

Filho DN, Lima S, do Nascimento Moura A (2021) Strong top–down effects of omnivorous fish and macroinvertebrates on periphytic algae and macrophytes in a tropical reservoir. Aquat Ecol 55(2):667–680

Francová K, Šumberová K, Janauer GA, Adámek Z (2019) Effects of fish farming on macrophytes in temperate carp ponds. Aquac Int 27(2):413–436

Gerbersdorf SU, Jancke T, Westrich B (2007) Sediment properties for assessing the erosion risk of contaminated riverine sites. An approach to evaluate sediment properties and their covariance patterns over depth in relation to erosion resistance. First investigations in natural sediments. J Soils Sediments 7(1):25–35. https://doi.org/10.1065/jss2006.11.190

Germ M, Janež V, Gaberš̌cik A, Zelnik I (2021) Diversity of macrophytes and environmental assessment of the Ljubljanica River (Slovenia). Diversity 13:278. https://doi.org/10.3390/d13060278

Getsinger KD, Turner EG, Madsen JD, Netherland MD (1997) Restoring native vegetation in a Eurasian water milfoil-dominated plant community using the herbicide triclopyr. Regul Rivers Res Manag 13:357–375

Gettys LA, Haller WT, Bellaud MD (2014) Biology and control of aquatic plants: a best management practices handbook. A best management practices handbook, 3rd edn. Aquatic Ecosystem Restoration Foundation, Marietta, GA

Ghirardi N, Bresciani M, Free G, Pinardi M, Bolpagni R, Giardino C (2022) Evaluation of Macrophyte community dynamics (2015–2020) in southern Lake Garda (Italy) from sentinel–2 data. Appl Sci 12(5):2693

Gillard M, Thiébaut G, Deleu C, Leroy B (2017) Present and future distribution of three aquatic plants taxa across the world: decrease in native and increase in invasive ranges. Biol Invasions 19:2159–2170

Gopal B, Goel U (1993) Competition and allelopathy in aquatic plant communities. Bot Rev 59: 155–210. https://doi.org/10.1007/BF02856599

Gross EM (2003) Allelopathy of aquatic autotrophs. Crit Rev Plant Sci 22:313–339. https://doi.org/10.1080/713610859

Gross EM, Hilt S, Lombardo P, Mulderij G (2007) Searching for allelopathic effects of submerged macrophytes on phytoplankton—state of the art and open questions. Hydrobiologia 584(1): 77–88. https://doi.org/10.1007/s10750-007-0591-z

Grutters BMC, Gross EM, Bakker ES (2016) Insect herbivory on native and exotic aquatic plants: phosphorus and nitrogen drive insect growth and nutrient release. Hydrobiologia 778:209–220. https://doi.org/10.1007/s10750-015-2448-1

Haller WT, Shireman JV, Durant DF (1980) Fish harvest resulting from mechanical control of hydrilla. Trans Am Fish Soc 109(5):517–520

Haroon AM (2022) Review on aquatic macrophytes in Lake Manzala, Egypt. Egypt J Aquat Res 48: 1–12. https://doi.org/10.1016/j.ejar.2022.02.002

Haroon AM, Abd Ellah R (2021) Variability response of aquatic macrophytes in inland lakes: a case study of Lake Nasser. Egypt J Aquat Res 47:245–252. https://doi.org/10.1016/j.ejar.2021.07.004

Harper JL (1977) Population biology of plants. Academic, New York

Hartmann DL, Klein Tank AMG, Rusticucci M, Alexander LV, Brönnimann S, Charabi Y, Dentener FJ, Dlugokencky EJ, Easterling DR, Kaplan A, Soden BJ, Thorne PW, Wild M, Zhai PM (2013) Observations: atmosphere and surface. In Climate change 2013: the physical science basis. In: Stocker TF, Qin D, Plattner GK, Tignor M, Allen SK, Boschung J, Nauels A, Xia Y, Bex V, Midgley PM (eds) Contribution of working group I to the fifth assessment report of the intergovernmental panel on climate Change. Cambridge University Press, Cambridge

Harvey RM, Pickett JR, Bates RD (1987) Environmental factors controlling the growth and distribution of submersed aquatic macrophytes in two South Carolina reservoirs. Lake Reserv Manag 3(1):243–255. https://doi.org/10.1080/07438148709354780

Havens KE, Sharfstein B, Brady MA, East TL, Harwell MC, Maki RP, Rodusky AJ (2005) Recovery of submerged plants from high water stress in a large subtropical lake in Florida, USA. Aquat Bot 78:67–82. https://doi.org/10.1007/s10750-004-4876-1
Hill MP, Coetzee J (2017) The biological control of aquatic weeds in South Africa: current status and future challenges. Bothalia 47(2):a2152. https://doi.org/10.4102/abc.v47i2.2152
Hill MP, Julien MH (2004) The transfer of appropriate technology; key to the successful biological control of five aquatic weeds in Africa. In: Cullen M, Briese DT, Kriticos DJ, Lonsdale WM, Morin L, Scott JK (eds) XI international symposium on biological control of weeds. CSIRO Entomology, Canberra, pp 370–374
Hill MP, Coetzee JA, Martin GD, Smith R, Strange EF (2020) Invasive alien aquatic plants in South African freshwater ecosystems. In: van Wilgen BW, Measey J, Richardson DM, Wilson JR, Zengeya TA (eds) Biological invasions in South Africa. Springer, Cham
Hilton J, O'Hare M, Bowes MJ, Jones JI (2006) How green is my river? A new paradigm of eutrophication in rivers. Sci Total Environ 365(1–3):66–83
Hintz WD, Schuler MS, Borrelli JJ, Eichler LW, Stoler AB, Moriarty VW, Ahrens EL, Boylen CW, Nierzwicki-Bauer S, Relyea R (2020) Concurrent improvement and deterioration of epilimnetic water quality in an oligotrophic lake over 37 years. Limnol Oceanogr 65(5):927–938. https://doi.org/10.1002/lno.11359
Houlahan JE, Findlay CS (2004) Estimating the 'critical'distance at which adjacent land-use degrades wetland water and sediment quality. Landsc Ecol 19(6):677–690
Howard-Williams C, Schwarz AM, Reid V (1996) Patterns of aquatic weed regrowth following mechanical harvesting in New Zealand hydro-lakes. Hydrobiologia 340:229–234
Howell AW, Richardson RJ (2019) Correlation of consumer grade hydroacoustic signature to submersed plant biomass. Aquat Bot 155:45–51
Hussner A, Hofstra D, Jahns P, Clayton J (2015) Response capacity to CO_2 depletion rather than temperature and light effects explain the growth success of three alien Hydrocharitaceae compared with native *Myriophyllum triphyllum* in New Zealand. Aquat Bot 120:205–211
Hussner A, Heidbüchel P, Coetzee J, Gross EM (2021) From introduction to nuisance growth: a review of traits of alien aquatic plants which contribute to their invasiveness. Hydrobiologia 848(9):2119–2151
IPCC (2018) Summary for policymakers. In: Masson-Delmotte V, Zhai P, Pörtner H-O, Roberts D, Skea J, Shukla PR, Pirani A, Moufouma-Okia W, Péan C, Pidcock R, Connors S, JBR M, Chen Y, Zhou X, Gomis MI, Lonnoy E, Maycock T, Tignor M, Waterfield T (eds) Global Warming of 1.5°C. An IPCC Special Report on the impacts of global warming of 1.5°C above pre-industrial levels and related global greenhouse gas emission pathways., in the context of strengthening the global response to the threat of climate change, sustainable development, and efforts to eradicate poverty. World Meteorological Organization, Geneva, Switzerland
Jones PE, Consuegra S, Börger L, Jones J, Garcia de Leaniz C (2020) Impacts of artificial barriers on the connectivity and dispersal of vascular macrophytes in rivers: a critical review. Freshw Biol 65(6):1165–1180
Joye GF (1990) Biocontrol of Hydrilla verticillata with the endemic fungus Macrophomina phaseolina. Plant Dis 74(12):1035–1036
Kading J, Xu L (2021) Documenting macrophytes and their habitat preferences in southeastern South Dakota. Proc South Dakota Academy Sci 100:37–52
Kagami M, Nishihiro J, Yoshida T (2019) Ecological and limnological bases for management of overgrown macrophytes: introduction to a special feature. Limnology 20:1–2. https://doi.org/10.1007/s10201-018-0565-z
Kang CX, Li J, Liu Y, Tong Z (2022) Effects of high nitrogen concentration and low water level on the growth of the submerged macrophyte Vallisneria spinulosa. J Freshw Ecol 37(1):161–172
Karst-Riddoch TL, Pisaric MFJ, Smol JP (2005) Diatom responses to 20th century climate-related environmental changes in high-elevation mountain lakes of the northern Canadian cordillera. J Paleolimnol 33:265–282. https://doi.org/10.1007/s10933-004-5334-9

Kirschbaum MUF (2004) Direct and indirect climate change effects on photosynthesis and transpiration. Plant Biol 6(3):242–253. https://doi.org/10.1055/s-2004-820883

Kolada A (2021) Charophyte variation in sensitivity to eutrophication affects their potential for the trophic and ecological status indication. Knowl Manag Aquat Ecosys 422:30. https://doi.org/10.1051/kmae/2021030

Kõrs A, Vilbaste S, Käiro K, Pall P, Piirsoo K, Truu J, Viik M (2012) Temporal changes in the composition of macrophyte communities and environmental factors governing the distribution of aquatic plants in an unregulated lowland river (Emajõgi, Estonia). Boreal Environ Res 17:460–472

Kosten S, Kamarainen AMY, Jeppesen E, van Nes EH, Peeters ET, Mazzeo N, Sass L, Hauxwell J, Hansel-Welch N, Lauridsen TL, Søndergaard M (2009) Climate-related differences in the dominance of submerged macrophytes in shallow lakes. Glob Chang Biol 15(10):2503–2517

Kriticos DJ, Brunel S (2016) Assessing and managing the current and future pest risk from water hyacinth, (*Eichhornia crassipes*), an invasive aquatic plant threatening the environment and water security. PloS one 11(8):e0120054

Kundzewicz ZW, Mata LJ, Arnell NW, Döll P, Jimenez B, Miller K, Oki T, Şen Z, Shiklomanov I (2008) The implications of projected climate change for freshwater resources and their management. Hydrol Sci J 53(1):3–10. https://doi.org/10.1623/hysj.53.1.3

Lacoul P, Freedman B (2006) Environmental influences on aquatic plants in freshwater ecosystems. Environ Rev 14(2):89. https://doi.org/10.1139/a06-00

Ladislas S, El-Mufleh A, Gérente C, Chazarenc F, Andrès Y, Béchet B (2011) Potential of aquatic macrophytes as bioindicators of heavy metal pollution in urban stormwater runoff. Water Air Soil Pollut 223(2):877–888. https://doi.org/10.1007/s11270-011-0909-3

Lind L, Eckstein RL, Relyea RA (2022) Direct and indirect effects of climate change on distribution and community composition of macrophytes in lentic systems. Biol Rev 97:1677. https://doi.org/10.1111/brv.12858

Liu Y, He L, Hilt S, Wang R, Zhang H, Ge G (2021) Shallow lakes at risk: nutrient enrichment enhances top-down control of macrophytes by invasive herbivorous snails. Freshw Biol 66(3):436–446

Lodge DM, Lorman JG (1987) Reductions in submersed macrophyte biomass and species richness by the crayfish *Orconectes rusticus*. Can J Fish Aquat Sci 44(3):591–597. https://doi.org/10.1139/f87-072

Lodge DM, Krabbenhoft DP, Striegl RG (1989) A positive relationship between groundwater velocity and submersed macrophyte biomass in Sparkling Lake Wisconsin. Limnol Oceanogr 34(1):235–239

Lodge DM, Kershner MW, Aloi JE, Covich AP (1994) Effects of an omnivorous crayfish (*Orconectes rusticus*) on a freshwater littoral food web. Ecology 75(5):1265–1281. https://doi.org/10.2307/1937452

Lougheed VL, Crosbie B, Chow-Fraser P (2001) Primary determinants of macrophyte community structure in 62 marshes across the Great Lakes basin: latitude, land use, and water quality effects. Can J Fish Aquat Sci 58:1603–1612. https://doi.org/10.1139/cjfas-58-8-1603

Lowery RS, Mendes AJ (1977) Procambarus clarkii in Lake Naivasha, Kenya, and its effects on established and potential fisheries. Aquaculture 11(2):111–121

Lu J, Wang Z, Xing W, Liu G (2013) Effects of substrate and shading on the growth of two submerged macrophytes. Hydrobiologia 700:157–167. https://doi.org/10.1007/s10750-012-1227-5

Madsen JD (1997) Methods for management of nonindigenous aquatic plants. In: Luken JO, Thieret JW (eds) Assessment and Management of Plant Invasions. Springer, New York, pp 145–171

Madsen JD, Adams MS (1989) The distribution of submerged aquatic macrophyte biomass in a eutrophic stream, Badfish Creek—the effect of environment. Hydrobiologia 171:111–119. https://doi.org/10.1007/BF00008171

Madsen TV, Cedergreen N (2002) Sources of nutrients to rooted submerged macrophytes growing in a nutrient-rich stream. Freshw Biol 47:283–291. https://doi.org/10.1046/j.1365-2427.2002.00802.x

Madzivanzira TC, South J, Wood LE, Nunes AL, Weyl OLF (2020) A review of freshwater crayfish introductions in continental Africa. Rev Fisheries Sci Aquacult 29(2):218–241. https://doi.org/10.1080/23308249.2020.1802405

Madzivanzira TC, Weyl OLF, South J (2022) Ecological and potential socioeconomic impacts of two globally-invasive crayfish. Neo Biota 72:25–43. https://doi.org/10.3897/neobiota.72.71868

Magadza CHD, Madzivanzira TC, Chifamba PC (2020) Decline of zooplankton food resources of *Limnothrissa miodon* fishery in Lake Kariba: global warming induced ecosystem disruption by *Cylindrospermopsis raciborskii*. Lakes Reserv Sci Policy Manag 25(2):117–132. https://doi.org/10.1111/lre.12318

Maltchik L, Rolon AS, Schott P (2007) Effects of hydrological variation on the aquatic plant community in a floodplain palustrine wetland of southern Brazil. Limnology 8:23–28

Maredová N, Altman J, Kaštovský J (2021) The effects of macrophytes on the growth of bloom-forming cyanobacteria: systematic review and experiment. Sci Total Environ 792:148413

Marufu L, Dalu T, Phiri C, Barson M, Simango R, Utete B, Nhiwatiwa T (2018) The diet of an invasive crayfish, *Cherax quadricarinatus* (Von Martens, 1868), in Lake Kariba, inferred using stomach content and stable isotope analyses. BioInvasion Re 7(2):121–132. https://doi.org/10.3391/bir.2018.7.2.03

Masese FO, Fuss T, Bistarelli LT, Buchen-Tschiskale C, Singer G (2022) Large herbivorous wildlife and livestock differentially influence the relative importance of different sources of energy for riverine food webs. Sci Total Environ 828:154452

McKee D, Hatton K, Eaton JW, Atkinson D, Atherton A, Harvey I, Moss B (2002) Effects of simulated climate warming on macrophytes in freshwater microcosm communities. Aquat Bot 74(1):71–83. https://doi.org/10.1016/s0304-3770(02)00048-7

Meehl GA, Stocker F, Collins WD, Friedlingstein P, Gaye AT, Gregory JM, Kitoh A, Knutti R, Murphy JM, Noda A, Raper SCB, Watterson IG, Weaver AJ, Zhao ZC (2007) Global climate projections. Climate change 2007. In: Solomon S, Qin D, Manning M, Chen Z, Marquis M, Averyt KB, Tignor M, Miller HL (eds) The physical science basis. Contribution of working group I to the fourth assessment report of the intergovernmental panel on climate Change. Cambridge University Press, Cambridge, pp 747–846

Mielecki M, Pieczynska E (2005) The influence of fragmentation on the growth of Elodea canadensis Michx. in different light conditions. Pol J Ecol 2(53):45–53

Mjelde M, Faafeng BA (1997) *Ceratophyllum demersum* hampers phytoplankton development in some small Norwegian lakes over a wide range of phosphorus concentrations and geographical latitude. Freshw Biol 37(2):355–365. https://doi.org/10.1046/j.1365-2427.1997.00159.x

Montes C, Bravo-Utrera MÁ, Baltanás Á, Duarte C, Gutiérrez-Yurrita JP (1993) Bases ecológicas para la integral del cangrejo de rojo de las marismas del Parque Nacional de Donãna. ICONA, Madrid, Espanã

Mormul RP, Thomaz SM, Jeppesen E (2020) Do interactions between eutrophication and CO2 enrichment increase the potential of elodeid invasion in tropical lakes? Biol Invasions 22:2787–2795

Moss B (1998) Ecology of freshwaters, man and medium, past to future, 3rd edn. Blackwell Science, Oxford

Murphy KJ (1988) Aquatic weed problems and their management: a review. II. Physical control measures. Crop Prot 7(5):283–302

Murphy K, Efremov A, Davidson TA, Molina-Navarro E, Fidanza K, Betiol TCC, Chambers P, Grimaldo JT, Martins SV, Springuel I, Kennedy M, Mormul RP, Dibble E, Hofstra D, Lukács BA, Gebler D, Baastrup-Spohr L, Urrutia-Estrada J (2019) World distribution, diversity and endemism of aquatic macrophytes. Aquat Bot 158:103127. https://doi.org/10.1016/j.aquabot.2019.06.006

Netshiongolwe NR, Cuthbert RN, Maenetje MM, Chari LD, Motitsoe SN, Wasserman RJ, Munyai LF, Dalu T (2020) Quantifying metal contamination and potential uptake by *Phragmites australis* Adans. (Poaceae) along a subtropical river system. Plan 9(7):846

Netten JJ, Van Zuidam J, Kosten S, Peeters ET (2011) Differential response to climatic variation of freefloating and submerged macrophytes in ditches. Freshw Biol 56(9):1761–1768

Newman RM (1991) Herbivory and detritivory on freshwater macrophytes by invertebrates: a review. J N Am Benthol Soc 10(2):89–114. https://doi.org/10.2307/1467571

Nichols SA (1991) The interaction between biology and the management of aquatic macrophytes. Aquat Bot 41:225–252

Nuñez MA, Kuebbing S, Dimarco RD, Simberloff D (2012) Invasive species: to eat or not to eat, that is the question. Conserv Lett 5:334–341. https://doi.org/10.1111/j.1755-263X.2012.00250.x

O'Farrell I, Izaguirre I, Chaparro G, Unrein F, Sinistro R, Pizarro H, Rodríguez P, Tezanos Pinto P, Lombard R, Tell G (2011) Water level as the main driver of the alternation between a free-floating plant and a phytoplankton dominated state: a long-term study in a floodplain lake. Aquat Sci 73:275–287. https://doi.org/10.1007/s00027-010-0175-2

O'Hare MT, Baattrup-Pedersen A, Baumgarte I, Freeman A, Gunn IDM, Lázár AN, Sinclair R, Wade AJ, Bowes MJ (2018) Responses of aquatic plants to eutrophication in rivers: a revised conceptual model. Front Plant Sci 9:451. https://doi.org/10.3389/fpls.2018.00451

Ortiz MF, Nissen SJ, Thum R, Heilman MA, Dayan FE (2020) Current status and future prospects of herbicide for aquatic weed management. Outlook Pest Manag 31(6):270–275. https://doi.org/10.1564/v31_dec_07

Penuelas J, Sardans J, Llusià J, Owen SM, Carnicer J, Giambelluca TW, Rezende EL, Waite M, Niinemets Ü (2010) Faster returns on 'leaf economics' and different biogeochemical niche in invasive compared with native plant species. Glob Chang Biol 16:2171–2185. https://doi.org/10.1111/j.1365-2486.2009.02054.x

Phillips GL, Eminson D, Moss B (1978) A mechanism to account for macrophyte decline in progressively eutrophicated freshwaters. Aquat Bot 4:103–126

Phiri G, Navarro L, Bashir M, Cilliers C, Diop O, Fayad Y, Khattab A (2000) Water-hyacinth management capability in AME: key issues emerging from surveys and case studies. In: Navarro L, Phiri G (eds) Water hyacinth in Africa and the middle east. International Development Research Centre, Ottawa, pp 45–64

Piedade MTF, Parolin P, Junk W, Schöngart J, Wittmann F, Demarchi LO et al (2022) Vegetation. In: Dalu T, Wasserman RJ (eds) Fundamentals of tropical freshwater wetlands: from ecology to conservation management. Elsevier, Cambridge

Pieterse AH (1990) Introduction. In: Pieterse AH, Murphy KJ (eds) Aquatic weeds. Oxford University Press, Oxford, pp 3–16

Pinder A, Harman A, Bird C, Quinlan K, Angel F, Cowan M, Lewis L, Thillainath E (2019) Spread of the nonnative redclaw crayfish *Cherax quadricarinatus* (von Martens, 1868) into natural waters of the Pilbara region of Western Australia, with observations on potential adverse ecological effects. BioInvasion Rec 8(4):882–897. https://doi.org/10.3391/bir.2019.8.4.17

Poikane S, Portielje R, Denys L, Elferts D, Kelly M, Kolada A, Mäemets H, Phillips G, Søndergaard M, Willby N, van den Berg MS (2018) Macrophyte assessment in European lakes: diverse approaches but convergent views of 'good' ecological status. Ecol Indic 94:185–197. https://doi.org/10.1016/j.ecolind.2018.06.056

Portielje R, Roijackers RMM (1995) Primary succession of aquatic macrophytes in experimental ditches in relation to nutrient input. Aquat Bot 50:127–140

Prajapati M, van Bruggen JJ, Dalu T, Malla R (2017) Assessing the effectiveness of pollutant removal by macrophytes in a floating wetland for wastewater treatment. Appl Water Sci 7(8):4801–4809

Preiner S, Dai Y, Pucher M, Reitsema RE, Schoelynck J, Meire P, Hein T (2020) Effects of macrophytes on ecosystem metabolism and net nutrient uptake in a groundwater fed lowland river. Sci Total Environ 721:137620. https://doi.org/10.1016/j.scitotenv.2020.137620

Puche E, Rojo C, Ramos-Jiliberto R, Rodrigo MA (2020) Structure and vulnerability of the multi-interaction network in macrophyte-dominated lakes. Oikos 129(1):35–48

Rao Q, Su H, Deng X, Xia W, Wang L, Cui W, Ruan L, Chen J, Xie P (2020) Carbon, nitrogen, and phosphorus allocation strategy among organs in submerged macrophytes is altered by eutrophication. Front Plant Sci 11. https://doi.org/10.3389/fpls.2020.524450

Reid AJ, Carlson AK, Creed IF, Eliason EJ, Gell PA, Johnson PTJ, Kidd KA, MacCormack TJ, Olden JD, Ormerod SJ, Smol JP, Taylor WW, Tockner K, Vermaire JC, Dudgeon D, Cooke SJ (2019) Emerging threats and persistent conservation challenges for freshwater biodiversity. Biol Rev 94:849–873. https://doi.org/10.1111/brv.12480

Robertson PA, Mill A, Novoa A, Jeschke JM, Essl F, Gallardo B, Geist J, Jarić I, Lambin X, Musseau C, Pergl J, Pyšek P, Rabitsch W, von Schmalensee M, Shirley M, Strayer DL, Stefansson RA, Smith K, Booy O (2020) A proposed unified framework to describe the management of biological invasions. Biol Invasions 22:2633. https://doi.org/10.1007/s10530-020-02298-2

Rosenzweig C, Casassa G, Karoly DJ, Imeson A, Liu C, Menzel A, Rawlins S, Root TL, Seguin B, Tryjanowski P (2007) Assessment of observed changes and responses in natural and managed systems. In: Parry ML, Canziani OF, Palutikof JP, Van Der Linden PJ, Hanson CE (eds) Climate Change 2007: impacts, adaptation and vulnerability contribution of working group II to the fourth assessment report of the intergovernmental panel on climate Change. Cambridge University Press, Cambridge, pp 79–73

Saintilan N, Rogers K, Kelleway J, Ens E, Sloane D (2018) Climate change impacts on the coastal wetlands of Australia. Wetlands 39(5):1145–1154. https://doi.org/10.1007/s13157-018-1016-7

Sala OE, Chapin FS III, Armesto JJ, Berlow E, Bloomfield J, Dirzo R, Huber-Sanwald E, Huenneke LF, Jackson RB, Kinzig A, Leemans R, Lodge DM, Mooney HA, Oesterheld M, Poff NL, Sykes MT, Walker BH, Walker M, Wall DH (2000) Global biodiversity scenarios for the year 2100. Science 287(5459):1770–1774. https://doi.org/10.1126/science.287.5459.1770

Santamaría L (2002) Why are most aquatic plants widely distributed? Dispersal, clonal growth and small-scale heterogeneity in a stressful environment. Acta Oecol 23(3):137–154. https://doi.org/10.1016/s1146609x(02)01146-3

Scheffer M, van Nes EH (2007) Shallow lakes theory revisited: various alternative regimes driven by climate, nutrients, depth and lake size. Hydrobiologia 584:455–466. https://doi.org/10.1007/s10750-007-0616-7

Scheffer M, Hosper SH, Meijer ML, Moss B, Jeppesen E (1993) Alternative equilibria in shallow lakes. Trends Ecol Evol 8:275–279. https://doi.org/10.1016/0169-5347(93)90254-m

Schmid S, Wersal R, Fleming J (2021) Abiotic factors that affect the distribution of aquatic macrophytes in shallow lakes located in Sibley County, Minnesota, USA: a spatial modelling approach. Res Square. https://doi.org/10.21203/rs.3.rs-667792/v1

Shabana YM (2005) The use of oil emulsions for improving the efficacy of *Alternaria eichhorniae* as a mycoherbicide for waterhyacinth (*Eichhornia crassipes*). Biol Control 32(1):78–89

Shearer JF (1994) Potential role of plant pathogens in declines of submersed macrophytes. Lakes Reserv Manag 10(1):9–12

Shearer JF (2010) A historical perspective of pathogen biological control of aquatic plants. Weed Technol 24(2):202–207

Shields EC, Moore KA (2016) Effects of sediment and salinity on the growth and competitive abilities of three submersed macrophytes. Aquat Bot 132:24–29

Silva AF, Cruz C, Pitelli RLCM, Pitelli RA (2014) Use of grass carp (*Ctenopharyngodon idella*) as a biological control agent for submerged aquatic macrophytes. Planta Daninha 32(4):765–773. https://doi.org/10.1590/s0100-83582014000400011

Silveira MJ, Thiébaut G (2017) Impact of climate warming on plant growth varied according to the season. Limnologica 65:4–9. https://doi.org/10.1016/j.limno.2017.05.003

Simberloff D (2006) Invasional meltdown 6 years later: important phenomenon, unfortunate metaphor, or both? Ecol Lett 9(8):912–919. https://doi.org/10.1111/j.1461-0248.2006.00939.x

Smith MJ, Ough KM, Scroggie MP, Schreiber ESG, Kohout M (2009) Assessing changes in macrophyte assemblages with salinity in non-riverine wetlands: a Bayesian approach. Aquat Bot 90(2):137–142
Smolders AJ, Lucassen ECHET, Roelofs JG (2002) The isoetid environment: biogeochemistry and threats. Aquat Bot 73(4):325–350
Špoljar M, Perić MS, Wang H, Zhang C, Kuczyńska-Kippen N, Fressl J, Ercegovac Z (2021) Does the size structure of the littoral community reflect water level fluctuations in shallow waterbodies? Ecol Indic 132:108330
Swe T, Lombardo P, Ballot A, Thrane JE, Sample J, Eriksen TE, Mjelde M (2021) The importance of aquatic macrophytes in a eutrophic tropical shallow lake. Limnologica 90:125910. https://doi.org/10.1016/j.limno.2021.125910
Tait LW, Schiel DR (2013) Impacts of temperature on primary productivity and respiration in naturally structured macroalgal assemblages. PLoS One 8(9):e74413. https://doi.org/10.1371/journal.pone.0074413
Tan CO, Beklioglu M (2005) Catastrophic-like shifts in shallow Turkish lakes: a modeling approach. Ecol Model 183(4):425–434
Tattersdill K (2017) Exotic invaders in boreal lakes: Assessing impact on biodiversity and ecosystem functioning. PhD thesis, Swedish University of Agricultural Sciences, Uppsala
Thiébaut G, Muller S (1998) The impact of eutrophication on aquatic macrophyte diversity in weakly mineralized streams in the northern Vosges mountains (NE France). Biodivers Conserv 7:1051–1068. https://doi.org/10.1023/A:1008809131487
Thiébaut G, Thouvenot L, Rodríguez-Pérez H (2018) Allelopathic effect of the invasive *Ludwigia hexapetala* on growth of three macrophyte species. Front Plant Sci 9:1835. https://doi.org/10.3389/fpls.2018.01835
Thiemer K, Schneider SC, Demars BO (2021) Mechanical removal of macrophytes in freshwater ecosystems: Implications for ecosystem structure and function. Sci Total Environ 782:146671
Thomaz SM (2021) Ecosystem services provided by freshwater macrophytes, vol 850. Hydrobiologia, pp 1–21
Thorpe AS, Thele GC, Diaconu A, Callaway RM (2009) Root exudate is allelopathic in invaded community but not in native community: field evidence for the novel weapons hypothesis. J Ecol 97:641–645. https://doi.org/10.1007/s10886-011-0005-6
Thouvenot L, Haury J, Thiebaut G (2013) A success story: water primroses, aquatic plant pests. Aquat Conserv Mar Freshwat Ecosyst 23:790–803. https://doi.org/10.1002/aqc.2387
Tobiessen P, Swart J, Benjamin S (1992) Dredging to control curly-leaf pondweed: a decade later. J Aquat Plant Manag 30:71–72
Toivonen H, Huttunen P (1995) Aquatic macrophytes and ecological gradients in 57 small lakes in southern Finland. Aquat Bot 51(3–4):197–221
Ueckermann C, Hill MP (2001) Impact of herbicides used in water hyacinth control on naturalenemies released against the weed for biologicalcontrol. WRC report No. 915/1/01. Water ResearchCommission, Pretoria, South Africa
Vaithiyanathan P, Richardson CJ (1999) Macrophyte species changes in the Everglades: examination along an eutrophication gradient. J Environ Qual 28(4):1347–1358. https://doi.org/10.2134/jeq1999.00472425002800040040x
Van den Berg MS, Scheffer M, Van Nes E, Coops H (1999) Dynamics and stability of *Chara* sp. and *Potamogeton pectinatus* in a shallow lake changing in eutrophication level. Hydrobiologia 408:335–342. https://doi.org/10.1023/A:1017074211970
Van Donk E, Otte A (1996) Effects of grazing by fish and waterfowl on the biomass and species composition of submerged macrophytes. Hydrobiologia 340:285–290
van Wyk E, van Wilgen BW (2002) The cost of water hyacinth control in South Africa: a case study of three options. Afr J Aquat Sci 27:141–149
Wagner T, Falter CM (2002) Response of an aquatic macrophyte community to fluctuating water levels in an oligotrophic lake. Lake Reserv Manag 18(1):52–65. https://doi.org/10.1080/07438140209353929

Waltham NJ, Fixler S (2017) Aerial herbicide spray to control invasive water hyacinth (*Eichhornia crassipes*): water quality concerns fronting fish occupying a tropical floodplain wetland. Tropical Conserv Sci 10:1–10. https://doi.org/10.1177/1940082917741592

Weyl OLF, Nunes AL, Ellender BR, Weyl PSR, Chilala AC, Jacobs FG, Murray-Hudson M, Douthwaite RJ (2017) Why suggesting Australian redclaw crayfish *Cherax quadricarinatus* as biological control agents for snails is a bad idea. Afr J Aquat Sci 42(4):325–327. https://doi.org/10.2989/16085914.2017.1414685

Wijewardene L, Wu N, Fohrer N, Riis T (2022) Epiphytic biofilms in freshwater and interactions with macrophytes: current understanding and future directions. Aquat Bot 176:103467

Wu H, Hao B, Jo H, Cai Y (2021) Seasonality and species specificity of submerged macrophyte biomass in shallow lakes under the influence of climate warming and eutrophication. Front Plant Sci 12:678259. https://doi.org/10.3389/fpls.2021.678259

Zambezi River Authority (ZRA) (1999) Report on the Control of Water Hyacinth (Eichornia Crassipes) on Lake Kariba. Zambezi River Authority, Kariba

Zhang Y, Jeppesen E, Liu X, Qin B, Shi K, Zhou Y, Thomaz SM, Deng J (2017) Global loss of aquatic vegetation in lakes. Earth Sci Rev 173:259–265. https://doi.org/10.1016/j.earscirev.2017.08.013

Zhao Y, Zhang L, Wang C, Xie C (2020) Biology and ecology of grass carp in China: a review and synthesis. N Am J Fish Manag 40:1379–1399. https://doi.org/10.1002/nafm.10512

Zhao F, Fang X, Zhao Z, Chai X (2021) Effects of water level fluctuations on the growth characteristics and community succession of submerged macrophytes: A case study of Yilong Lake, China. Water 13(20):2900

Zhu G, Li W, Zhang M, Ni L, Wang S (2012) Adaptation of submerged macrophytes to both water depth and flood intensity as revealed by their mechanical resistance. Hydrobiologia 696(1): 77–93. https://doi.org/10.1007/s10750-012-1185-y

3 Aquatic Macrophytes: Ecology, Function, and Services in Niger Delta, Nigeria

H. E. Dienye

Abstract

Aquatic macrophytes are crucial within the inland waters of the Niger Delta as they enhance the surrounding, act as a dwelling region for minute water creatures, and make a contribution considerably to fisheries productivity. Aquatic macrophytes also provide substrates, meal, and habitat for aquatic animals, in addition to enhancing habitat physical structure and organic complexity, which will increase biodiversity in our water bodies. Macrophytes have crucial characteristics in our water bodies. Nonetheless, rather mild attention is paid to their protection, and if they are not properly managed, they can become out of control and cause issues. This article looks at the ecology, benefits, and drawbacks of common aquatic macrophytes in Nigeria's Niger Delta, in addition to great strategies for controlling the macrophytes in the inland waters of the Niger Delta. Managing aquatic macrophytes in this region is to accomplish stability in the environment by controlling extreme foray of plant species. A thorough assessment of the nature, scope, and potential of aquatic macrophyte problems is required before implementing control measures. Management actions in this region must raise awareness among the local population.

Keywords

Aquatic macrophyte · Management · Classification · Water bodies · Ecology

H. E. Dienye (✉)
Department of Fisheries, Faculty of Agriculture, University of Port Harcourt, Choba, Harcourt, Nigeria
e-mail: henry.dienye@uniport.edu.ng

S. Kumar et al. (eds.), *Aquatic Macrophytes: Ecology, Functions and Services*,
https://doi.org/10.1007/978-981-99-3822-3_3

3.1 Introduction

Inland water basins in Nigeria offer crucial habitat for a variety of flora and aquatic fauna, which support the areas they surround. However, activities (synthetic and native) that caused ecological concerns have recently damaged the connected ecological system, affecting the provided natural resources. Despite this, not much is established as regards aquatic bodies with plants and wildlife, stock-taking, socioeconomic, and conservation in Nigeria (Daddy et al. 1993). Nigeria's Niger Delta is among the significant deltas with natural resources such as crude oil, gas, animals, beneficial plants, and other resources abound in the region. It is the world's sixth largest producer and exporter of crude oil. It encompasses a diverse range of natural zones, including sand ridge barriers, brackish mangroves, and freshwater swamps (Udo 1987). The Niger Delta spans 20,000 km^2 and is surrounded by 70,000 km^2 of natural wetlands. The physically generated flood plains cover 7.5% of the 923,800 km^2 entire surface. This incredibly well-endowed ecological system harbors great biodiversity around the sphere. It also sustains abundant species, with additional species of freshwater organisms than any other ecological system in West Africa (Akinbode 2005; Vida 2010).

3.1.1 Aquatic Macrophyte Ecology

Macro-flora with roots that grow constantly or intermittently in aquatic environments is usually referred to as "aquatic macrophytes." They are a category of big, macroscopic photosynthetic organisms that grow in an aquatic environment (Jones et al. 2012). They are floras that grow in the presence of standing water that is at or above the soil's surface. Different water bodies and culture systems are examples of standing water. Plants having photosynthetic components that are always or occasionally underwater or detached in water and evident to the naked eye are known as macrophytes (Cook 1990). Macrophytes are significant parts of the brook environment since they boost the structure of habitats and increase biodiversity (Wetzel 2001; Pelicice et al. 2008). Furthermore, both animate and inanimate aquatic macrophytes can serve as food sources for other creatures (Lopes et al. 2007). They are important in the hydro-environment because they provide a spawning substratum for species such as fin fish, insects, and plankton, and they also help as fish diet (Ratusshnyale 2008). Excessive macrophyte growth might have detrimental consequences in most rivers and lakes (Bini et al. 2005).

Many individuals are unaware of the relevance of macrophytes in our aquatic environment. Macrophytes have an important function in water bodies and pastoral populations. Sadly, minute attention is given to their conservation. Aquatic macrophytes have not been given attention, and this is regrettable as fluctuations in macrophyte assemblage could be particularly predictive key urban stress classes. Water quality is thought to be influenced by the health and structure of macrophyte populations (Suren 2000; Balanson et al. 2005).

Despite the current focus on fisheries research and development in Nigerian waters, little attention has been paid to the non-fish resources that go with them (aquatic macrophytes). Aquatic plants, particularly in freshwater ecology, have a scarcity of information. The current trend of fully destroying these resources without first gaining a thorough understanding of their ecology, population dynamics, and socioeconomic significance could signal doom for other aquatic resources that rely on them. They offer recreational and medical value in a well-balanced environment. However, with a case study of the Niger Delta region and knowledge of the ecological characteristics and possible uses of these resources, better management, protection, and conservation of aquatic macrophytes in Nigerian water bodies would be required (Ita et al. 1985). This article looks at the ecology, species, distribution, and abundance of macrophytes in the Niger Delta region, Nigeria.

3.2 Macrophyte Taxonomy Groups

Macrophytes are a wide range collecion of taxonomic groups that come in a variety of forms and dimensions, in a particular, totally submerged, and others drift on the water's surface. Despite the fact that they are vital to our aquatic ecology, many still don't value them. The location of the plant in relation to the surface and substrate. Macrophytes are frequently divided into four categories: floating unattached, floating attached, submerged, and emergent (Puijalon et al. 2008). Aquatic macrophytes are aesthetically beautiful and environmentally beneficial when used in moderation. They are described as essential components of a river's aging process. Though they can be found in deep, clean lakes and rivers, their presence is not guaranteed. An abundance of aquatic macrophytes represents a sign of "middle" or "old" age. In large quantities, they can interfere positively or negatively with some water uses (Okaeme et al. 1999) (Fig. 3.1) (Table 3.1).

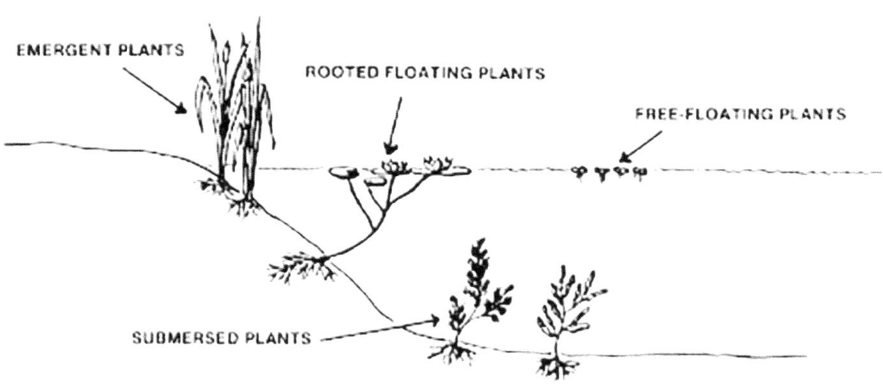

Fig. 3.1 Macrophyte categories based on their habitat of growth (Davidson et al. 2005)

Table 3.1 Taxonomy and zonation of aquatic flora in Niger Delta

Family	Species	Common name	Emergent	Floating
Araceae	*Pistia stratiotes*	Water lettuce	+	
	Lemna minor	Duckweed	+	
Asteraceae	*Aspilia africana*	Wild sunflower	+	
Athyriaceae	*Diplazium sammatti*		+	
Commelinaceae	*Aneilema beniniense*	Aneilema	+	
Cyperaceae	*Cyperus difformis*	Small flower umbrella	+	+
	Rhynchospora corymbosa	Matamat	+	
	Cyperus iria	Rice flat sedge	+	
Lamiaceae	*Platostoma africanum*	Asirisiri	+	
Nymphaeaceae	*Nymphaea lotus*	Water lily	+	
Onagraceae	*Ludwigia decurrens*	Water primrose	+	
Poaceae	*Sacciolepis africana*	Wild rice	+	
Pontederiaceae	*Eichhornia crassipes*	Water hyacinth	+	
Tiliaceae	*Triumfetta cordifolia*	Burweed		

Source: Dienye (2015)

3.3 Ecological Functions of Aquatic Macrophytes

Various forms of aquatic ecosystems rely heavily on macrophyte plants. Aquatic macrophytes are present in a variety of aquatic habitats, and their occurrence is of benefit to fisheries and pastoralism in the basins. Macrophytes are important for not just the biological community but also the natural processes, which take place in the aquatic environment. There are benefits to macrophytes' performance in an aquatic ecological system. The commonly found macrophytes in the Niger delta region is shown in Plates 3.1, 3.2, 3.3, 3.4, 3.5, 3.6, 3.7, 3.8, 3.9, 3.10, 3.11, 3.12.

3.3.1 Fisheries Production

Aquatic plants are regarded as undervalued components of the aquatic ecosystem. It does, however, play an important part in the fishing industry. Herbivorous fishes like *Tilapia zillii*, locally farmed, rely on aquatic plants as a food source. Some fishes consume *Lemna paucicostata* species (Mbagwu and Adeniji 1988). According to known information, 37 freshwater herbivorous fish species feed on macrophytes and belong to 24 families (Opuszynski and Shireman 1995). More specifically, periphytic algae that grow on the surface of aquatic plants serve as food for some fish species. Oreochromis eats coarser things, such as macrophytes, than other members of the genus (Ezeri et al. 2003). Macrophytes have been shown to provide breeding sites and refuge for fish. Carnivorous fish fingerlings feed on aquatic plants till their intestines mature to take animals, according to Agbogidi et al. (2000).

3.3.2 Habitat for Water Organisms

Smaller animals use aquatic macrophytes as a home. These little animals play an important role in ecology because they feed fish. Studies have shown that vegetated regions harbor fewer organisms than non-vegetated ones (Agbogidi et al. 2000). Many fishes find shelter, spawning substrates, and nursery sites under the leaves of *Ceratophyllum demersum* and *Myriophyllum spicatum.* Aquatic macrophytes provide cover for juvenile fish from predatory fish, making them a vital nursery for baby fish (juveniles). Water lettuce provides a safe haven for fish and crustaceans from predatory fish (ICAAE 1992). *Heterotis niloticus* makes its nest out of aquatic macrophytes, but *Gymnarchus niloticus* spawns in stagnant waters containing macrophytes and then migrates to flowing waters (Meske 1985). Freshwater and marine plants have an impact on animal and aquatic organism communities through a series of habitat-related mechanisms, such as providing nurseries, dwelling spaces, and feeding areas (Hyndes et al. 2018).

3.3.3 Healthy Ecosystem/Nutrient Cycling

In the aquatic environment, some macrophytes perform a dynamic function in a healthy ecosystem that produces oxygen via photosynthesis and provide a substrate and cover for various species. They also help stabilize the sediments together. This helps increase water clarity and reduce the volume of pollution released into the environment through sediment erosion (Kumar et al. 2020).

Agbogidi et al. (2000) reported that the sewage is channeled through macrophytes in order to assimilate and decrease the nutrient concentrations prior to discharge inside the water body. Macrophytes are exploited in bio-manipulation to boost fish culture (Dar et al. 2011). They absorb large amounts of nutrients as a means of removing nutrients from effluent (Uka et al. 2009). Macrophytes are used in phytoremediation procedures of polluted water bodies and in engineered structures known as "constructed swamps" for the treatment and decontamination of wastes (Vymazal 2013). They are also an indicator of water quality by absorbing excess nutrients (Petre 1990).

3.3.4 Source of Alternative Medicine

Traditional communities also use a variety of macrophytes in healing therapy. As a result, a significant portion of these ethnobotanical materials may yield molecules that could be employed as modern medicine and pharmaceuticals (Olayide 1981). *Polygonum senegalense* is mashed with soda ash and used for rheumatoid arthritis, according to Kio and Ola-Adams (1987). Water lettuce is also employed for treating "flu," according to Obot and Ayeni et al. (1999). According to Bubayero (1986), most Nigerians patronize traditional healers. Most of these macrophytes produce chemicals that are extremely promising for application in current medications and pharmaceuticals. The species fever in youngsters and is used as a dewormer and eye ointments. It's regularly used with clay in Ghana to prevent abortion. The root of *Ethulia conyzoides* is used to relieve constipation when blended with red pepper.

3.3.5 Industrial Uses

Aquatic macrophytes variety of resources that could be beneficial in industries, construction, matting, bedding, and pulp or paper. In Northern Nigeria, the dry root of *C. maculatus* is used for perfume, and the ripe silky inflorescences of *T. australis* are utilized in padding pillows (Ita 1993). The leaves of *C. asticulatus* have mosquito resistance, and their stems are utilized in multicolored mats (Kio and Ola-Adams 1990). *Vossia cuspidata*, *Cyperus papyrus*, and *Eichornia crassipes* have monetary value for pulp, paper, and fiber. *Raphia vinifera* is used as a raw material for brushes, brooms, and mats (Okojie 1995).

3.3.6 Sources of Energy

Aquatic flora as a source of energy, according to, Edewor (1998), is primarily used as a fuel for fish smoking and residential energy. Aquatic floras can become liquid, gaseous, or stable fuels through bio-methanation, fermentation, and pyrolysis, in line with reviews from different growing countries. *Eichornia crassipes* is digested without delay in China and India to make biogas, which is used to generate electricity to rural regions at a low cost. Stems of *Aeschynomene crassicaulis* and *Cyperus papyrus* are used as fuel for domestic cooking and fish smoking (Kio and Ola-Adams 1987).

3.4 Aquatic Macrophytes as Nuisance

Macrophytes produce an explosively excessive population, when the environment changes as a result of pollution. Aquatic macrophytes play an essential function in maintaining the richness and role of the aquatic environment. Several of these macrophytes can be unsafe when in abundance. When non-native species are purposely or by accident brought into places where they have no natural enemies to limit their growth, they are able to produce massive, uncontrollable populations. Plants developing in an aquatic environment can become dense (Chambers et al. 2008). The nuisance macrophytes cause:

3.4.1 Effect on Water Body

A floating mat of macrophyte vegetation can hinder sunlight from reaching the surface of water, which results in low natural food, eventually affecting fish production. The bloom of macrophyte vegetation causes enormous fish mortality due to the excessive oxygen requirement and contest for available nutrients. These invasive aquatic macrophytes have a negative effect on water condition and biodiversity (Uka et al. 2009). Submerged macrophytes degrade breeding grounds (particularly almonds). Dense macrophytes can cause a huge variation in oxygen, putting several fish species at risk. Similarly, when photosynthesis is lower than respiration, fish death may occur.

3.4.2 Hindrance to Navigation

Towering macrophytes above and submerged in water prevent entry, impede navigation, and damage hydroelectric infrastructure, and floating mats obstruct watercraft transportation routes. The lifestyle of a floating mat makes the aquatic surrounding insecure due to the hazard of craft, the penetration of massive predatory aquatic animals, and additional mechanical problems. It additionally has an effect on fish nets within the surrounding. Macrophytes halt boats through the means of winding round their propellers. Macrophyte mats, inclusive of water hyacinth, may even block a ship (Mandal 2007).

3.4.3 Habitat for Spread of Diseases

While certain aquatic macrophytes prevent disease-supporting organisms, others create the best surroundings for them. Most individual ailments are spread through transitional hosts, which are reliant on certain macrophytes for completion of their cycle. Blocked waterways as a result of floral vegetation or infestation of *Pistia stratiotes* harbor *schistosomiasis* (African sleeping sickness). An aquatic snail that dwells among flora serves as the intermediate host. Anything that brings this deoxygenated water to the surface (such as high wind) reduces the oxygen level in the water column, resulting in fish fatalities. As a result of this, even if the fish does not die, continuous low oxygen levels weaken the fish, and it grows to be extra susceptible to illnesses. The tranquil aquatic surrounding that macrophyte growth can create is optimal for mosquito larvae development (Bromilow 2010).

3.5 Interaction Between Macrophytes and Environmental Variables

This interplay is a considerable function of the aquatic surrounding that is vital for aquatic movement and ecological function (Xia et al. 2010). Though increase and spread are regular occurrences in water bodies, actions like agriculture, building, and development initiatives have increased concern about aquatic macrophytes and water quality in recent years (Wang et al. 2009). According to Dienye et al. (2017), the interplay confirmed that as pH and dissolved oxygen decreased through the wet season, the extent of macrophyte abundance increased while salinity increased. Dense macrophytes were slightly influenced by salinity in the Niger Delta vicinity. As the temperature rises, fewer species become abundant, while macrophytes in the vicinity decline in abundance. Chemical oxygen demand was negatively correlated with all species, and biological oxygen demand was positively correlated with all species of macrophytes. BOD increases due to dead organic matter, which supports macrophyte abundance. Therefore, pH, dissolved oxygen, and chemical oxygen demand affect the distribution and abundance in the region.

3.5.1 Methods of Managing Macrophytes

The number one goal of coping with flora in the Niger Delta is to create a balance in the ecological system through monitoring the intense invasion of various macrophyte species. The ways of controlling aquatic macrophytes include:

3.5.2 Precautionary Control

There are special ways by which macrophytes get into our waters: fishermen's nets, boats, ballast waters, wind, birds, and more. However, precautionary measures begin with either reduction or total eradication of the sources. Plants should never be rinsed into aquatic surrounding wherever they can develop and regrow.

3.5.3 Machine-Driven Control

It involves uprooting or raking the macrophytes out of the soil. Some aquatic floras are recurrent and possess roots that could resprout, and harvesting growth beneath is vital for efficient management. Mechanical weed harvesters with submerged blades are beneficial for larger bodies of water, according to McComas (1993). The principle of the operation these harvesters used is like that of lawn mowing. The macrophytes will not be eliminated but cannot get to the surface and cause problems. Crop the harvested flora and discard it appropriately, so it won't be re-introduced into the water while mechanically regulating the aquatic macrophytes. If left to float in a body of water, the harvested flora fragments can sprout new ones.

3.5.4 Biological Control

For biological management, lots of strange and natural organisms have been used. Beneficial organisms are used to prevent the spread of macrophytes in this approach. People could also rely on introducing animal or microbe that can feed on poisonous floras. Nevertheless, when the incorrect type of management is implemented, this process could have terrible effect for the environment (Gallagher and Haller 1990). Biological control measures are:

3.5.4.1 Water Plants (Macrophytes)

Introduction of certain desirable aquatic flora has the capacity to eradicate aquatic nuisance species. Native macrophytes, on the alternative, are generally desirable since they have more control with the surrounding ecosystem. Invasive species can effortlessly displace desirable macrophytes, and this approach can be tough, but when the invasive plants are eliminated, this method works well.

3.5.4.2 Herbivorous Fish Species

Herbivores can help to keep aquatic flora abundance under check; to help decrease aquatic flora, grass carp is adopted. They've been advanced genetically to stop breeding to consume the flora. When the surroundings are advantageous, the Chinese grass carp (*Ctenopharyngodon idella*) will eat up aquatic macrophytes, and this is temperature dependent. Their activities have minute result below 16 °C but attain a peak at 25 °C. Grass carp are choosy eaters, desiring soft flora over fibrous ones (Wells and Clayton 2005).

3.5.4.3 Microorganisms (Bacteria)

Bacteria and fungus are also used to regulate macrophyte richness. Those that live on different floras can be exploited to control the flora selectively. These macrophytes die due to contamination when the microorganisms are introduced, while the more needed plants will be spared.

3.5.5 Chemical Control

When rightly applied, herbicides overpower aquatic macrophyte plants without inflicting damage to fish or wildlife. In few instances, herbicides can be employed to manage definite floral species while sparing others. It can function as part of an aquatic plant control approach when treating few vegetative areas leaving others untreated. "Contact herbicides" kill the contact plant part. Translocated herbicides do no longer kill plant as swiftly but alternatively enter the plant itself. Generally, only the latter groups are efficient for decreasing perennial floral regrowth. They are separated as selective (killing definite plants) and non-selective (killing all plants).

Herbicide remedy can be expensive and may only afford temporary result from the fundamental problem, which is habitually enriched waterways. One should also note that when pesticides destroy aquatic flora, they decompose and discharge their stored nutrients into the aquatic surrounding, and it encourages successive growth of aquatic plants, which repeatedly necessitates additional medications. The following are examples of treatment:

3.5.6 Diquat Herbicide

This is a contact herbicide that is principally active at managing aquatic weeds and algae in a short period of time, and it's normally sprayed on aquatic plants. This treatment will make the treated flora to quickly die and turn brown. REWARD is a standard diquat herbicide logo that works effectively on floating macrophytes and is absolutely safe to use. When the water is muddy, diquat herbicide should no longer be used as soil debris absorb it in the water (Netherland et al. 1997).

3.5.7 Fluridone Herbicide

The herbicide fluid-like non-touch herbicide that is more gradual in method than diquat. Fluridone principally controls submerged flora, and its brand tag is SONAR. It works for 30–90 days. When it is applied, it demonstrates signs for 1–2 weeks. The plants lose its green color and change to white.

3.5.8 Glyphosate

Glyphosate is available only as liquid and it controls plants above water. It is not effective for submerged plants. AquaMaster, AquaPro and Rodeo are trade names. It is prohibited to use glyphosate chemicals not explicitly branded for aquatic use (Getsinger 1998).

3.6 Commonly Found Macrophytes in the Niger Delta Region, Nigeria

Plates 3.1, 3.2, 3.3, 3.4, 3.5, 3.6, 3.7, 3.8, 3.9, 3.10, 3.11, and 3.12

3.7 Conclusion and Recommendations

In spite of the opinion about aquatic plants causing a nuisance to the surrounding, they can be ecologically welcoming, when combined with mechanical method of control, which permits riparian communities to sustain dependable but long-term aquatic flora control at a reduced cost and with added economic benefits. Industrial activities in the Niger Delta vicinity should be thoroughly monitored since they have an effect on macrophytes, which are important for fish production. They offer substrates, food, and habitat for aquatic animals, in addition to enhanced habitat physical structure and biological complexity, which increases biodiversity in water bodies. The plants can only be managed and kept within tolerable limits if they are controlled properly, with some used on a long-term basis. Aquatic macrophyte utilization can only be successful on a long-term basis if their habitat is correctly

Plate 3.1 *Cyperus difformis* spp. (Source: Dienye 2014)

Plate 3.2 *Nymphaea lotus* spp. (Source: Dienye 2014)

handled, and this necessitates preservation of the environment. Management initiatives, on the other hand, should create awareness among the local populace. A thorough assessment of the nature, scope, and potential of aquatic macrophyte problems is required before implementing control measures.

Plate 3.3 *Sacciolepis africana* spp. (Source: (Dienye 2014)

Plate 3.4 *Rhynchospora corymbosa* spp. (Source: Dienye 2014)

Plate 3.5 *Aneilema beniniense* spp. (Source: Dienye 2014)

Plate 3.6 *Cyperus iria* spp. (Source: Dienye 2014)

Plate 3.7 *Platostoma africanum* spp. (Source: Dienye 2014)

Plate 3.8 *Aspilia africana* spp. (Source: Dienye 2014)

Plate 3.9 *Triumfetta cordifolia* spp. (Source: Dienye 2014)

Plate 3.10 *Eichhornia crassipes* spp. (Source: Dienye 2014)

Plate 3.11 *Ludwigia decurrens* spp. (Source: Dienye 2014)

Plate 3.12 *Diplazium sammatti* spp. (Source: Dienye 2014)

References

Agbogidi OM, Bamidele JE, Ekokotu PA, Olele NF (2000) The role and management of aquatic macrophytes in fisheries and aquaculture. Issues Anim sci 10:221–235

Akinbode A (2005) Introductory environmental resource management. Daybis Limited, Ibadan

Ayeni JSO, Obot EA and Daddy F (1999) Aspects of the biology, conservation and management of aquatic vascular plant resources of Nigerian wetland based on the Kainji Lake experience. Proceedings of a workshop on sustainable management and conservation of fisheries and other aquatic resources of Lake Chad and the arid zone of Nigeria, Jan 16-17, Maiduguri. pp 144–145

Balanson PC, Mac MJ, Smith SB (2005) Challenges in the development and use of ecological indicators. Eco ind 1:3–10

Bini LM, Oliveria LG, Souza DC, Cavalho P, Pinto MP (2005) Patterns of the aquatic macrophytes cover in *Cachoeira dourada* Reservoir *Brazil*. J Fish Biol 65(1):19–24

Bromilow C (2010) Problem plants and alien weeds of South Africa. Briza Publication, Pretoria, p 243

Bubayero AM (1986) Traditional medicine in the science of man. In: Sofowara A (ed) The state of medicinal plants research in Nigeria, pp 129–133

Chambers PA, Lacoul P, Murphy KJ, Thomaz SM (2008) Global diversity of aquatic macrophytes in freshwater. Hydrobiologia 198:9–26. https://doi.org/10.1007/978-1-4020-8259-7_2

Cook CDK (1990) Aquatic plant book. SBP Academic, The Hague, pp 110–112

Daddy F, Adesina GO, Bankole NO, Isah U and Owotunse S (1993) Flora and fauna resources associated with Kainji, Jebba, Shriroro and Wuya water bodies in. NIFFR annual report. pp 25–27

Dar SH, Kumawat DM, Singh N, Wani KA (2011) Sewage treatment potential of water hyacinth (*Eichhornia crassipes*). Res J Environmental Science 5:377–385

Davidson NC, D'Cruz R, Finlayson CM (2005) Ecosystems and human Well-being: wetlands and water synthesis: a report of the millennium ecosystem assessment. World Resources Institute, Washington, DC, p 68

Dienye HE (2014) Seasonal distribution and abundance of Macrophyte of the new Calabar River Niger Delta. M.Sc. thesis, Department of Animal And Environmental Biology, Faculty of Science, University of Port Harcourt) (Unpublished). p 112

Dienye HE (2015) Species diversity of Macrophytes of the new Calabar River, Niger Delta, Nigeria. Int J Fisheries Aquatic Studies 3(1):409–413

Dienye HE, Olopade OA, Woke GN (2017) The Relationship between aquatic Macrophytes and water quality in new Calabar River Niger Delta Nigeria. Int J Res Agric Forestr 4(10):1–6

Edewor JO (1998) Developing water hyacinth from menace status to national profitable level. The proceedings of the international workshop/seminar on water hyacinth. pp 15–17

Ezeri GNO, Gabriel, Ashade OO (2003) Effects of partial shading by water lettuce (*Pistia stratiotes*) on growth of tank cultured *Oreochromis niloticus*. Fortschr Zool 2:29–38

Gallagher JE, Haller WT (1990) History and development of aquatic weed control in the United States. Rev Weed Sci 5:115–192

Getsinger KD (1998) Chemical control research in the Corps of Engineers. J Aquat Plant Manage 36:61–64

Hyndes GA, Francour P, Guidetti P, Heck KL Jr, Jenkins G (2018) The roles of seagrasses in structuring associated fish assemblages and fisheries. In: Larkum AWD et al (eds) Seagrasses of Australia. Springer, New York, pp 589–627. https://doi.org/10.1007/978-3-319-71354-0_18

ICAAE (1992) Introduction to fish culture in ponds international Centre for aquaculture and aquatic environments. Auburn University, Alabama

Ita EO (1993) Inland fishery resources of Nigeria. CIFA occasional paper no. 20 CIFA/OP 20FAO of the UN Rome. pp 120

Ita EO, Sado EK, Balogun JK, Pandogari A and Ibitoye B (1985) Inventory survey of Nigeria inland waters and their fisheries resources in: preliminary checklist of inland water bodies in Nigeria with special reference to ponds, lakes, reservoir and major rivers. Kainji Lake research institute technical report series. pp 14–51

Jones JI, Collins AL, Naden PS, Sear DA (2012) The relationship between fine sediment and macrophytes in rivers. River Res Appl 28:1006–1018

Kio PRO, Ola-Adams BA (1987) Economic importance of aquatic macropytes. In: Iloba C (ed) Ecological implications in the development of water bodies in Nigeria. National Institute for Fresh water Fisheries Research, New Bussa, pp 112–157

Kio PRO, Ola-Adams BA (1990) Utilization and development of wetlands. In: Akpata TVA, Okali DUU (eds) Nigerian wetlands, pp 41–53

Kumar V, Kumar P, Singh J, Kumar P (2020) Potential of water fern (*Azolla pinnata* R. Br.) in phytoremediation of integrated industrial effluent of SIIDCUL, Haridwar, India: removal of physicochemical and heavy metal pollutants. Int J Phytoremediation 22(4):392–403

Lopes CA, Benedito-Cecilio E, Martinelli LA (2007) Variability in the carbon isotope signature of *Prochilodus lineatus* (*Prochilodontidae, Characiformes*) a bottom-feeding fish of the Neotropical region. J Fish Biol 70:1649–1659
Mandal RC (2007) Weed Weediade and weed control: principles and practice, 1st edn. Agrobios, Delhi, pp 128–154
Mbagwu IG, Adeniji HA (1988) The nutritional content of duckweed *Lemna paucicostata* (Hegelm) in the Kainji lake area, Nigeria. Aquat Bot 29:357–366
McComas S (1993) Lake smarts: the first Lake maintenance handbook. Terrene Institute, Washington DC
Meske C (1985) Fish aquaculture: technology and experiments. Pergamon press, New York, p 33
Netherland MD, Getsinger KD, Skogerboe JD (1997) Mesocosm evaluation of the species-selective potential of fluridone. J Aquat Plant Manage 35:41–50
Okaeme AN, Olatunde AA and Ayeni JOS (1999) Lake Chad and arid zones fisheries proceeding and conservation of fisheries and other aquatic resources of Lake Chad and arid zone of Nigeria. New Bussa NIFFR. pp 202
Okojie AO (1995) Strategy for sustainable development and utilization of aquatic and wetland resources in Nigeria. In Sustainable utilization of aquatic/wetland resources. $9^{th}/10^{th}$ Annual Conference of National Association of Aquatic Science (NAAS). pp 25–32
Olayide SO (1981) Scientific research and the Nigerian economy. Ibadan University Press, Ibadan, p 83
Opuszynski K, Shireman JV (1995) Herbivorous fishes: culture and use for weed management. CRC Press, Boca Raton
Pelicice FM, Thomas SM, Agostinhno AA (2008) Simple relationships to predict attributes of fish assemblages in patches of submerged macrophytes. Neotrop Khthyol 6:543–550
Petre T (1990) Fish, fisheries aquatic macrophytes and water quality in inland waters. Water Qual Bull 12:103–106
Puijalon S, Boumo TJ, Van Groenedael J, Bormette G (2008) Clonal plasticity of aquatic plants species submitted to mechanical stress: escape versus resistance strategy. Ann Bot 102:989–996
Ratusshnyale AA (2008) The role of aquatic macrophytes in hydro ecosystem of the Kuibysheu reservoir. J Agric Environ Sci 4(1):01–08
Suren L (2000) Urban drainage impacts a receiving waters quality water. Sci Technol 27:151–158
Udo RK (1987) A comprehensive geography of West Africa. Heinemann Educational Books, Ibadan
Uka UN, Mohammed HA, Ovie SI (2009) Current diversity of aquatic macrophytes in Nigerian freshwater ecosystem. Brazil J Aquat Sci Technol 13:9–15
Vida J (2010) Nigeria's agony dwarfs the Gulf oil spill: the US and Europe ignore it. https://www.theguardian.com/world/2010/may/30/oil-spills-nigeria-niger-delta-shell
Vymazal J (2013) Emergent plants used in free water surface constructed wetlands: a review. Ecol Eng 61:582–592. https://doi.org/10.1016/j.ecoleng.2013.06.023
Wang SR, Jin XC, Jiao LX, WU F.C. (2009) Response in root morphology and nutrient contents of Myriophyllum, spicatum to sediment type. Ecol Eng 35:1264–1270
Wells RDS, Clayton JS (2005) Mechanical and chemical control of aquatic weeds; cost and benefits. In: encyclopedia of pest management. Taylor and Francis, London. https://doi.org/10.1081/EPM-120024643
Wetzel RG (2001) Limnology: lake and river ecosystems. Academic, San diego
Xia C, Wang X, Xia J, Liu G (2010) The effect of temperature, water level on seed germination of *Myriophyllum spicatom* and *Potamogeton malaianus*. Aqaut Bot 92:28–32

Environmental and Ecological Importance of Indian Aquatic Macrophytes

4

Vimlesh Rawat and Amit Kumar Singh

Abstract

This chapter is the study of various aquatic macrophytes (AMs) in their natural surroundings and their type and ecological importance. AMs are commonly known as hydrophytes and are the key components of aquatic and wetland ecosystems. Such ecosystems include river, lakes, streams, ponds, canals, water-logged/muddy/marshy areas, wetlands, etc. AMs play a significant role in balancing the quality of aquatic ecosystem very efficiently as they possess an effective root system that helps them in absorption of dissolved nutrients, chemical constituents, heavy metals, etc. Thus, it contributes in bioremediation/ phytoremediation process. Bioaccumulation of metals in macrophytic plant body through roots from water and soil regulates their concentration in aquatic body; however, the death and decay of the macrophytes releases it again in the same system, which may lead to nutrient enrichment or eutrophication of the aquatic body or ecosystem. Macrophytes primarily affect the floral and faunal diversity of any aquatic ecosystem. It also affects the ecosystem functioning and balancing by other such services that are important for the ecosystem equilibrium in many ways.

Keywords

Macrophytes · Hydrophytes · Wetlands · Ecosystem functioning · Phytoremediation

V. Rawat (✉) · A. K. Singh
Department of Botany, University of Lucknow, Lucknow, Uttar Pradesh, India

S. Kumar et al. (eds.), *Aquatic Macrophytes: Ecology, Functions and Services*, https://doi.org/10.1007/978-981-99-3822-3_4

4.1 Introduction

The term "macrophytes" (Greek: macro = big or large sized, phytes = plants) is used to represent "groups of larger sized plants," which are easily visible and distinguishable to the naked eyes. The term "aquatic macrophytes" is often used to designate "macrophytes" of all aquatic habitat plants, which consist of thallus large enough to be seen by the naked eyes and can be easily distinguishable by their characteristics of adaptations to various moistened or submerged habitat (Holmes and Whitton 1977). Their thallus organizations are so diversified and have various adaptations to facilitate photosynthesis very efficiently either in submerged or floating conditions, either permanently or at least for several months in a year on the water surface. Aquatic macrophytes include not only spermatophytes or flowering plants but also ferns, bryophytes, and macrophytic algae (Crowder and Painter 1991).

4.2 Classification of Aquatic Macrophytes (AMs)

The aquatic macrophytes (AMs) may occur in different types of aquatic habitats and are able to develop self-sufficient ecosystem, which encourages the growth of a varied range of flora. The natural freshwater AMs are well noted in ponds, lakes, rivers, canals, marshy places, wetlands, or other such waterlogged areas. As the macrophytes are belonging to various kinds of plant groups and having different adaptations, their classification was found to be difficult by various authors. According to the Danish botanist Raunkiaer (1934) system of plant classification, the land plants that are under the category of macrophytes are kept in "mesophyte," and the aquatic plants are classified in groups as helophyte and hydrophyte, under subdivision "cryptophyte" and "geophyte" as aquatic "hemicryptophyte" and aquatic "geophytes." According to Reid (1961), hydrophytes are plants "whose seeds germinate in either the water phase or the substrate of a body of water and which must spend part of their life cycle in water." AMs include multicellular green algae and certain non-vascular and vascular plants.

4.2.1 Rooted Aquatic Macrophytes

4.2.1.1 Emergent Aquatic Macrophytes

These are particularly the most productive macrophytes among all other types as their roots reside in the sediments under water and upper or leafy part floats above water in the air (Westlake 1963). These macrophytes usually occur at a depth of about 1–1.5 m, the littoral zone of any aquatic body. The root and rhizome of such emergents are often suitable for permanent anaerobic sediments and have airborne reproductive organs. Emergent aquatic macrophytes can be further grouped into two:

(a) **Erect Emergent**

These plants grow straight through the water column, extending straight out of the water. These plants often occur on the coasts, e.g., cattails, *Sagittaria*, *Scirpus*, *Carex*, *Phragmites*, *Typha*, *Butomus*, etc.

(b) **Creeping Emergent**

These plants are of creeping nature and show rapid growth and nitrogen accumulation, e.g., *Nasturtium aquaticum*, *Ludwigia* spp., *Hydrocotyle* spp., *Alternanthera philoxeroides*, *Myriophyllum aquaticum*, etc.

4.2.1.2 Floating-Leaved Aquatic Macrophytes

These AMs are normally rooted in the sediment at water depth of about 0.5–3 m with leaves floating on the water surface. They possess long and slender petioles with broad leaves that are adapted to deal with mechanical stress. Reproductive bodies are found floating above the water surface, e.g., *Hydrilla*, *Nymphaea*, *Polygonum*, *Nymphoides*, *Potamogeton*, etc.

4.2.1.3 Submerged Aquatic Macrophytes

These AMs grow entirely submerged and are anchored in sediments. Their benthic occurrence of angiosperms is usually restricted to approximately 10 m. However, submerged macrophytes of other taxonomic groups may be found at any depth of the photic zone. Such macrophytes usually possess narrow, ribbon-shaped, or dissected leaves to deal with water flow/velocity within water bodies. The reproductive organs of submerged plants are found in floating conditions, and submerged occurrence is rare. Certain algae such as *Chara* and *Nitella*, mosses such as *Fontinalis*, and angiosperms such as *Myriophyllum*, *Elodea*, *Potamogeton*, etc. are the example of submerged macrophytes.

4.2.2 Free-Floating Aquatic Macrophytes

Free-floating macrophytes are found floating on or under water surface. Leaves and reproductive bodies are found floating in the open air. Such macrophytes have comparatively less developed root system, which is not attached with the bottom sediments. They absorb nutrients directly and completely through water, e.g., *Lemna*, *Salvinia*, *Eichhornia*, *Pistia*, *Ceratophyllum*, *Hydrocharis*, *Azolla*, *Salvinia*, etc.

4.2.3 Other Associated Life Forms

(a) **Amphibious plants:** Such plants complete most of their living activities in saturated soil, but water is not necessary for them, e.g., *Polygonum* spp.

(b) **Epiphytes/epiphytic plants:** They are also known as "air plants" as they grow above the aquatic macrophytes and extract their food from them or other sources, e.g., *Oxycarium cubense*.

4.3 Characteristic Features of Aquatic Macrophytes

Aquatic macrophytes possess a thin cuticle, which usually prevents water loss. Most of the hydrophytes do not require cuticles. Availability of sufficient water discourages the closing of stomata, and they remain open most of the time. It shows the inactive role of guard cells in aquatic macrophytes. Stomatal count is comparatively high in aquatic plants, which may be present on both sides of the leaves. It generally endures the pressure of water.

The flat, horizontal, and comparatively large leaf structure of aquatic macrophytes makes them suitable for surface flotation. Such leaves contain a large number of air sacs that support flotation. Smaller roots can directly diffuse water into leaves. Ranaticafus flabellaris (buttercup) floats slightly submerged in water and has highly dissected submerged leaves while broad and lobed emerged leaves that helps them to spread over a larger surface area and increases its buoyant nature. It also enhances the surface for mineral absorption and gaseous exchange (Table 4.1).

4.4 Aquatic Adaptations in Macrophytes

According to the nature and requirement, aquatic macrophytes acquire certain specific adaptations. Aerenchyma tissues are one of them, which are a spongy network of cells that creates air spaces in the plant. This plays an important role in gaseous exchange under water. With the increasing depth of water, the level of oxygen usually drops; in that case, the air spaces act as tunnels, allowing plants to pass oxygen from the upper parts to lower parts. The carbon dioxide produced as a byproduct of respiration is transported back through the aerenchyma to be used in photosynthesis by some aquatic macrophytes. Aerenchyma makes plants more buoyant. Due to the presence of gases in the air spaces of leaves and stems, they act like flotation device and give them structure and support.

Different kinds of hydrophytes have different nature to survive in particular habitats as floating hydrophytes require more sunlight than submerged hydrophytes, and the presence of aerenchyma supports floating plants to receive more sunlight, while submerged hydrophytes receive less sunlight as light energy decreases as they travel through a water column. Certain floating macrophytes have hairs on their leaf surface; their function is to trap air. Some plants like duckweed, water cabbage, etc. have chloroplast in the upper leaf surface, which is coated with a thick wax layer that repels water and keeps stomata clear and open. Some aquatic plants especially partially submerged plants have two types of leaf structures; submerged leaves are linear ribbon-shaped or highly dissected, and the leaves above the water surface are broad, circular, or slightly lobed. This may be helpful in reducing quantitative transpiration. Larger leaves on water surface provide shade to submerged leaves, which requires comparatively less light. The presence of air spaces within the tissues of partially submerged plants keeps them buoyant that help their leaves to reach at the top of the water body in order to receive the sufficient amount of sunlight.

Table 4.1 Characteristics of different types of aquatic macrophytes

S. no.	Floating macrophytes	Submerged macrophytes	Amphibious macrophytes
1.	Lower plant body specially remains in the water, while the others, specially the leaves and reproductive parts, float on the surface of water	The entire plant remains submerged in the water	Some plant parts commonly root and initial stem are found within aquatic system, and upper parts like petioles, leaves, reproductive organs, etc. are found in aerial conditions
2.	Aerial parts of the plants take up CO_2 and O_2 from air and submerged parts from water	Whole plant takes up CO_2 and O_2 from water	Leaves absorb CO_2 and O_2 from atmosphere through stomata
3.	Heterophylly may be present in some plants	Heterophylly is not present	Heterophylly is present
4.	Floating leaves are broad, entire, and comparatively larger in size, while submerged leaves are tapered, dissected, and smaller in size	Leaf size is greatly reduced. It may be thin, narrow, and dissected	Large and entire leaves that seem tough and laminated. Leaf and reproductive organs are found above the water surface
5.	A protective thin wax layer is present on the whole plant body to cope with direct water current	Cuticle, suberin, and epidermal hairs are not found. Whole plant body is adapted to absorption functions. A brief mucilaginous layer may be found in some submerged macrophytes	Cuticle layer is present on the upper (aerial) parts of the plant body
6.	Stomatal presence is usually restricted to the upper surface of floating leaves	Stomata are absent or may be present in highly reduced form	Stomatal presence is common on both sides of the floating leaves
7.	Palisade cells are slightly developed, and spongy aerenchyma is present	No clear differentiation between palisade and spongy parenchyma is present. Large air chambers are present in mesophyll to form aerenchyma	Well-developed spongy parenchyma and large air spaces partitioned by diaphragm cells are found Mesophyll is clearly differentiated into spongy and palisade tissues
8.	Well-developed stems while reduced vascular system is present Aerenchyma and air spaces are well developed	Stem is slender and reduced. It may be either creeping or in the rhizomatous form Vascular bundles are not much developed Well-developed aerenchyma and air chambers properly	Well-developed stem is present. In some cases, it may be found as rhizome Stem is clearly differentiated into epidermis, cortex, and stellar zones. Cortex may be differentiated into outer aerenchymatous parts having large air spaces and

(continued)

Table 4.1 (continued)

S. no.	Floating macrophytes	Submerged macrophytes	Amphibious macrophytes
		separated by diaphragms are present	inner thick-walled cortex with pith
9.	Roots may be absent in free-floating macrophytes, while well-developed roots are present in rooted floating plants, and root hairs are also present Roots show poor differentiation of internal tissues	Poorly developed, unbranched roots are present. Root hairs absent Root system shows poor internal organization	Well-developed roots having properly differentiated internal tissues. Epidermal root hairs are present

Hydrilla shows rapid growth to reach and branches out on water surface to access more light. It lacks xylem tubes or well-developed vascular bundles.

Submerged hydrophytes have usually small and narrow leaves. Plants like *Utricularia*, *Myriophyllum*, and *Ceratophyllum* have finely dissected leaves. Aerenchyma is well developed in the leaves and stem of submerged plants. Both the respiratory gases carbon dioxide, a byproduct of respiration, is present in the air chambers and is used in the photosynthesis, while oxygen, which is produced during the process of photosynthesis, is also present in the air chambers and is used in respiration. A one- or two-cell-thick chlorenchymatous partition is present between two cavities. Stomata are completely absent in submerged plants; however, vestigial and functionless stomata have been found in some cases. Gaseous exchange directly takes place through cell walls. Conducting tissue shows greatest reduction, are poorly developed absorption of water and nutrients takes place through the entire surface of submerged parts. Amphibious hydrophytes have tougher leaves in comparison to other hydrophytes. They show much resemblance to mesophytic characteristics. The lower part especially roots and some parts of stems and leaves may be submerged in water, but some foliage, branches, and flowering shoots grow well above the surface of water. The aerial parts of such plants show mesophytic features, while submerged parts show hydrophytic characteristics, e.g., some varieties of rice (*Oryza sativa*), *Marsilea*, *Jussiaea, Phragmites*, *Sagittaria*, *Alisma*, *Enhydra fluctuans*, *Neptuma*, *Commelina*, *Polygonum*, *Ranunculus aquatilis*, etc. The shoots of such plants are completely exposed to air as in land plants, but the roots are buried in water-lodged soil, and these are called marsh plants, e.g., *Cyperus*, *Typha*, *Scirpus*, *Rumex*, etc.

4.5 Effects of Macrophytes on Aquatic Fauna

Macrophytes acquire a larger part of any aquatic body and have a profound effect on the whole aquatic ecosystem and its functioning. Their presence affects the physiochemical properties of water and influences nutrient cycle and their availability to other aquatic organisms. Many invertebrates and vertebrates extract their food from macrophytes either from living parts or dead biomass. They also provide living space to smaller organisms such as epiphytes or other smaller ones. The presence of macrophytes and other such plants supports the growth and number of various zooplankton, insect, and fish population that has a diverse effect on the food chain. The diversity among species reduces the predation pressure over a particular species. The trophic-level diversity of organisms enriches the food web of any aquatic ecosystem. Species richness also defines the wellness of any ecosystem. Macrophytes positively affect the water clarity as some of them have phytoremediation ability by absorbing excess amount of minerals from the water bodies. The luxuriant growth of macrophytes enhances denitrification of soil, which may affect the growth of other phytoplanktons. They also suppress the growth of phytoplanktons by releasing certain "allelopathic" substances. There are various pathways by which submerged macrophytes positively affect the resource base of epiphytic and benthic macro-invertebrates.

Significant growth of macrophytes in lakes and reservoirs can alter food interactions among different trophic levels. The abundance of submerged macrophytes negatively affects the planktonic algal biomass. It may be due to competition for nutrients, space, light, dissolved oxygen, or other such things. Macrophytes support the growth of epiphytes that survive on macrophytes. Productivity of the epiphyte complex may reach up to 93% of host macrophyte productivity (Jones 1984). The negative relation of macrophyte and planktonic algae directly affects the growth and number of certain aquatic fauna that feeds on planktonic algae. Aquatic macrophytes can directly influence the prey-predator interaction by altering the composition and abundance of zooplankton and macro-invertebrates, which are directly associated with fish population. They may aggravate the correlation by favoring the reproductive success of certain predators (Maceina et al. 1991).

4.6 Role of Aquatic Macrophytes in Improving Water Quality

Rapidly increasing population and its pressure on urbanization and extraction of natural resources have aggravated the current scenario of water bodies or aquatic habitats. Surface runoff and sewage flow into aquatic bodies are the most common sources by which various chemicals or nutrients enter into the aquatic system, which ultimately damages the natural functioning of the whole ecosystem (Sudhira and Kumar 2000). Increasing water pollution by laundry workers, domestic usages, sewage, etc. is enhancing pollution problems which leads water bodies toward eutrophication that results in massive macrophytic and weed growth and decay of plant bodies that again release nutrients into the aquatic system, which is a matter of serious concern. However, eutrophication signifies the aging of any aquatic system; with the passage of time, it accumulates silt, sediments, organic matter, and nutrients. The self-purification of water bodies is naturally performed by aquatic macrophytes. They have the ability to absorb nutrients and cleanse the aquatic system. Their ability to absorb nutrients in large quantities makes them more useful in wastewater treatment.

Free-floating aquatic macrophyte *Eichhornia crassipes* has tremendous capacity to absorb nutrients and other such constituents through roots. It has the ability to remove nitrogen, phosphorus, organic carbons, phenols, pesticides, suspended solids, and certain heavy metals from the polluted water. Thus, it lays down the pollution load and decreases the BOD and COD of the aquatic system (Gupta 1982). *Hydrilla verticillata, a submerged macrophyte* efficiently absorbs the nutrients but requires oxygenated water for better growth; thus, it is not useful for the treatment of polluted water, which has high BOD value. Such plants have comparatively more space to attach denitrifying bacteria than emerging macrophytes (Weisner et al. 1994). Aquatic macrophytes are present in almost all the water resources. They may be rooted in coastline, free-floating, or submerged in nature. They act as natural purifiers of any aquatic body or ecosystem.

4.7 Effects of Macrophytes on Biogeochemical Cycles

Aquatic macrophytes serve as "pumps for essential and non-essential elements" (Odum 1998). Ever-increasing urbanization, industrialization, agrochemicals, and vehicular load have increased the flow and dump of metals, trace metals, and pollutants toward aquatic resources. It may be through direct discharge, surface runoff, or washout of atmospheric pollutants through rain water. Such flow toward aquatic systems contains high amount of contaminants such as surfactants, heavy metals (lead, nickel, cadmium, zinc, copper, cadmium, etc.), oil or grease residues, fertilizer and pesticide residues, and human or animal wastes. Undefined pollutant sources add several primary and secondary pollutants in water bodies, which may have undefined effects on the aquatic flora and fauna. Sediment formation in aquatic ecosystems provides base, space, and nutrients to develop aquatic life. Vitality of any ecosystem is defined by its producer community that makes the nutrients available to flow from a trophic level to another or the one who begins the cycling flow of nutrients (Prasad et al. 2001). Thus, aquatic macrophytes are the most important components of any aquatic system as they provide habitat, food, and shelter to other organisms (Chilton II 1990). The well-being of macrophytes directly affects the wellness of other associated life forms and nutrients' flow in them. The importance and effect of submerged macrophytes on storage and cycling of nutrients in aquatic system can be clearly observed by understanding their ubiquitous nature, their growth patterns, and their ability to concentrate particular elements. Submerged macrophytes affect the metal cycling within any aquatic system by absorbing or accumulating them. The growth of submerged plants is season dependent and observed maximum in the end of the summer season (St-Cyr et al. 1994).

Floating macrophytes greatly depend on roots for heavy metal uptake, while submerged plants show less dependency as their leaves and shoots also act as a site of metal uptake. Heavy metal absorption through leaves is by passive movement across the cuticle. The negative charges of the pectin and cutin polymers of the thin cuticle and the polygalacturonic acids of the cell walls create a suck inward. Due to the increase in the charge density inward, transport of positive metal ions takes place (Marschner 1995; Yamada et al. 1964). Flow reduction in macrophyte patches permits longer time duration for chemical and biological processes to take place. This includes absorption and respiration by macrophytes, as well as by the closely attached community of epiphytic algae, fungi, and bacteria. That is the reason of higher metabolic rates within macrophyte patches than in adjacent non-vegetated patches (O'Brien et al. 2014).

4.8 Economic Importance of Aquatic Macrophytes

4.8.1 Used as Human Food

Aquatic macrophytes such as wild rice (*Zizania*), few varieties of rice (*Oryza*), Chinese water chestnut (*Eleocharis dulcis*), Indian lotus (*Nelumbo nucifera*), watercress (*Rorippa nasturtium-aquaticum*), water mimosa (*Neptunia natans*), taro (*Colocasia esculenta*), water pepper (*Polygonum hydropiper*), totora (*Scirpus californicus*), wasabi (*Wasabia japonica*), water caltrop (*Trapa natans*), bullrush, cattail, *Typha*, etc. are used as food by humans. Uncooked hydrophytes may cause fasciolopsiasis. It is suggested to cook aquatic plants well before eating to prevent such infections.

4.8.2 Used as Animal Food

Macrophytes provide food to many fishes and other small insects or organisms in any aquatic ecosystem. Amaranthaceae, Araceae, and Typhaceae are the major macrophyte families that provide food. The fishes belonging to the Cyprinidae and Cichlidae family directly feed on macrophytes. This use of macrophytes has given livelihood to many rural peoples who are involved in fish feed preparation as aquaculture is one of the fastest growing fields.

4.8.3 Medicinal Uses of Various Aquatic Macrophytes

Herbal medicine is one of the most widely accepted medicinal types. Some aquatic plants are used for treatment of serious diseases like cancer and genetic disorders. Traditional remedies indicate the use of some water plants even to induce polyploidy (Tables 4.2 and 4.3).

Table 4.2 Uses of some dicotyledon aquatic macrophytes

S. no.	Botanical name	Family	Parts used	Usage/applications
1.	*Alternanthera Sessilis* (L.) DC	Amaranthaceae	Whole plant	The leaf juice is useful in blood purification and cures night blindness and snakebites Decoction of its leaves and stem cures blood vomiting. It also improves the lactation in lactating mother Its leafy part is eaten as vegetable as well
2.	*Asteracantha longifolia* (L.) Nees	Acanthaceae	Whole plant	The whole plant is useful in the treatment of rheumatism, anemia, and jaundice. It prevents bleeding and boosts sexuality Seed paste cures tubercular fistula, while root paste is applied externally in rheumatism Leaf decoction is helpful in dropsy
3.	*Bacopa monnieri* (L.) Pennell	Scrophulariaceae	Whole plant	This plant is used as potent cardiotonic, diuretic, bronchodilator, and laxative Powder of dried leaves is useful in weakness and nerve disorders All plant parts are useful in the treatment of skin diseases, epilepsy, ulcer, and leprosy It also cures cold and cough
4.	*Centella asiatica* (L.) Urb.	Apiaceae	Leaves	It has the wound healing ability Helps in memory improvement Cures mental fatigue, bronchitis, asthma, kidney trouble, and dysentery Leaf decoction cures leprosy, nerve disorders, and heart problems Direct chewing of soft raw leaves is useful in acidity and peptic ulcers
5.	*Ceratophyllum demersum* L.	Ceratophyllaceae	Leaves	Decoction of leaf is used to regulate bile secretion. It is also used as cardiotonic and antipyretic Leaf paste is applied externally to cure scorpion sting
6.	*Dentella repens* (L.) Forst.	Rubiaceae	Leaves	Leaves are useful in blood purification, improve eyesight, and heal wounds
7.		Asteraceae		

(continued)

Table 4.2 (continued)

S. no.	Botanical name	Family	Parts used	Usage/applications
	Eclipta prostrata L.		Whole plant	Plant juice is useful in skin ailments and asthma Leaf extract (boiled) with coconut oil improves hair growth
8.	*Ipomoea aquatica* Forssk.	Convolvulaceae	Leaves, shoot	Boiled leaf and stem extract is used to cure arsenic toxicity and opium poisoning Its decoction has blood purification ability
9.	*Jussiaea repens* L.	Onagraceae	Whole plant	Plant powder (dried) is useful in various skin diseases. Applied externally on skin
10.	*Nelumbo nucifera* Gaertn.	Nelumbonaceae	Whole plant	Decoction of flower petals is used as tonic to cure liver and cardio problems. It cures diarrhea. Foliar paste with lime is applicable as plaster on bone fractures Seed paste has cooling effect in skin diseases Root/rhizome paste cures ringworms
11.	*Nymphaea nouchali* Burm. f.	Nymphaeaceae	Whole plant	Its rhizome is used to stop excessive menstrual flow Rhizome is also used to treat goiter, dyspepsia, diarrhea, dysentery, and burn wounds Decant water of overnight soaked flower and leaves cures heart problems
12.	*Polygonum barbatum* L.	Polygonaceae	Leaves, young shoots	Plant extract is taken to cure pneumonia. Leaf juice is taken with honey to cure fever and colic pain Growing shoots and roots are cooked with vegetables
13.	*Ranunculus sceleratus* L.	Ranunculaceae	Leaves, stem, seeds, and root	Its rhizome powder is used to cure diarrhea and dysentery Daily sip of leaf extract juice removes ringworms Seeds cure kidney disorders
14.	*Trapa bispinosa* Roxb.	Trapaceae	Fruits and leaves	Leaf paste is used as cooling agent Used as aphrodisiac Cures leprosy and inflammation

Table 4.3 Uses of some aquatic monocotyledon macrophytes

S. no.	Botanical name	Family	Parts used	Usage/applications
1.	*Colocasia esculenta* (L.) Schleid	Araceae	Leaves, seeds, and young shoots	The leaves and rhizome are useful in constipation Peel of outer skin of petiole is tied to cure cracked feet
2.	*Eichhornia crassipes* (Mart.) Solms	Pontederiaceae	Leaves, petiole, and flower	Contains good amount of antioxidant. Carotene-rich vegetables Dried plant is used as compost
3.	*Hydrilla verticillata* (L.f.) Royle	Hydrocharitaceae	Leaves	The boiled leaf extract is used to treat cuts and wounds
4.	*Monochoria hastata* (L.) Solms	Pontederiaceae	Leaves and stalk	Leaf extract is applied on boils Stalk and leaves are used to cure insanity and also used as cooling agent and tonic
5.	*Pistia stratiotes* L.	Araceae	Leaves and root	These are effective in leprosy, eczema, piles, and ulcer
6.	*Sagittaria sagittifolia* L.	Alismataceae	Leaves and tuber	Leaf paste cures skin problems. Root tuber is used as birth control
7.	*Spirodela polyrhiza* (L.) Schleid	Lemnaceae	Leaves	Leaf decoction regulates urination. Dried plant is used as compost

4.9 Conclusion

The aquatic macrophytes (AMs) are the key component of any aquatic body. It regulates the functionality of the whole aquatic ecosystem. It gives structural and mechanical support to aquatic habitat, provides food to primary consumers, gives shelter to epiphytes, and regulates nutrient cycle. Luxuriant growth of macrophytes contributes in species richness and diversity of particular water body. The reason behind animal abundance is the availability of food, shelter, and living assistance. Complexity of any habitat is defined by the number of organisms present at per trophic level. Competition among organisms regulates the wellness of any ecosystem. Bioaccumulation or absorption of mineral constituents cleanses the water and assures the wellness of other organisms and water body too. Macrophytes are of high economic value as they provide food and medicinal products to humans and other animals too. A lot of diseases are cured by using plant parts of various macrophytes. Fresh or dried, dead, or alive, in every condition, they add value to the life. Certain macrophytes are used as green manure, which is an urgent requirement of present time. Thus, macrophytes are of great importance as they function as the center of biodiversity, clarify water pollution by bioremediation and bioaccumulation, and regulate nutrient cycle. Some aquatic macrophytes have high aesthetic values, while some have medicinal or economic importance.

References

Chilton EW II (1990) Macroinvertebrate communities associated with three aquatic macrophytes Ceratophyllum demersum, Myriophyllum spicatum, and Vallisneria americana in Lake Onalaska, Wisconsin. J Freshw Ecol 5:455

Crowder LB, Painter DS (1991) Submersed macrophytes in Lake Ontario: current knowledge, importance, threats to stability and needed studies. Can J Fish Aquat Sci 48:1539–1545

Gupta GC (1982) Use of water hyacinth in wastewater treatment. J Environ Health 43(2):80–82

Holmes NTH, Whitton BA (1977) Macrophytes vegetation of the river swale, Yorkshire. Freshw Biol 7:545–558

Jones RC (1984) Application of a primary production model to epiphytic algae in a shallow eutrophic lake. Ecology 65:1895–1903

Maceina MJ, Bettoli PW, Klussmann WG, Betsill RK, Noble RL (1991) Effect of aquatic macrophyte removal on recruitment and growth of black crappies and white crappies in Lake Conroe, Texas. N Am JFisheries Manag 11:556–563

Marschner H (1995) Mineral nutrition in higher plants. Academic, New York

O'Brien JM, Lessard JL, Plew D, Graham SE, McIntosh AR (2014) Aquatic macrophytes alter metabolism and nutrient cycling in lowland streams. Ecosystems 17:405–417. https://doi.org/10.1007/s10021-013-9730-8

Odum WE (1998) Comparative ecology of tidal freshwater and salt marshes. Annu Rev Ecol Syst 19:147

Prasad MNV, Greger M, Smith BN (2001) Aquatic macrophytes. In: Prasad MNV (ed) Metals in the environment: analysis by biodiversity. Marcel Dekker Inc., New York, p 259

Raunkiær C (1934) The Life Forms of Plants and Statistical Plant Geography, being the collected papers of C. Raunkiær. Oxford University Press, Oxford. Translated by H. Gilbert-Carter, A. Fausbøll, and A. G. Tansley. Reprinted 1978 (ed. by Frank N. Egerton), Ayer Co Pub., in the "History of Ecology Series"

Reid GK (1961) Ecology of Inland Waters and Estuaries. Reinfold Publication Corporation, New York

St-Cyr L, Campbell PGC, Guertin K (1994) Evaluation of the role of submerged plant beds in the metal budget of fluvial lake. Hydrobiologia 291:141

Sudhira HS, Kumar VS (2000) Monitoring of lake water quality in Mysore city, proceedings of Lake 2000. International symposium on restoration of lakes and wetlands, 27–29 Nov. 2000. CSIC Auditorium, Indian Institute of Science, Bangalore. pp 1–10

Weisner SE, Eriksson PG, Graneli W, Leonardson L (1994) Influence of macrophytes on nitrogen removal in wetlands. Ambio 23(6):363–366

Westlake DF (1963) Comparisons of plant productivity. Biol Rev Camb Philos Soc 38:385–425. https://doi.org/10.1111/j.1469-185X.1963.tb00788.x

Yamada Y, Bucovac MJ, Wittwer SH (1964) Ion binding by surfaces of isolated cuticular membranes. Plant Physiol 39:978

Macrophytes Used as Complementary Medicines for Curing Human Ailments to Facilitate Livelihood Opportunities

5

R. Bar and R. N. Mandal

Abstract

A total of 16 macrophytes comprising aquatic and semi-aquatic plants are documented as used as complementary medicine for therapy of a number of human ailments through traditional knowledge. This traditional practice of therapy using macrophytes has close proximity to the modern medicinal values. A great many people of India make their dietary satisfaction with these macrophytes by means of leafy vegetables being potentially nutritious, apart from the benefit of complementary medicine. These macrophytes are eaten in local food cultures as a practice of therapeutic purpose. This study has identified 59 human ailments and 20 affected human organs/systems for which these macrophytes are used. Gastrointestinal disorder is found to be the highest recorded ailment, and the skin is the most affected organ for which such macrophytes are administered for remedy. Some workers consider these macrophytes as 'famine plants' because of their availability during scarcity of usual vegetables. Also, the present study considers these macrophytes as 'poor man medicine' for their substantial use in both curative and preventive measures to human ailments. Despite their phenomenal service to the society, they are neglected to the great extent by which their existence is found to be in the alarming state. No policy for their conservation is formulated. Here are mentioned some measures for their protection and propagation in relation to livelihood opportunities of common people.

R. Bar (✉)
Department of Botany, Bangabasi Morning College, Kolkata, India

R. N. Mandal
Regional Research Center, ICAR-CIFA, Kolkata, India

S. Kumar et al. (eds.), *Aquatic Macrophytes: Ecology, Functions and Services*,
https://doi.org/10.1007/978-981-99-3822-3_5

Keywords

Human ailment · Livelihood · Macrophytes · Medicine

5.1 Introduction

India is endowed with diverse water sources of various forms such as rivers, lakes, reservoirs, ponds, marshes, ditches, swamps, bogs, canals, karanjali, and sewage lagoons. Most of them are perennial, and others are seasonal, based on water availability. Besides, there are other forms of water bodies available but less frequent, which includes flowing water, streams, rivers, and springs. All these diverse forms of water bodies are habitats of different aquatic life forms; aquatic macrophytes among others are important ones—an integral part of the aquatic habitats.

5.1.1 Aquatic and Semi-aquatic Macrophytes as Important Bio-resources

A simple understanding on the definition of macrophytes refers to the plant communities that grow within water, inside water, in surface water, near water, or in saturated soil. They are of diverse life forms such as submerged, floating, rooted floating, and emergent (Sculthorpe 1967). However, the term 'macro' literally means larger one than 'micro' as a smaller one. Rationally, macrophytes comprise those aquatic plants which are visible in the naked eye in the sense that their morphology may be identified scientifically and of which one species may be differentiated from another. On the contrary, the identification of microphytes requires the microscopic instrument to observe their morphology. We use to mention a distinct terminology as 'aquatic macrophytes' and 'semi-aquatic macrophytes'. Aquatic macrophytes refer to those plant communities which require exclusive water sources for their growth, development, and completion of the life cycle; else they never exist. On the other hand, semi-aquatic macrophytes prefer growing near water bodies, in marshy areas, or in the saturated soil. They may withstand dry conditions for a short while but do not survive long. Whatever life forms, life cycles, and habits they may have, the existence of both aquatic and semi-aquatic plants requires water sources. Like terrestrial plants, macrophytes are also important bio-resources for medicinal purpose, but unlike terrestrial ones, their potential as medicinal plants is yet to be explored.

Let's consider the macrophytes as important bio-resources. The definition of 'bio-resources' is understood to mean 'an organism as a whole or part/organ of an organism or active ingredient/compound of an organism identified to be used for the specific purpose of human benefit' (Mandal and Mukhopadhyay 2012). Given this definition to ascertain the specific purpose of all biological entities, we discuss some aquatic and semi-aquatic plants for their medicinal values used for therapy of a variety of human ailments.

5.1.2 Different Uses of Aquatic and Semi-aquatic Macrophytes in Human Welfare

The aquatic and semi-aquatic plants which are considered as the valuable plant bio-resources have multifaceted roles in human welfare, as catering to the society several basic needs, namely, food crops, cash crops, horticulture crops, bio-fertilizers, feeds, fodders, green manure, leafy vegetables, ornamental plants, integrated crops, wastewater treatment, and therapeutic substances (Mandal and Saha 2007; Mandal and Jayasankar 2014). Despite having immense values, the utility of these aquatic plant resources is usually overlooked; otherwise, they could be much more useful for various beneficial purposes of human societies. Most of these aquatic and semi-aquatic plants still grow wildly and are utilized traditionally rather than being cultivated following the modern agricultural methods blended with scientific know-how (Mandal et al. 2012a; Mandal and Jayasankar 2014). Importantly, a substantial number of populations in rural India utilize these bio-resources for their livelihood support. Below is highlighted a glimpse of the potentials of these aquatic bio-resources.

Macrophyte bio-resources comprise a diverse group of plant communities, which have incredible treasures of nutritional properties (Gopalan et al. 2007) to support human as well as animals' health benefits. Such bio-resources are used as foods, fodders, feeds, and vegetables and thus contribute substantially to promote and progress human society and its livestock (Mandal and Jayasankar 2014). Macrophytes are considered as the 'foodstuffs' coined primarily in the sense of both satisfaction of hunger and maintenance of good health through their intake with appropriate nutritional requirements. Many common aquatic plants used as foodstuff include water chestnut, makhana, lotus, water lily, and *Colocasia* (Mandal and Saha 2007, 2012; Mandal et al. 2010a,d, 2012a; Mandal and Bar 2013; Mandal and Jayasankar 2014), used as feed constituting a variety of items given to livestock and fish for maintaining commercial husbandry. Feed is prepared by using different ingredients obtained from macrophytes such as *Azolla*, duckweed (*Spirodela polyrhiza*, *Lemna minor*, and *Wolffia arrhiza*), water spinach, lotus, and water chestnut; all these are familiar to the rural people (Mandal et al. 2007, 2008, 2010b,c, 2012b; Mandal and Jayasankar 2014). The term 'fodder' which refers to green plants used directly or after processing for feeding livestock includes macrophytes such as *Azolla*, duckweed, water hyacinth, dhaincha (*Sesbania bispinosa*), water lily, and water spinach (Mandal et al. 2011; Mandal and Jayasankar 2014). Several wild aquatic plants represent leafy vegetables utilized as human food for their unique nutrient properties, apart from important therapeutic substances (Mandal and Mukhopadhyay 2010). Three species of duckweeds and *Azolla*, which contain 33% protein as weight/weight basis, are treated as the valuable food used directly to fish or indirectly used as ingredients for preparing fish feeds (Kalita et al. 2007; Mandal et al. 2010b,c, 2012b).

5.1.3 Relevance of Aquatic and Semi-aquatic Macrophytes as Drugs in the Present Time

Systematic investigation of drugs used in indigenous medicine was started during the early period of the twentieth century (Chopra et al. 1986). Some aquatic and semi-aquatic plants that grow in the backyard of households, which were prescribed by 'kavirajas' and 'hakims' to cure human ailments, are now investigated by modern science to explore their active compounds useful to remedy the particular human disease. These common plant bio-resources, which have been treated by rural people as their reliable medicinal drugs, make them satisfied and assured them from curing common diseases. These bio-resources have important drug constituents by scientific examination as to have explored their pharmacological action of the active principles. All these scientific validities are initially tested by animal experimentation, and then prepared drugs are applied to humans (Chopra et al. 1986). When only a thorough investigation confirms the merit of a particular drug useful for the remedy of a particular disease, such a bio-resource is only recommended as a proven natural drug. Substantial aquatic and semi-aquatic plants have been examined to have shown valuable active compounds used as substitutes of chemical commercial drugs.

The present-day scientific investigation ascertains the following:

- How many numbers of different active compounds are present in the particular macrophyte?
- How much amount of specific active compound is available in the macrophyte?
- If all the active compounds of the particular macrophytes are used to the particular human ailment following the crude method, whether any side effect may occur to the patient for which it is used?
- Which organ/part (root, rhizome, leaf, stem, fruit, and seed) of the macrophytes possesses the particular active compound in maximal amount?
- Which stage of the maturity (seedling, sapling, young plant, full-grown plant) of the particular macrophytes possesses the maximal amount of the active compound?

When the scientific investigation explores all these information accurately and reliably with sufficient evidences through repeated analyses and examinations, the credential of macrophytes to cure the particular human ailment is established to the great extent, and the confidence of common people to use such macrophytes as a remedial drug is assured. On the contrary, if a particular macrophyte does not have the reliable active compound for which people use it to cure ailment based on traditional belief, the macrophyte needs to be known as useless to the public. However, all the scientific investigations need to be published in the refereed journals, so that noted experts can check their scientific validity and recommend them for publication; otherwise, false information published in the fishy journals may harm the reputation of indigenous macrophytes used traditionally.

5.2 Understanding of Complementary Medicines

5.2.1 Definition of Complementary Medicine

Wetlands are known repositories of many unique biotic resources useful to human welfare (Mandal 2011). Use of some macrophytes as a therapy to several human ailments is one important purpose among others. As complementary/alternative medicine, their use inherits through generations and has been prevalent in rural areas (Mandal and Mukhopadhyay 2010; Azam et al. 2014). The term 'complementary' medicine refers to diagnosis, therapies, and preventive procedures excluded from mainstream medicine (Ernst 1993). The 'complementary' medicine is also known as 'alterative' medicine that means a herb said to alter a disease, purify blood, increase appetite, improve digestion, and eliminate toxins (Med Dict 2007). As a practice of the traditional therapy (Azam et al. 2014; Mandal and Mukhopadhyay 2010), the herbs are eaten in local food cultures (Guinard and Lemessa 2000; Islam et al. 2011; Mandal et al. 2008, 2012a; Mandal and Bar 2013; Mandal and Jayasankar 2014; Ong et al. 2011; Roy et al. 2013; Soni and Singh 2012), an inseparable item of both food and medicine. However, all the herbs fall beyond the scope of any range of conventional medicines but are seen used alongside conventional medicines by rural folks in the treatment of disease and ill health.

5.2.2 Meaning of Curative and Preventive Measures

Rural masses who are unable to afford costly drugs get complementary medicinal benefits using the easily available macrophyte resources for both 'curative' and 'preventive' measures. 'Curative' means to cure disease and 'preventive' to increase immunity as defined by Ayurveda therapy (Bhattacharya 1993). Common people use all such macrophytes for both curative and preventive purposes, not knowing exactly their mode of action, either curative or preventive. These macrophytes are supposed to be 'future foods' (Mandal and Jayasankar 2014), as they are important sources of the valuable nutrient substances (Gopalan et al. 2007) protecting human health through their consumption (Borah et al. 2009; Mandal and Jayasankar 2014; Siddique et al. 2011).

5.2.3 Use of Macrophytes for Therapy to Human Ailments: Traditional Methods

Apparently, common people of rural India are familiar with the medicinal value of these macrophytes with respective human ailments to cure. People who use these herbs as remedial measures of ailments/ill health follow the standard methods as much as traditional while using them for therapies (Table 5.1). Decoction of leaves/stems/entire plants obtained by thrashing or by squeezing is a common practice. Probably, this traditional practice of collecting decoction is unique since medicinal properties (secondary metabolites) along with essential nutrients (vitamins and minerals) contained in it are protected as such without degeneration of medicinal properties. Sometimes, mild heating is provided, or other ingredients like honey or cumin or both are added, whenever necessary as administered by traditional therapy.

Despite seminal usefulness, they are neglected. Their disappearance from natural habitats is found alarming due to wetland destruction (Mandal and Mukhopadhyay 2010). Also considered to be famine plants eaten and utilized for centuries (Chopak 2000; Paul et al. 2011; Rahamatullah et al. 2011), they are capable of growing under adverse climatic conditions. Compared to the terrestrial medicinal plants, they are poorly documented. The present review aims to document the usefulness of such macrophytes, which have therapeutic uses in a number of human ailments.

5.3 Aquatic and Semi-aquatic Macrophytes vis-à-vis Modern Approach to Human Ailments

Since long macrophytes have been useful medicinal plants for human health benefits. Their medicinal use is of a potential relevance despite a remarkable progress of synthetic drugs in the modern health therapy. A large section of Indian people, the rural mass in particular, rely on their efficacy of curing human ailments by means of direct use as the extraction of decoction or a method of specific application through indigenous knowledge or daily consumption as leafy vegetable. Our approach is to explore how all such macrophytes benefit human health pertaining to their constituent drug properties in relation to human ailments and affected organs.

5.3.1 Macrophytes with Their Constituent Compounds Used as Drugs

We compile all the necessary information regarding active ingredients or constituent drugs the respective macrophytes possess (Table 5.2). However, the information about constituent drugs of the respective macrophytes mentioned here is very limited as compared to their vast treasure of medicinal properties. For instance, the use of the sacred lotus (*Nelumbo nucifera*) as medicinal plant by human is as old as the human civilization; Buddhist monks of China used lotus for their daily use as a health

Table 5.1 Macrophytes used as complementary medicine, with scientific name, parts used, status, and TK

Scientific name (family)	Part (s) used	Status	Traditional knowledge (TK)	
			Therapy	Method of use
Alternanthera philoxeroides Griseb. (Amaranthaceae)	Stem and leaf	Wild	Remedy of constipation, clearing bowel; check vomiting with cough	Stem and leaf made parboiling and fried with little amount of edible oil before use
Alternanthera sessilis (L.) R.Br. (Amaranthaceae)	Entire plant, stem, and leaf	Wild	Remedy of constipation, clearing bowel; leaves used against eczema	Stem and leaf made parboiling and fried with little amount of edible oil before use
Arum trilobatum Linn. (Araceae)	Leafy twig and petiole	Wild	Used for secretion of saliva, increase appetite, acting as cooling agent for remedy of boil and insect sting	Leaf made semisolid dough and mixed with black cumin, followed by mild heating before use
Asteracantha longifolia Nees. (Acanthaceae)	Leaf, root, and seeds	Wild and partially cultivated	Use for remedy of anaemia, particularly for female and mothers giving immediate birth of child	1. Decoction of stem and leaf used after addition of honey 2. Leaf made semisolid dough, followed by mild heating before use
Bacopa monnieri (L.) Pennell (Plantaginaceae)	Leafy twig, stem, and whole plant	Wild	Improve memory, especially children, regain memory power	1. Decoction of stem and leaf used after addition of honey 2. Leaf mixed with ghee made fried before use
Centella asiatica L. (Apiaceae)	Leaf	Wild	Remedy of dysentery, diarrhoea, and skin disease and purification of blood Improve digestion	Decoction of leaf used as raw or used after addition of honey
Colocasia esculenta (L.) Schott. (Araceae)	Leaf, petiole, and rhizome	Wild and cultivated	Remedy for bowel complaint and constipation, indigestion, rheumatism	1. Leaf and stalk made dough, followed by mild heating before use2. Rhizome eaten through parboiling
Commelina benghalensis L. (Commelinaceae)	Whole plant and young leaf	Wild	Bowel clearance, remedy of constipation	Stem and leaf made parboiling and fried with little amount of oil before use

(continued)

Table 5.1 (continued)

Scientific name (family)	Part (s) used	Status	Traditional knowledge (TK)	
			Therapy	Method of use
Enhydra fluctuans Lour. (Asteraceae)	Stem and leaf	Wild	Remedy of skin disease and purification of blood	1. Decoction of leaf used as raw 2. Stem and leaf made parboiling and fried with little amount of oil before use
Ipomoea aquatica Forssk. (Convolvulaceae)	Stem and leaf	Wild and partially cultivated	Remedy of constipation and nervous diseases and improve tastiness	Leaf made parboiling and fried with little amount of oil before use
Marsilea minuta L. (Marsileaceae)	Foliages	Wild	Remedy of insomnia	1. Leaf used after being flattened 2. Leaf made parboiling and fried with little amount of oil before use
Mollugo cerviana Ser. (Molluginaceae)	Leafy twig	Wild	Remedy of skin disease and purification of blood and liver disease	Stem and leaf made parboiling and fried with little amount of oil before use
Nelumbo nucifera Gaertn. (Nelumbonaceae)	Leaf, flower, fruit, seeds, and roots	Wild and partially cultivated	Leaf used in liver disease, cataract, and blood piles	Rhizome used after parboiling Seeds eaten raw after grinding
Nymphaea alba L. (Nymphaeaceae)	Leaf stalk, fruit	Wild	Remedy of diabetic problem to check frequent urination	Inflorescence stalk used after parboiling
Nymphaea rubra Roxb. (Nymphaeaceae)	Leaf stalk, fruit	Wild	Remedy of diabetic problem to check frequent urination	Inflorescence stalk used after parboiling
Polygonum plebeium R. Br. (Polygonaceae)	Leafy twig	Wild	Remedy of bowel complaints	Entire plants parboiling and mixed with juice of boiled rice before use

therapy apart from treating it as a symbol of sanctity (Mandal and Bar 2013). Likewise, the medicinal wealth of each macrophyte is rich, but this article just mentions a few. We expect that other researchers will investigate and make them useful for human welfare.

Table 5.2 Macrophytes possessing active compounds and their medicinal use with references

Scientific name (family)	Constituent drugs	Medicinal use with references
Alternanthera philoxeroides Griseb. (Amaranthaceae)	Alkaloids, flavonoids, amino acids, carbohydrates, phenols, steroids, terpenoids, saponins, and glycosides—from leaf extracts	Fungal and bacterial infections (Sivakumar and Sunnathi 2016) Inflammation and arthritis (Sunmathi et al. 2016)
Alternanthera sessilis (L.) R.Br. (Amaranthaceae)	Alkaloids, flavonoids, amino acids, carbohydrates, phenols, steroids, terpenoids, saponins, and glycosides—from leaf extracts	Fungal and bacterial infections (Arollado and Osi 2010; Sivakumar and Sunnathi 2016) Inflammation and arthritis (Subhashimi et al. 2010; Sunmathi et al. 2016)
Arum trilobatum Linn. (Araceae)	Alkaloids, flavonoids, amino acids, carbohydrates, phenols, steroids, terpenoids, saponins, and glycosides—from leaf extracts	Pain, inflammation, and diarrhoea (Ali et al. 2012)
Asteracantha longifolia Nees. (Acanthaceae)	Alkaloids, flavonoids, amino acids, carbohydrates, phenols, steroids, terpenoids, saponins, and glycosides—from leaf extracts	Pain, inflammation (Al Amin et al. 2012)
Bacopa monnieri (L.) Pennell (Plantaginaceae)	Alkaloids, flavonoids, bacosides, bacopasides, and bacopa saponins	Anxiety; depression; Alzheimer's disease; cardiovascular, gastrointestinal, hepatic, neurological, and respiratory disorders; cancer; diabetes; inflammation; microbial infection; and oxidative damage (Charoenphon et al. 2016)
Centella asiatica L. (Apiaceae)	Triterpene, volatile and fatty oil, hydrocotylin, asiaticoside, madecassoside, centelloside, thankuniside, isothankuniside, indocentelloside, brahminoside, flavonoids, carotenoids, tannins	Insomnia, gastric ulcer, wound healing, memory improvements, protecting neurons, cardiac damage, liver damage, oxidative damage, improves immunity, inflammation, reduces protein deficiency, prevents against radiation damage, vascular diseases, depression, filariasis, fertility disorder, protozoan infection, spasm of involuntary muscle (Jamil et al. 2007)
Colocasia esculenta (L.) Schott. (Araceae)	Soluble extracts of plant containing cyanoglucoside and dietary fibre	Diabetes, fungal infection, colon cancer, liver disease, inflammation, anxiety, depression, thiopental-induced sleeping time, and rotarod performance (Prajapati et al. 2011)
Commelina benghalensis L. (Commelinaceae)	Alkaloids, flavonoids, coumarins, triterpenoids, steroids, resins, carbohydrates, phenols, tannins, amino acids, quinones, oils and	Burns, sore throats, headache, leprosy, fever, snake bite and jaundice, microbial infection, allergy, diarrhoea, fertility

(continued)

Table 5.2 (continued)

Scientific name (family)	Constituent drugs	Medicinal use with references
	fats, saponins, salicylic acid, chlorogenic acid, 8-hydroxyquinoline, caffeic acid, quinol, resorcinol, catechol, anthocyanin, beta-amyrin, lutein, zeaxanthin, violaxanthin, carotenoids, nutraceutical like vitamin C, proteins, calcium, iron, and wax	problem, cancer, psychosis, insanity and epilepsy, oxidative damage (Kansagara and Pandya 2019)
Enhydra fluctuans Lour. (Asteraceae)	Polyphenolic compounds especially flavonoids from crude extracts	Liver diseases (Sannigrahi et al. 2009) Oxidative damage (Sannigrahi et al. 2010)
	Alkaloids, tannins, flavonoids, saponins, carbohydrates, steroids, triterpenoids, and anthraquinone and glycosides from extracts of aerial parts	Anxiety and CNS active disorders (Roy et al. 2011)
Ipomoea aquatica Forssk. (Convolvulaceae)	Fibres, carotenes, and flavonoids	Flatulence, inflammation, fever, jaundice, biliousness, bronchitis, liver complaints (Manvar and Desai 2013)
Marsilea minuta L. (Marsileaceae)	Steroids, flavonoids, alkaloids, and saponins from whole plant extract termed as 'marsiline'	Stress (Tiwari et al. 2009) Tumour (Sarker et al. 2011)
	Steroid/triterpenoidal and flavonoid glycosides and saponins	Liver and oxidative damage (Praneetha et al. 2011)
	Flavonoids, tannins	Diabetes (Madhu et al. 2012)
Mollugo cerviana Ser. (Molluginaceae)	Carbohydrates, saponins, tannins, terpenoids, flavonoids, steroids, phenols, proteins, alkaloids, mollugenol A and B	Fungal and microbial infection, inflammation, oxidative damage, liver damage, skin damage by UV radiation (Aglin 2018)
Nelumbo nucifera Gaertn. (Nelumbonaceae)	Dauricine, lotusine, pronuciferine, armepavine, gallic acid, roemerine, and nuciferine from fruit and seeds Procyanidin, anonaine, coclaurine, norcoclaurine, liriodenine, dehydroemetine, dehydronuciferin, dehydroanonaine, nelumboside, remerine, quercetin 3-O-β-D-glucuronide, asimilobine, lirinidine, N-methylcoclaurine, quercetin, rutin, hyperoside, leucocyanidin, leucodelphinidin, isoquercitrin, catechin, and astragalin from leaves	Ischaemic heart, oxidative damage, liver damage, inflammation, fertility control, arrhythmic heart disease, lung tissue damage, herpes virus infection, cell cycle progression, immunomodulatory activity, diarrhoea, diabetes, urine formation, fever, bacterial infection, improvement of sperm health, prevention of blood clotting, obesity, cardiovascular activity (Mehta et al. 2013)

(continued)

Table 5.2 (continued)

Scientific name (family)	Constituent drugs	Medicinal use with references
	Kaempferol and its derivatives, arbutin, nelumboroside, quercetin 3-O-β-D-glucopyranoside, and β-sitosterol glucopyranoside from flower Betulinic acid from rhizome	
Nymphaea alba L. (Nymphaeaceae)	Flower extract	Check proliferation of human cervical and breast carcinoma (Selvakumari et al. 2012)
Nymphaea rubra Roxb. (Nymphaeaceae)	Polysaccharide	Diarrhoea, piles, and cough (Alhazmi et al. 2021) Immunity-related diseases (Cheng et al. 2012)
Polygonum plebeium R. Br. (Polygonaceae)	Flavonoids, alkaloids, glycosides, steroids, saponins, and carbohydrates	Oxidative damage (Hasan et al. 2015)

5.3.2 Affected Organs/Systems of Human Body

The study correlates three factors: human ailments, affected organ/system, and use of respective macrophytes as a remedial measure. Fifty-nine human ailments are identified in which the respective macrophytes by any means are used (Table 5.3). Twenty human organs/systems are identified, for which these herbs are administered for therapy. Gastrointestinal disorder that includes related organs/systems such as the liver, stomach, intestine, colon, and anus is found to be the highest recorded ailment for which maximum of such herbal remedies are administered. The skin is the most affected organ among others recorded, for which these herbs are used.

5.3.3 Human Ailments, Causative Factors, and Use of Macrophytes as Drugs

With a particular reference, we mention that geo-climatic conditions of Bengal are favourable for gastrointestinal disorder, which includes infectious diseases like dysentery, amoebiasis, diarrhoea, bowel complaint, intestinal pain, acidity, piles, and fistula prevalent due to unhygienic condition and poor sanitation (Azam et al. 2014). Skin disease becomes dominant due to fungal infection, which occurs under humid weather as well as wet conditions (Azam et al. 2014). People believe that impurity of blood invites all those diseases such as ringworm, eczema, itching, abscess, swelling, burning sensation, and fungal infection; oral intake of decoction of respective plants is considered to be a great herbal remedy (Boddupalli et al. 2012; Haque et al. 2009; Siddique et al. 2011). Out of 16 herbs, *Centella asiatica* is recorded for maximum uses for therapy, followed by others such as *A. longifolia*,

Table 5.3 List of human ailments and affected organs/systems in relation to macrophytes used as complementary medicines

Human ailments	Organs/ systems	Macrophytes							
		Alternanthera philoxeroides	*Alternanthera sessilis*	*Arum trilobatum*	*Asteracantha longifolia*	*Bacopa monnieri*	*Centella asiatica*	*Colocasia esculenta*	*Commelina benghalensis*
Abscess	Skin	...	...	●	●		●		●
Alopecia	Hair follicle	...	...	...	...	...	●	●	...
Amnesia	Brain	...	...	...	...	●	●	...	...
Anaemia	Blood	...	●	...	●	...	...	●	...
Aphrodisia	Nerve	...	...	...	...	...	...	...	...
Asthma	Lungs	...	...	...	...	●	...	...	...
Biliousness	Liver	●	...	...	...	...	...	...	...
Cancer	Any organ	...	...	...	...	...	...	●	...
Cardio trouble	Heart	...	...	...	...	...	●	...	...
Cataract	Eye	...	...	...	...	...	...	...	...
Constipation	Colon	●	●	●	...	●	...	●	●
Cough	Pulmonary system	...	...	...	...	●	...	...	...
Depression	Nerve	...	...	...	...	...	●	●	...
Diabetes	Blood	...	●	...	...	...	...	...	...
Diaphoresis	Skin	...	...	...	...	...	●	...	...
Diarrhoea	Intestines	...	...	...	...	...	●	...	...
Distastefulness	Tongue	...	...	...	...	...	...	...	...
Diuresis	Kidney	...	...	...	●	●	...	...	...
Dropsy	Lower limb	...	...	...	●	...	...	...	...
Drowsiness	Nerve	●	...	...	...	...	...	...	...
Dysentery	Colon	...	...	...	...	...	●	...	...

(continued)

Table 5.3 (continued)

		Macrophytes							
Human ailments	Organs/ systems	*Alternanthera philoxeroides*	*Alternanthera sessilis*	*Arum trilobatum*	*Asteracantha longifolia*	*Bacopa monnieri*	*Centella asiatica*	*Colocasia esculenta*	*Commelina benghalensis*
Dysmenorrhoea	Vaginal tract	...	...	...	...	...	●	...	...
Dyspepsia	Stomach	●	...	●	...	...	●	●	●
Ear ache	Ear	...	...	...	...	...	...	...	●
Emesis	Stomach	●	...	...	...	...	●	...	...
Encephalitis	Brain	...	...	...	...	...	...	...	...
Epilepsy	Nerve	...	...	...	...	●	...	...	...
General debility	Genital organ	●	●	...	●	...	...	...	...
Gonorrhoea	Genital organ	...	...	...	...	...	...	...	...
Helminthiasis	Intestines	...	●	...	...	...	...	●	...
Haemorrhage	Blood	...	...	...	...	...	...	●	...
Hepatitis	Liver	...	...	...	●	...	...	...	...
Herpes	Mouth	...	...	...	●	...	...	...	...
Hoarseness	Vocal chord	...	...	...	...	●	...	...	...
Hypoglycaemia	Blood	...	...	...	...	...	...	●	...
Infertility	Genital organ	...	...	...	...	...	...	...	...
Inflammation	Intestine	●	●	●	...	...	●	...	●
Insanity	Brain	...	...	...	...	●	...	...	...
Insomnia	Brain/ nerve	...	...	...	●	...	...	...	...

(continued)

Table 5.3 (continued)

Human ailments	Organs/ systems	Macrophytes							
		Alternanthera philoxeroides	*Alternanthera sessilis*	*Arum trilobatum*	*Asteracantha longifolia*	*Bacopa monnieri*	*Centella asiatica*	*Colocasia esculenta*	*Commelina benghalensis*
Ischaemia	Blood vessels	…	…	…	…	…	…	…	…
Kidney stone	Kidney	…	…	…	•	…	…	…	…
Lactation	Mammary gland	…	…	…	…	…	…	…	…
Leprosy	Skin	…	…	…	…	…	•	…	•
Leukaemia	Blood	…	…	…	…	…	…	…	…
Lochial discharges	Uterus/ trachea	…	…	…	…	…	•	…	…
Lustre	Skin	…	…	…	•	…	•	…	…
Menorrhagia	Uterus	…	…	…	…	…	…	…	…
Microbial infection	Any part of the body	…	•	…	…	…	…	•	…
Pain	Nerve	…	…	…	…	…	•	…	…
Piles	Anus	…	…	•	…	…	…	…	…
Pneumonia	Lungs	…	…	…	…	…	…	…	…
Poisoning	Blood/ nerve	•	…	•	…	•	•	•	…
Redness of skin	Skin	…	…	…	…	…	…	•	…
Rheumatism	Limb joints	…	…	…	•	•	…	•	•
Skin irritation	Skin	…	…	…	…	…	•	…	•
Spermatorrhoea	Genital organ	…	…	…	•	…	…	…	…
Syphilis	Genital organ	…	…	…	…	…	•	…	…

(continued)

Table 5.3 (continued)

Human ailments	Organs/ systems	Macrophytes							
		Alternanthera philoxeroides	*Alternanthera sessilis*	*Arum trilobatum*	*Asteracantha longifolia*	*Bacopa monnieri*	*Centella asiatica*	*Colocasia esculenta*	*Commelina benghalensis*
Tumour	Any part of the body	...	...	...	...	...	...	...	...
Ulcer	Mouth	...	●	...	...	...	●	...	...

Human ailments	Organs/ systems	Macrophytes							
		Enhydra fluctuans	*Ipomoea aquatica*	*Marsilea minuta*	*Mollugo cerviana*	*Nelumbo nucifera*	*Nymphaea alba*	*Nymphaea rubra*	*Polygonum plebeium*
Abscess	Skin	...	...	...	...	...	...	...	...
Alopecia	Hair follicle	...	...	...	...	...	...	...	...
Amnesia	Brain	...	...	●	...	...	...	...	...
Anaemia	Blood	...	...	...	...	...	...	...	...
Aphrodisia	Nerve	...	...	...	...	...	●	●	...
Asthma	Lungs	...	...	●	...	...	...	...	...
Biliousness	Liver	●	...	...	...	...	...	...	...
Cancer	Any organ	...	●	...	...	●	●	●	...
Cardio trouble	Heart	...	...	●	...	●	●	●	...
Cataract	Eye	...	...	...	...	●	...	...	...
Constipation	Colon	●	●	...	...	...	●	...	●
Cough	Pulmonary system	...	...	...	...	...	...	...	...
Depression	Nerve	●	●	...	...	...	...	...	...
Diabetes	Blood	●	●	●	...	...	●	●	...
Diaphoresis	Skin	...	...	...	...	...	●	●	...
Diarrhoea	Intestines	...	...	...	...	●	●	●	...

(continued)

Table 5.3 (continued)

Human ailments	Organs/ systems	Macrophytes *Enhydra fluctuans*	*Ipomoea aquatica*	*Marsilea minuta*	*Mollugo cerviana*	*Nelumbo nucifera*	*Nymphaea alba*	*Nymphaea rubra*	*Polygonum plebeium*
Distastefulness	Tongue	●	●	...	...	...	...	...	●
Diuresis	Kidney	...	...	...	...	●	...	...	...
Dropsy	Lower limb	...	...	...	...	...	●	●	...
Drowsiness	Nerve	...	...	...	...	...	...	...	...
Dysentery	Colon	...	...	...	...	●	●	●	...
Dysmenorrhoea	Vaginal tract	...	...	●	●	...	...	...	...
Dyspepsia	Stomach	●	...	...	●	●	●	●	...
Ear ache	Ear	...	...	...	...	...	...	...	...
Emesis	Stomach	...	●	...	...	●	...	...	...
Encephalitis	Brain	...	...	...	...	...	●	●	...
Epilepsy	Nerve	...	...	●	...	...	...	...	...
General debility	Genital organ	...	●	...	...	...	...	...	...
Gonorrhoea	Genital organ	...	●	...	●	...	...	...	...
Helminthiasis	Intestines	...	...	...	...	...	...	...	...
Haemorrhage	Blood	...	...	...	...	●	●	●	...
Hepatitis	Liver	●	...	●	●	●	●	●	...
Herpes	Mouth	...	...	...	...	...	...	...	...
Hoarseness	Vocal chord	...	...	...	...	...	...	...	...
Hypoglycaemia	Blood	...	...	...	...	...	...	...	...
Infertility	Genital organ	●	...	●	...	...	...	...	...
Inflammation	Intestine	...	...	...	●	●	...	...	...
Insanity	Brain	...	...	...	...	...	...	...	...
Insomnia	Brain/nerve	...	...	●	...	...	...	...	...

(continued)

Table 5.3 (continued)

Human ailments	Organs/ systems	Macrophytes *Enhydra fluctuans*	*Ipomoea aquatica*	*Marsilea minuta*	*Mollugo cerviana*	*Nelumbo nucifera*	*Nymphaea alba*	*Nymphaea rubra*	*Polygonum plebeium*
Ischaemia	Blood vessels	...	...	...	...	●	...	...	...
Kidney stone	Kidney	...	...	...	...	...	...	...	...
Lactation	Mammary gland	...	●	...	...	...	...	...	...
Leprosy	Skin	...	...	...	...	●	...	...	...
Leukaemia	Blood	...	...	...	...	...	●	...	...
Lochial discharges	Uterus/ trachea	...	...	...	●	...	...	...	...
Lustre	Skin	...	...	...	...	...	...	...	...
Menorrhagia	Uterus	...	...	...	...	●	...	...	...
Microbial infection	Any part of the body	...	●	...	...	●	...	...	...
Pain	Nerve	...	...	●	...	●	●	●	...
Piles	Anus	...	...	...	...	●	●	●	...
Pneumonia	Lungs	...	...	...	...	...	...	...	●
Poisoning	Blood/nerve	●	●	●	...	●	...	...	...
Redness of skin	Skin	...	...	...	...	...	...	...	...
Rheumatism	Limb joints	●	●	...	...	...	...	...	...
Skin irritation	Skin	●	...	●	●	●	●	●	...
Spermatorrhoea	Genital organ	...	...	...	...	...	...	...	...
Syphilis	Genital organ	...	...	...	...	...	...	...	...
Tumour	Any part of the body	...	...	●	...	...	...	...	...
Ulcer	Mouth	...	●	...	...	...	...	...	...

(References: Bhattacharya 1993; Chopra et al. 1986)
'●' represents the respective plants in the column used for curing the human ailments mentioned in the row
'...' indicates no role of the particular species against human ailments in the row

B. monnieri, *A. trilobatum*, *E. fluctuans*, *M. cerviana*, and *N. nucifera*. The present study considers these herbs as 'poor man medicine' for their substantial use in both curative and preventive measures of human ailments.

5.4 Status of Macrophytes Having Medicinal Values

5.4.1 Description of Macrophytes and Availability

There are several macrophytes (Naskar 1990) having medicinal values, but we mentioned few selective macrophytes, which are well known for their medicinal values and familiar to common people. Their morphological descriptions are given below so that users can identify them and collect them for specific treatment of human ailments.

Alternanthera philoxeroides **Griseb. (Amaranthaceae):** common name, alligator weed (Fig. 5.1a); perennial creeper with erect branching; leaves opposite decussate, lanceolate; inflorescence spikes; flower white, capitulate at axil. Flowering

Fig. 5.1 (**a**) *Alternanthera philoxeroides*, (**b**) *Alternanthera sessilis*, (**c**) *Arum trilobatum*, (**d**) *Asteracantha longifolia*, (**e**) *Bacopa monnieri*, (**f**) *Centella asiatica*, (**g**) *Colocasia esculenta*, (**h**) *Commelina benghalensis*, (**i**) *Enhydra fluctuans*, (**j**) *Ipomoea aquatica*, (**k**) *Marsilea minuta*, (**l**) *Mollugo cerviana*, (**m**) *Nelumbo nucifera*, (**n**) *Nymphaea alba*, (**o**) *Nymphaea rubra*, (**p**) *Polygonum plebeium*

period occurs from December to April; fruit appears to be an indehiscent reniform. This plant population grows near water body, pond dyke, and saturated soil and grows above water surface occasionally.

***Alternanthera sessilis* (L.) R.Br. (Amaranthaceae):** common name, sessile joyweed (Fig. 5.1b); perennial herb, prostrate purple stem; leaves opposite decussate, lanceolate; inflorescence spikes, sessile; flower white, clustered at axil; fruits utricle with winged margins; seed, suborbicular. This plant population grows near water body, pond dyke, and saturated soil and even can withstand dry summer for a short while.

***Arum trilobatum* Linn. (Araceae):** common name, Bengal Arum (Fig. 5.1c); stem underground, tuberous rhizome; petiole up to 30 cm long; leaf blade deeply three-lobed; spathe convolute at base, inside dark purple to reddish purple, spadix shorter than spathe; female zone slightly conic; ovary yellowish green; stigma sessile, dark or mid-purple, disciform; proximal half densely covered with staminodes, distal half naked; staminodes strongly curled but mostly directed downwards and covering most of female zone. This species grows in marshy area or in saturated soil under tree shade.

***Asteracantha longifolia* Nees. (Acanthaceae):** common name, marsh barbel (Fig. 5.1d); erect herb with spines at each node; stem quadrangular with thickened nodes; leaves are sessile and oblong-lanceolate; flowers are white-purple, clustered at whorls, 2–3 cm long, bilabiate; fruits two-celled, linear oblong, compressed capsule, 8 mm long, 4–8 seeded; seed ovate, flat, or compressed. This species grows in marshy places at dyke of paddy fields and pond embankment. This species is cultivated in different parts of Bengal and marketed as a leafy vegetable.

***Bacopa monnieri* (L.) Pennell (Plantaginaceae):** common name, water hyssop (Fig. 5.1e); prostrate stem with erect branches and succulent herb; leaves opposite, sessile, fleshy, lush green, oblong; flowers solitary, white with 4–5 petals; fruit capsule, ovoid or oblong, enclosed within calyx; seeds oblong, reticulate. This species grows in marshy places, saturated soil, and the areas where water drainage is available. This species is cultivated unconventionally and marketed as a leafy vegetable.

***Centella asiatica* L. (Apiaceae):** common name, Indian pennywort (Fig. 5.1f); creeper with roots developed from lower nodes; leaves orbicular, reniform, crenate or sub-entire, glabrous, nerves radiating; petiole to 12 cm long, sheathing at base; inflorescence umbel shaped, inconspicuous; flowers sessile, pink, in 2–5 flowered umbels; fruit ovoid, reticulate-rugose, 7–9-ribbed. This species grows in marshy places, saturated soil, and the areas where water drainage is available. This species is not cultivated but grows profusely and is marketed as a leafy vegetable.

***Colocasia esculenta* (L.) Schott (Araceae):** common name, elephant's ear (Fig. 5.1g); rhizomatous, rhizome stoloniferous; leaves triangular ovate with long petiole forming spathe at base; inflorescence spadix, 10 cm long, cylindrical, appendages terete, obtuse; male flowers above, 5–6 cm long of the spadix, stamens 6; female flowers on lower, 2 cm long of the spadix; ovary one-celled, neutral flowers many, peltate, lying between the female and male flowers; fruit an aggregate of berries, globose. This species grow wildly but is cultivated in shallow pond for vegetable and marketed.

***Commelina benghalensis* L. (Commelinaceae):** common name, tropical spiderwort (Fig. 5.1h); creeper, diffuse herb; rooting at lower nodes; leaves oblong; petiole long, sheath at base, apex with hairs, spathe funnel-shaped, truncate at apex; flowers cyme, bluish violet; petals blue, larger one broadly ovate; stamens 3; staminodes 2; ovary long; capsule long, ellipsoid, seeds pitted. This species grows in marshy areas and withstands dry weather for a short time.

***Enhydra fluctuans* Lour. (Asteraceae):** common name, watercress (Fig. 5.1i); perennial herb with profuse branching, trailing; rooting at nodes; leaves opposite decussate, sessile; inflorescence capitulate, axillary; flower sessile, white; terminal and small. This species grows above the surface water and is known as a free-floating macrophyte that is able to cover a substantial area of water surface.

***Ipomoea aquatica* Forssk. (Convolvulaceae):** common name, water spinach (Fig. 5.1j); creeping, or floating, a perennial herb with a trailing stem, stem hollow; rooting at the nodes, root poorly developed; leaves alternate, varies in form, usually oblong-lanceolate or narrowly triangular, base hastate, apex acute; petiole 6–10 cm long; flowers purplish white, solitary or few in cymes; sepals subequal, 6–8 mm long, oblong-lanceolate, membranous, glabrous; corolla funnel-form, pale purple to nearly white, tube 2 cm long, lobes obscure; stamens included, filaments unequal, hairy at the base; ovary glabrous; capsule globose; seeds, minutely pubescent. This species grows in a diverse habitats right from above water surface to marshy areas to saturated soil. When it grows in shallow water, it can cover whole water body. It grows wildly and is also cultivated and marketed as an important leafy vegetable.

***Marsilea minuta* L. (Marsileaceae):** common name, dwarf water clover (Fig. 5.1k); perennial fern, creeper with long creeping rhizome; fronds simple; stipe long, slender, glabrous or softly pubescent; lamina quadrifid, each lobe obovate or obtriangular, lobed to serrate along the outer margins; sporocarp oblongoid, black, produced in clusters. This species grows mainly in marshy places but occasionally is found growing on water surface. It is used as an important leafy vegetable and marketed, though it grows wildly.

***Mollugo cerviana* Ser. (Molluginaceae):** common name, thread-stem carpetweed (Fig. 5.1l); prostrate herb with erect branching; leaves opposite to whorled, lamina broadly linear; inflorescence terminal cyme; flower with whitish petal, long pedicel; stamens 6–8; ovary three-lobed; capsule ellipsoid; seeds many, ovoid. This species grows in marshy areas and can withstand dry summer for quite a while. It is used as an important leafy vegetable and marketed, though it grows wildly.

***Nelumbo nucifera* Gaertn. (Nelumbonaceae):** common name, sacred lotus (Fig. 5.1m); perennial rhizomatous herb; roots advantageous; leaves simple, solitary, petiolate; petiole may grow up to 2 m long, beset with scattered hard minute papillae; long stalk with round lamina; lamina, suborbicular, dark green above, paler beneath, glabrous and glaucous on both surfaces, coriaceous, shallowly notched and apiculate at one side, terminating in a simple vein, membranous when dry; flowers solitary on long peduncles, rose-pink or white, expanding and emerging above water; sepals linear-elliptic, concave, obtuse, with midrib distinct near at apex; greenish purple outside, purplish inside; corolla shorter than sepals, green outside, purplish inside, lobes broadly or narrowly oblong; stamens long, fertile, incurved; fruit berry; seeds globose, brown, spinulose. This species grows profusely in abandoned water bodies and the periphery of lake, reservoir, ponds, and tanks. All the parts of this species are used in one or other ways and marketed.

***Nymphaea alba* L. (Nymphaeaceae):** common name, white water lily (Fig. 5.1n); perennial rhizomatous herb, rhizomes erect or creeping, stoloniferous; rooted; leaves polymorphic, suborbicular to cordate, long stalk; lamina oval, floating, green; flower bisexual, long peduncle, receptacle cylindrical, white petals on long stalk; stamens numerous, distributed up to summit of ovary; carpels completely united, ovary superior, stigma flat; fruit berry; seeds ellipsoid, smooth. This species grows in abandoned water bodies, beel, jheel, karanjali, and shallow water bodies. This species grows wildly and is marked.

***Nymphaea rubra* Roxb. (Nymphaeaceae):** common name, red water lily (Fig. 5.1o); perennial rhizomatous herb, rhizomes erect or creeping, stoloniferous; rooted; leaves polymorphic, suborbicular to cordate, long stalk; lamina oval, floating, reddish green; flower bisexual, long peduncle, receptacle cylindrical, pink petals on long stalk; stamens numerous, distributed up to summit of ovary; carpels completely united, ovary superior, stigma flat; fruit berry, reddish black; seeds ellipsoid, smooth. This species grows in abandoned water bodies and shallow water bodies. This species grows wildly and is marked occasionally.

***Polygonum plebeium* R. Br. (Polygonaceae):** common name, not known (Fig. 5.1p); a seasonal prostrate semi-aquatic herb with diffuse branching; leaves sessile, elliptic; inflorescence axillary, flowers with white-pink tepals, axillary. This species grows in marshy areas and saturated muddy soil. This species grows wildly and marketed occasionally.

5.4.2 Threats to Macrophytes' Habitats and Ecology

Ecological system is typically structured and varies from one system to another. Further, any ecological system is governed by local scale pattern in relation to environmental factors. The regional scale pattern, by and large, determines species distributions (Cottenie 2005; Levin 1992). So the existence of any habitat that may be terrestrial or aquatic or marshy areas is greatly dependent on the ecological system in which the particular habitat exists. The ecological factors and the respective habitat co-exist in an integral manner; if one changes or alters, the other would be affected. Besides climatic factors (discussed in the later paragraph), the major cause that is alarming to be responsible to threaten wetlands or aquatic systems is anthropogenic pressure that causes to exploit aquatic system indiscriminately. So, there is a major concern about dwindling of aquatic system—a known repository of many worth biotic resources that serve human civilization since immemorial time. Already, the rapid increase of urbanization has engulfed many water bodies and wetlands, which are not traceable now. Those habitats are either destroyed or fragmented or disappeared due to inhabitations by tremendous increasing population pressure. Experts opine that loss of aquatic habitats might bring a number of inevitable threats towards the environment, one of which is loss of aquatic biodiversity. The following consequences of aquatic system may occur due to anthropogenic interference:

- Changes in aquatic species, communities, and ecosystems would certainly affect the sustainability of aquatic ecosystem health, causing tremendous alteration of aquatic biological diversity.
- If one species is exploited by human activities, another one may occupy its niche, so there would be an alternation of macrophytes' composition.
- Changes in macrophytes' composition may affect biological diversity that might impact overall ecosystem services, even if there may be mono-specific composition.
- Changes in aquatic biological diversity may impact land-use alterations, nitrogen deposition, and exotic species invasion to threaten the ecosystem functioning and biodiversity of aquatic ecosystems and aquatic-terrestrial ecotones.

Ironically, there is not so much strong hue and cry about the alteration of such aquatic habitats and their biodiversity (Mandal 2011). Such anthropogenic interference causes disturbances of livelihood support of common people, because a considerable population in rural areas depends upon such ecosystem services in one way or the other.

5.4.3 Vulnerability of Macrophytes due to Climate Change

Changing climate has a great impact on the earth ecosystem including its biological diversity, which has been felt tremendously. The components which are responsible for climatic changes include GHG (greenhouse gases) including carbon dioxide, methane, nitrous oxide, and fluorinated gases. Greenhouse gases cause to increase the global surface temperatures by 1.4–5.8 °C predicted in the next 100 years (IPCC 2007). Increasing surface temperature due to climate change may bring the following consequences on macrophytes:

- Climate changes cause to alter the water chemistry and hydrological regimes that affect the structure and function of aquatic ecosystems (Alahuhta 2015; Alahuhta et al. 2011; Heino et al. 2009; Knutti and Sedlacek 2013; Nielsen 2003; Poff et al. 2002; Rahel and Olden 2008). Alteration of water characteristics along with components including high nutrient loading causes shifting from oligosaline or mesosaline conditions to eutrophication (Beklioglu et al. 2011; Trenberth et al. 2014; Wrona et al. 2006), leading to develop high competition among macrophytes and phytoplankton. Eventually, the ecosystem turns to be the house of nuisance aquatic macrophytes for sheltering dangerous aquatic animals in the place of a vibrant aquatic ecosystem.
- Climate changes affect air temperature, precipitation, and other stressors that cause to alter the physical, chemical, and biological characteristics of freshwater

ecosystems and their biological communities, especially macrophytes (Alahuhta 2015; Ejankowski and Lenard 2015; Wrona et al. 2006). Temperature is one such factor that regulates plant growth as an increase in shoot length, plant height, leaf surface area, and biomass production. The increased photosynthetic rate supported height development and production of root, rhizome, and leaf biomass. Over and above, temperature changes are related to thermal tolerance of individual species, causing to affect phenology—budding, leaf growth, flowering, fruiting, and seed production (Thackeray et al. 2010).

- Rising temperature causes to change CO_2 concentration and alteration of precipitation, which impact directly and indirectly growth, productivity, and distribution of aquatic vegetation (Heino et al. 2009; Kankaala et al. 2000; Pearson and Dawson 2003; Peeters et al. 2013; Tokoro et al. 2014; Wrona et al. 2006). In general, net photosynthesis of macrophytes increases with temperature up to an optimum value and then decreases dramatically. Very high temperatures have an overall negative impact on the net primary productivity (NPP) of plants as the rate of respiration increases at a greater pace than photosynthesis, leading to a steady decrease in the photosynthesis to respiration (P:R) ratio. Eventually, warming will favour the growth of few species; hence, the diversity and species richness of macrophytes will decrease (Erwin 2009). The overall aquatic ecosystem will be affected, and maybe some macrophytes disappear, while other nuisance macrophytes occupy their place due to inter-species competition.
- Temperature changes affect nutrient uptake of individual species, and the consequences of that lead to intra- and inter-species competition (Mooij et al. 2007; Wrona et al. 2006). Also, changing temperature causes to affect metabolic events such as photosynthesis and respiration and enzyme-mediated processes including primary productivity. Nutrient enrichment or alteration is crucial for the growth of macrophytes that affect the composition of aquatic plant communities, particularly free-floating and rooted macrophytes (Madsen and Cedergreen 2002).

In a nutshell, climate change leads to subtle shift of macrophytic population that impacts the overall aquatic ecosystem, biodiversity, and its functioning. Mitigating this challenge needs to maintain, conserve, and protect aquatic ecosystem and its floral community in a sustainable manner; otherwise, we lose this precious ecosystem that harbours plant bio-resources—the potential repository of a variety of gene pool of food, fodder, feed, and vegetable crops, apart from a substantial number of known medicinal macrophytes.

5.4.4 Common People and Use of Macrophytes as Drugs in the Past, Present, and Future

5.4.4.1 In the Past

The history of medicine by using plant bio-resources in India may be traced to the remote past (Chopra et al. 1986), which bears the rich heritage of Indian medicinal wisdom. Dated back between 4500 and 1600 B.C., the earliest mention of use of plants is recorded in the *Rig Veda*—the oldest repository of human knowledge documented on medicine (Chopra et al. 1986). The 'Ayurveda', which means to increase the longevity of human life, is known to be the repository of the information of various drugs mentioned in detail, mainly the usage of various plants' materials as a drug. The remarkable book *Sushruta Samhita*, which was written less than 1000 B. C., contains comprehensive chapters on therapeutics. Another book *Charaka Samhita* written at the same time as *Sushruta Samhita* provides a comprehensive description of the 'materia medica' as known to ancient Hindus (Chopra et al. 1986). During the Buddhist period, medicinal plants were widely used as household drugs, and their cultivation was practised to a noticeable extent. There was knowledge outside India, which also contributed to enrich Indian 'materia medica', which contains a number of aquatic plants used for treatments of human ailments.

5.4.4.2 In the Present

There are many references dealing with plants including macrophytes used as a drug for remedy of human ailments. In India, many states are still practising 'Ayurveda' to cure human ill health; Kerala is one of the foremost states among others in south India, and Uttarakhand is the pioneering state in the north and Bengal in the eastern region. Here, we name one or two remarkable works that are the valuable documents of plants used as drugs. Shibkali Bhattacharyya, a renowned Ayurvedic doctor, documented all common medicinal herbs comprising 500 plants (approx.) and their use in human ailments in his book *Chiranjib Banousadhi* written in Bengali vernacular language. In his documentary, a substantial number of macrophytes and their usage along with therapeutic procedures are well documented. Prof. Asima Chatterjee, University of Calcutta, made a remarkable progress to analyse all the organic compounds from plant materials used drugs.

5.4.4.3 In the Future

Despite dependency on modern medical therapy, a large number of populations in rural India still rely on 'Ayurvedic' therapy using a variety of plant materials. 'Baidyanath', 'Dabur', and the recently manufactured 'Patanjali' drugs, which are made with plant-based materials, carry forward the past legacy of Indian ancient therapy of plant-based medicine and are popular to common people across rural and semi-urban and urban India. Unfortunately, all these plants containing medicinal properties are vanishing rapidly due to human interference. Macrophytes are the most victims among other vegetation since macrophytes are easily destroyed without monitoring. Forest department can check large trees, prevent offences, and catch the culprits for punishment, but for herbs or macrophytes, there is no checkpoint of monitoring. Fortunately, there are people who maintain semi-aquatic macrophytes in their backyard for their daily use as household medicine through inherited knowledge. Even there are small ponds and water bodies in which several common macrophytes are kept growing for medicinal purposes as a part of vegetables. Besides, many macrophytes are cultivated, and their cultivation techniques are well established, which are consumed as leafy vegetables. So it is expected that as long as human civilization exists, the importance of macrophytes as drugs will remain nurtured for the benefit of human health.

5.5 Utility of Macrophytes to Facilitate Livelihood Opportunities

5.5.1 Awareness Drive for Protection of Macrophytes About Their Medicinal Values

There are macrophytes having potential medicinal values and used as leafy vegetables by local people who consume them for two basic purposes such as food crops (Chandra 1997) in one way and health protective vegetable of medicinal values in the other (Mandal and Mukhopadhyay 2010). However, their importance as medicinal properties is not much cared for so much so that they are exploited tremendously—their habitats are destroyed, fragmented, encroached, and utilized for other purposes. Indiscriminate destruction of *Centella asiatica*, *Asteracantha longifolia*, *Bacopa monnieri*, *Enhydra fluctuans*, *Marsilea minuta*, and *Mollugo cerviana* is a well-established fact in rural India (Mandal and Mukhopadhyay 2010; Mandal and Jayasankar 2014). Irony is that people who destroy them from natural habitats are found purchasing them at a high cost when they are prepared as a tonic packed in a bottle by the reputed drug companies. It is not a scattered phenomenon but is a well-known fact. Even these macrophytes are collected by locals ignorantly, and they supply them to pharmaceutical agencies for a minimum price. It is just like a committed suicide because one day will come when these valuable macrophytes are not to be available in our environment. It is the time to aware local people about the medicinal values of such drug plants and to guide them for their conservation religiously.

5.5.2 Encouragement of Research to Explore Medicinal Potentials of Macrophytes

Probably, there is a gap of consciousness between traditional eaters and those who think about their conservation. Traditional eaters are not in a position of thinking about the disappearance of these wild leafy vegetables. They think if these are available, they will collect them. They are not much aware of the impact of their disappearance. It is not expected so! Environmental conservationists know about their utility, but probably they suffer from the lack of information about the data of proper record pertaining to the account of their use by rural population: how many peoples per annum feed upon these wild leafy vegetables and how much amount of them are utilized annually. People use them as a non-conventional vegetable and to some extent know about their medicinal importance through traditional knowledge (Mandal and Mukhopadhyay 2010). There are suggestions which may be implemented under the research programme. Encouragement of scientific research in place of encouragement of research is necessary to have a thorough survey essentially required to generate data about their record of use: (a) total area of their coverage, (b) locality where they grow, (c) climatic condition they grow in, (d) phenological record, (e) method of reproduction, (f) varieties of usage, (g) their amount of exploitation/annum and ways of exploitation, (h) status in red data book, (i) details on medicinal and nutritive value, and (j) present status in 'Ayurveda' therapeutic use.

5.5.3 Priority on Conservation of Macrophytes: An Hour of Need

Conservation of freshwater is one of the priorities worldwide (Mandu 1995; Martin 1995) Freshwater in the near future can seriously be in crisis because of global warming and climate change. In this context, freshwater conservation needs to be considered as an utmost priority for the protection of freshwater bodies and their resources including macrophytes. Conservation of macrophytes may bring the following advantages:

- Maintenance of water bodies along with their biodiversity with particular reference to macrophytes may conserve freshwater from the loss due to excessive evaporation.
- Macrophytes' coverage hinders light penetration inside the water, which is thus protected from its loss. Such a conservation effort through macrophytes has advantages like storage of water and maintenance of aquatic biodiversity.
- Macrophytes' coverage provides a suitable refuge for a variety of wild life including fish species, which are now threatened, and thus conservation of macrophytes means protection of diverse aquatic lives.
- A variety of wild fish species which take shelter, breed, hatch spawn, rear fry, and grow adult complete the life cycle—a biological existence is possible in water bodies in association with macrophytes. It is evident that few species lay eggs which remain adhered with roots of macrophytes—an ecological significance of mutual interest between macrophytes and fish; else a gene pool of diverse aquatic species might get lost from this earth.
- Presence of macrophytes in water bodies is a natural phenomenon that makes an ecosystem vibrant. The submerged portion of macrophytes acts as suitable substrata around which a number of periphytons grow on which aquatic animals graze to feed them, particularly fish. So a definite food chain is established that further enhances to make food web of an aquatic system. Besides, a huge amount of periphyton assemblage enhances productivity of water bodies by addition of O_2 through photosynthesis.

It may be perceived that if one-tenth area of the total fallow water bodies and wetlands are utilized in the purpose of conservation, it can surely bring about a remarkable change in the scenario of protecting medicinal macrophytes. It may also be another advantage for revenue generation by exploiting such medicinal macrophytes to the great extent. The country has tremendous human resources of invigorating energy who need to be engaged in such a case—a concerted effort is required to harness the potential of both conservation of water bodies and their medicinal macrophytes' bio-resources.

5.5.4 Low-Cost Medicinal Drugs Curing Human Ailments: An Alternative to Synthetic Drugs

Rural masses are acquainted with the medicinal as well as the nutritional value of macrophytes, which are mostly seasonal, and a few are annual and perennial. They have dense population while growing wildly, thus occupying a major portion of any aquatic vegetation. Not that they are merely leafy vegetables eaten but that they are eaten as their nutritional and medicinal values are familiar to rural masses to the great extent (Borah et al. 2009; Deb et al. 2013; Fleuret 1993; Kumarasamyraja et al. 2012; Mandal et al. 2008; Mandal and Bar 2013; Mandal and Jayasankar 2014; Ong et al. 2011; Roy et al. 2013; Soni and Singh 2012; Weng and Chen 1996). Probably, therefore, anybody could find them as one of the daily items available in the full meal in rural areas. Without manuring and with less care, they grow wildly in fallow water bodies and wetlands. Treated as non-conventional leafy vegetables, they are more organic as well as nutritious than a number of vegetables cultivated. They are also treated as 'poor man drug' because population across economic status in rural areas consumes them for their taste as well as health benefits (Chen 2000; Mandal and Mukhopadhyay 2010; Rodale and Mcgrath 1991). On the contrary, synthetic drugs are so costly that poor masses may not afford them, except in emergency requirements (Mandal and Mukhopadhyay 2010). In such case, these aquatic and semi-aquatic macrophytes serve as an alternative drug to care and protect for human health instead of costly synthetic drugs.

5.5.5 Development of Cultivation Protocols of Macrophytes with Package of Practice

Cultivation practice which requires strategic methods to develop crops in the fields includes quality seeds, sowing seeds, preparation of seed bed through ploughing, transplanting of seedlings, manuring, irrigation, nurturing crops with fertilization, and harvesting. These are some important aspects of agriculture required to develop when medicinal macrophytes are to be brought under cultivation practices. Each crop requires a specific technique to grow because individual species has its own ecological cycle, and accordingly, its cultivation protocol needs to be developed. For example, *Centella asiatica* contains *asiaticoside* as an active ingredient used for treatment of leprosy (Chopra et al. 1986). The question is that which life stage or maturity stage of *Centella asiatica* or its leaf has produced the maximal amount of *asiaticoside*. If it is known that a particular phase of leaf maturity produces the optimal amount of *asiaticoside*, the leaves are to be harvested at that particular stage for medicinal purposes. It is only possible when plant is cultivated with the distinct package of practice. *Ipomoea aquatica* is widely cultivated and for which a distinct package of practice is available to farmers. *Asteracantha longifolia* and *Nelumbo nucifera* are brought under cultivation practice, though partially because an established package of practice is yet to be standardized for them. There are many more medicinal macrophytes which need to be brought under cultivation practice, and there are places outside India where such medicinal plants are brought under cultivation practice (Moreno-Black et al. 1996). Their incorporation in cultivation will serve many advantages:

1. New crops to be introduced in the cultivation system.
2. When these are medicinal crops, production of herbal drugs is assured.
3. Employment of generations created, when medicinal plants are undertaken in cultivation.
4. Conservation of such threatened macrophytes assured.
5. Maximal production of active ingredient from particular macrophyte assured.
6. Trading of such macrophytes to encourage people to adopt macrophytes' cultivation.
7. Indigenous system of therapy using herbs to be spread among common people.

5.5.6 People Engagement to Proliferate Macrophytes in Production of Drugs

Rural people need to be engaged to protect and nurture their environment and resources. Nurturing macrophytes as medicinal crops needs 3Ps: protection, propagation, and proliferation (Mandal and Jayasankar 2014). Protection includes both maintenance of natural habitats of aquatic plants and conservation of its wild genetic stocks. Propagation requires production and development of aquatic plants of cultivable varieties through application of advanced biotechnology. Proliferation certainly indicates cultivation of aquatic plants with the aim to bring as much as fallow water bodies and wetlands for their sustainable utilization (Mandal and Mukhopadhyay 2010). Proliferation of medicinal macrophytes can surely bring about a remarkable change in the scenario of future crops in India. It may also augment the revenue generation compared to the present context; else a huge amount of nutrients which aquatic plants contain will get lost without proper utilization.

5.5.7 Use of Mobile App to Spread the Drug Value of Macrophytes to Common Users

'App' is an abbreviated form of the word 'application'. It is designed in the specific software program to perform specific information directly catered for the user or, in some cases, for another application program. App is run on a mobile device such as a phone, tablet, or watch. Most mobile devices are sold with several apps bundled as pre-installed software, such as a web browser, email client, calendar, mapping program, and an app for buying music, other media, or more apps. Some pre-installed apps can be removed by an ordinary uninstall process, thus leaving more storage space for desired ones. Where the software does not allow this, some devices can be rooted to eliminate the undesired apps. However, several other apps which are not preinstalled in the device are usually available through distribution platforms called app stores. These may be operated by the owner of the device's mobile operating system, such as the App Store (iOS) or Google Play Store; by the device manufacturers, such as the Galaxy Store and Huawei AppGallery; or by third parties, such as the Amazon Appstore and F-Droid.

Maybe many software companies are interested to develop apps on payment basis. To make advertisement the spread of medicinal herbs with their importance, macrophytes in particular, 'apps' may be designed to store all the information regarding specific macrophytes possessing a certain drug to cure the specific human ailment. App is to be designed in such a way that is attractive to users and caters essential information so that users feel the necessity to buy medicinal macrophytes as a household item and perceive them as an alternative to synthetic drugs in case of selective common human ailments. Besides, app needs to store all the information of macrophytes in detail as follows:

- Name of the macrophytes
- Local name in different regions
- Diagrammatic view of macrophyte for easy identification
- Seasonal description of availability
- Available plant constituent/compound used as drug
- Procedure of application of drug
- How it benefits more as compared to synthetic commercial drug
- Cultivation procedure of macrophytes
- Harvesting of macrophytes and extraction of drug

5.5.8 Incorporation of Macrophytes into School Curriculum

Our young minds who are nurtured in school need to be educated about their environment, its biota, and their importance to human welfare. In such case, medicinal plants may be introduced in the school curriculum. They are to be educated in such a way so that students feel interested to have inquisitiveness to know all those plants which are an integral part of their daily life. The study curriculum needs to be formulated in such a way that it is to be survey-oriented. School students have practical knowledge about them and their status in fields. A questionnaire may be framed in such a way that may include the following: (1) common name of the herb, (2) parts used, (3) why to use/which human ailment it is used to, (4) how to use, (5) source of knowledge, and (6) present status of the herbs.

5.5.9 Trade of Macrophytes as Ayurvedic Drugs

There are now many establishments of business-related management courses. Trade of medicinal macrophytes may be thought to be incorporated in such a management course so that their trades get momentum. People of other countries may get interested towards such medicinal plants and their worthiness of therapeutic usages. Once they are introduced in trading, a business chain is automatically developed—right from producers to consumers via a series of middleman involvement. Proper advertisement that highlights the values of medicinal macrophytes and their utility to cure human ailments without side effects is the need of the hour. India is a large country, and its water resources is enormous and so its aquatic bio-resources. Only one enthusiastic mind is required to initiate the spread of their potentials in a proper direction, and the rest is assured to bring a fruit of success.

All the above information related to macrophytes certainly manifests livelihood opportunities through engagement of a large number of people and thus creates employment opportunities to the great extent.

5.6 Epilogue

Alarmingly, availability of these plants is becoming scarce as surveyed. No wonder that these macrophytes are reported as important sources of drug for human ailments as practised by local people, even in other parts of the world, Bangladesh in particular. Considering their disappearance rapidly, a conservation effort is necessary to protect them as practised in different parts of India by individual or organization levels such as some NGOs. We feel that our indigenous plants' wealth is worthy enough to protect us from suffering various diseases. Our care for such macrophytes will transcend our hope of medicinal prospect in diverse ways crossing the limited boundaries. People should take care of their precious health through consumption of potential macrophytes which provides both nutrition and drug as a means of food.

References

Aglin AA (2018) Medicinal effects of Mollugo cerviana–a review. Int J Sci Res in Multidisciplinary Studies 4:9

Al Amin M, Chowdhury IA, Mahbub KMM, Sattar M, Shahriar M, Kuddus MR, Rashid MA (2012) Anti-inflammatory and analgesic activities of Asteracantha longifolia Nees. Bangladesh Pharm J 15(2):171–176

Alahuhta J (2015) Geographic patterns of lake macrophyte communities and species richness at regional scale. J Veg Sci 26(3):564–575

Alahuhta J, Heino J, Luoto M (2011) Climate change and the future distributions of aquatic macrophytes across boreal catchments. J Biogeogr 38(2):383–393

Alhazmi HA, Najmi A, Javed SA, Sultana S, Al Bratty M, Makeen HA, Meraya AM, Ahsan W, Mohan S, Taha MM, Khalid A (2021) Medicinal plants and isolated molecules demonstrating

immunomodulation activity as potential alternative therapies for viral diseases including COVID-19. Front Immunol 12:1721
Ali K, Ashraf A, Biswas NN (2012) Analgesic, anti-inflammatory and anti-diarrheal activities of ethanolic leaf extract of Typhonium trilobatum L. Schott Asian Pacific J Tropical Biomed 2(9): 722–726
Arollado EC, Osi MO (2010) Hematinic activity of *Alternanthera sessilis* (L.) R.BR. (Amaranthaceae) in mice and rats. E-Int Sci Res J 12:110–117
Azam FMS, Biswas A, Mannan A, Afsana NA, Jahan R, Rahamatullah M (2014) Are famine plants also ethnomedicinal appraisal of famine food plants of two districts of Bangladesh. Evid Based Complementary Altern Med 2014:1–28
Beklioglu M, Meerfhoff M, Sondergaard M, Jeppesen E (2011) Eutrophication and restoration of Shallow Lakes from a cold temperate to a warm Mediterranean and a (sub) tropical climate. In: Ansari AA, Gill GS, Lanza GR, Rast W (eds) Eutrophication: causes, consequences and control, 1st edn. Springer, New York, pp 91–108
Bhattacharya S (1993) Chiranjib Banousadhi (in Bengali). Ananda Publishers Pvt. Ltd., Kolkata, p 6
Boddupalli A, Kumanan R, Elumalai A, Eswaraiah MC, Veldi N, Pendern A (2012) A twelve-monthly review on antidiabetic plants: Jan–Dec – 2011. Int Res J Pharm 3:77–80
Borah S, Am B, Das AK, Borah J (2009) Determination of mineral content in commonly consumed leafy vegetables. Food Anal Methods 2(3):226–230
Chandra RK (1997) Nutrition and the immune system: an introduction. Am J Clin Nutr 66:4605–4635
Charoenphon N, Anandsongvit N, Kosai P, Sirisidthi K, Kangwanrangsan N, Jiraungkoorskul W (2016) Brahmi (Bacopa monnieri): up-to-date of memory boosting medicinal plant: a review. Indian J Agric Res 50:1–7
Chen J (2000) Regulatory control of functional food in China. Scand J Nutr 44(3):130–131
Cheng JH, Lee SY, Lien YY, Lee MS, Sheu SC (2012) Immunomodulating activity of Nymphaea rubra Roxb. Extracts: activation of rat dendritic cells and improvement of the TH1 immune response. Int J Mol Sci 13(9):10722–10735
Chopak C (2000) Famine early warning system project, early warning primer: an overview of monitoring and reporting. United States Agency for International Development
Chopra RN, Nayar SL, Chopra IC (1986) Glossary of Indian medicinal plants. CSIR, New Delhi
Cottenie K (2005) Integrating environmental and spatial processes in ecological community dynamics. Ecol Lett 8(11):1175–1182
Deb D, Sarkar A, Barma BD, Datta BK, Majumdar K (2013) Wild edible plants and their utilization in traditional recipes of Tripura, Northeast India. Adv Biol Res 7:203–211
Ejankowski W, Lenard T (2015) Climate driven changes in the submerged macrophyte and phytoplankton community in a hard water lake. Limnologica-Ecol Manage Inland Waters 52: 59–66
Ernst E (1993) Complementary medicine—scrutinizing the alternatives. Lancet 341:1626
Erwin KL (2009) Wetlands and global climate change: the role of wetland restoration in a changing world. Wetl Ecol Manag 17:71–84
Fleuret A (1993) Dietary and therapeutic uses of fruit in three Taita communities. In: Etkin N (ed) Plants in indigenous medicine and diet, Redgrave Bedford Hills, NY, pp 151–170
Gopalan G, Rama Shatri BV, Balasubramanian SC (2007) In: BSN R, Deosthale YG, Pant KC (eds) Nutritive value of Indian foods. National Institute of Nutrition, ICMR, Hyderabad
Guinard Y, Lemessa D (2000) Wild-food plants in Southern Ethiopia, reflections on the role of "famine foods" at a time of drought. United Nations Development Programme Emergencies Unit for Ethiopia
Haque R, Mondal D, Karim A, Molla IH, Rahim A, Faruque ASG, Ahmad N, Kirkpatrick BD, Houpt E, Snider C, Petri WA Jr (2009) Prospective case-control study of the association between common enteric protozoal parasites and diarrhea in Bangladesh. Clin Infect Dis 48(9):1191–1197

Hasan AN, Roy P, Bristy NJ, Paul SK, Wahed TB, Alam MN (2015) Evaluation of in vitro antioxidant and brine shrimp lethality bioassay of different extracts of Polygonum plebeium R Br. Int J 3(12):97–107

Heino J, Virkkala R, Toivonen H (2009) Climate change and freshwater biodiversity: detected patterns, future trends and adaptations in northern regions. Biol Rev 84(1):39–54

IPCC (2007) Climate change 2007: synthesis report: contribution of working groups I, II and III to the fourth assessment report of the intergovernmental panel on climate change. IPCC, Geneva, pp 1–104

Islam MT, Das PR, Salehin BMAFM, Hasan ME, Jahan FI, Seraj S, Islam F, Khatun Z, Chowdhury AR, Rahman MA, Rahmatullah M (2011) A survey of non-conventional plant items consumed during food scarcity in two randomly selected villages of Kurigram district, Bangladesh. Am-Eur J Sustain Agric 5(2):233–239

Jamil SS, Nizami Q, Salam M (2007) *Centella asiatica* (Linn.) urban—a review. Nat Prod Radiance 6(2):158–170

Kalita P, Mukhopadhyay PK, Mukherjee AK (2007) Evaluation of nutritional quality of four unexplored aquatic weeds of Northeast India for the formulation of cost effective fish feed. Food Chem 103:204–209

Kankaala P, Ojala A, Tulonen T, Haapamaki J, Arvola L (2000) Response of littoral vegetation on climate warming in the boreal zone; an experimental simulation. Aquat Ecol 34(4):433–444

Kansagara PA, Pandya DJ (2019) A complete review on medicinally active herbal weed: Commelina benghalensis L. (Commelinaceae). J Pharm Sci Res 11(4):1165–1171

Knutti R, Sedlacek J (2013) Robustness and uncertainties in the new CMIP5 climate model projections. Nat Clim Chang 3:369–373

Kumarasamyraja D, Jeganathan NS, Manavalan R (2012) A review on medicinal plants with potential wound healing activity. Int J Pharm Sci 2:105–111

Levin SA (1992) The problem of pattern and scale in ecology. Ecology 73(6):1943–1967

Madhu S, Kannabirran V, Frank PR, Reddy MS, Gnanasekher N (2012) Evaluation of anti-diabetic activity of *Marsilea minuta* Linn. against alloxan-induced diabetes in albino rats. Int Res J Pharm 3(8):223–225

Madsen TV, Cedergreen N (2002) Sources of nutrients to rooted submerged macrophytes growing in a nutrient-rich stream. Freshw Biol 47(2):283–291

Mandal RN (2011) Nurturing wetlands. Science Reporter, pp 45–48

Mandal RN, Bar R (2013) The sacred lotus–an incredible wealth of wetlands. Resonance 18:732–737

Mandal RN, Jayasankar P (2014) Aquatic weeds as potential future foods. In: Peter KV (ed) Future crops, vol 2. Daya Publishing House, New Delhi-2, pp 17–50

Mandal RN, Mukhopadhyay PK (2010) Little known treasures of wetlands need conservation. Science Reporter, pp 27–29

Mandal RN, Mukhopadhyay PK (2012) Diversity of aquatic bio-resources and their role aquaculture. Abstract, national seminar on 'biodiversity and sustainability vis-à-vis economic development in the northern parts of West Bengal', Raiganj Surendranath Mahavidyalaya, during 26-27th august 2012, p 81

Mandal RN, Saha GS (2007) Aquatic vegetation—a potential support for rural economy. Agric Situation India LXIV:251–255

Mandal RN, Saha GS (2012) Producing Makhana (*Euryale ferox* Salisbery) for its diverse benefits to depressed economic regions of India. Water Garden J 27:13–16

Mandal RN, Saha GS, Mukhopadhaya PK (2007) Lotus—an aquatic plant of versatile qualities. Aquaculture Asia X11(1):11–13

Mandal RN, Saha GS, Kalita P, Mukhopadhyay PK (2008) *Ipomoea aquatica*—an aquaculture friendly macrophytes. Aquaculture Asia XIII(2):12–13

Mandal RN, Chattopadhyay DN, Saha GS, Mukhopadhyay PK (2010a) Water chestnut cum fish culture: a viable integration for rural aquaculture. Agrovet Buzz 3(5):53–56

Mandal RN, Chattopadhyay DN, Saha GS, Mukhopadhyay PK (2010b) Duck weeds-Nature's potential treasures of nourishment source for production of herbivorous fish. Agrovet Buzz 3(2):26–31

Mandal RN, Datta AK, Sarangi N, Mukhopadhyay PK (2010c) Diversity of aquatic Macrophytes as food and feed components to herbivorous fishes—a review. Indian J Fish 57:65–73

Mandal RN, Saha GS, Sarangi N (2010d) Harvest and processing of Makhana (Euryale ferox Salisb.)—an unique assemblage of traditional knowledge. Indian J Tradit Knowl 9(4):684–688

Mandal RN, Bar R, Chattopadhyay DN, Datta AK (2011) Assessment of impact of water hyacinth (*Eichhornia crassipes*) on ecology and socio-economy. In: Roy AK (ed) Evaluation and impact assessment of technologies and developmental activities in agriculture. Fisheries and Allied Fields. New India Publishing Agency, New Delhi, pp 291–312

Mandal RN, Chattopadhyay DN, Mukhopadhyay PK (2012a) Potentials of aquatic plant bioresources. Water Garden J 27:12–19

Mandal RN, Pandey BK, Chattopadhyay DN, Mukhopadhyay PK (2012b) Azolla—an aquatic fern of significance to small-scale aquaculture. Aquaculture Asia **17**(1):11–15

Mandu P (1995) Methodology for collecting and sharing indigenous knowledge: a case study. Indigenous Knowl Dev Monitor 3:3–5

Manvar MN, Desai TR (2013) Phytochemical and pharmacological profile of *Ipomoea aquatica*. Indian J Med Sci 67(3–4):49–60

Martin GJ (1995) Ethnobotany: a "people and plants" conservation manual. Chapman and Hall, London

Dictionary AH The American heritage medical dictionary, Med Dict 2007 32: 909

Mehta NR, Ekta P, Pragnesh P, Patani V, Shah B (2013) *Nelumbo nucifera* (lotus): a review on ethnobotany, phytochemistry and pharmacology. Indian J Pharm Biol Res 1(4):152–167

Mooij WM, Janse JH, de Senerpont Domis LN, Hulsmann S, Ibelings BW (2007) Predicting the effect of climate change on temperate shallow lakes with the ecosystem model PCLake. Hydrobiologia 584:443–454

Moreno-Black G, Somnasang P, Thamathawan S (1996) Cultivating continuity and creating change: women's home garden practices in northeastern Thailand. Agric Hum Values 13(3): 3–11

Naskar KR (1990) Aquatic and semi-aquatic plants of the lower ganga delta. Daya Publishing House, New Delhi

Nielsen KJ (2003) Nutrient loading and consumers: agents of change in open-coast macrophytes assemblages. Proc Natl Acad Sci U S A 100(13):7660–7665

Ong HC, Mojiun PFJ, Milow P (2011) Traditional knowledge of edible plants among the Temuan villagers in Kampung Guntor, Negeri Sembilan, Malaysia. Afr J Agric Res 6(8):1962–1965

Paul AK, Chakma P, Nahar N, Akber M, Ferdausi D, Ahsan S, Nasrin D, Jamal F, Ahmed R, Rahmatullah M (2011) A survey of non-conventional plant items consumed during times of food scarcity by the Chakma people of Hatimara village of Rangamati district, Bangladesh. Am-Eur J Sustain Agric 5(1):87–91

Pearson RG, Dawson TP (2003) Predicting the impacts of climate change on the distribution of species: are bio-climate envelope models useful? Glob Ecol Biogeogr 12(5):361–371

Peeters ETHM, van Zuidam JP, van Zuidam BG, van Nes EH, Kosten S, Heuts PGM, Roijackers RMM, Netten JJC, Scheffer M (2013) Changing weather conditions and floating plants in temperate drainage ditches. J Appl Ecol 50(3):585–593

Poff NL, Brinson MM, Day JW (2002) Aquatic ecosystems and global climate change. Potential impacts on inland freshwater and coastal wetland ecosystems in the United States. Pew center on global climate Change, Arlington

Prajapati R, Kalariya M, Umbarkar R, Parmar S, Sheth N (2011) *Colocasia esculenta*: a potent indigenous plant. Int J Nutr Pharmacol Neurol Dis 1:90–96

Praneetha P, Rani VS, Kumar VR (2011) Hepatoprotective activity of methanolic extract of leaves of *Marsilea minuta* Linn. against CCl4 induced hepatic damage in rats. Global J Pharmacology 5(3):164–171

Rahamatullah M, Jahan FI, Seraj S, Khatun Z, Islam F, Sattar MM, Khan T, Ishika T, Rahman M, Jaha R (2011) Correlation between non-conventional plants consumed during food scarcity and their folk medicinal usages: a case study in two villages of Kurigram district, Bangladesh. Am-Eur J Sustain Agric 5(2):240–246

Rahel FJ, Olden JD (2008) Assessing the effects of climate change on aquatic invasive species. Conserv Biol 22(3):521–533

Rodale R, Mcgrath M (1991) Foods that defy drought. In: Rodale R, Mcgrath M (eds) save three lives: a plan for famine prevention. Sierra Club Books, San Francisco

Roy SK, Mazumder UK, Islam A (2011) Pharmacological evaluation of *Enhydra fluctuans* aerial parts for central nervous system depressant activity. Pharmacology 1:632–643

Roy DK, Talukdar AD, Choudhary MD, Sinha BK (2013) Less known uses of *Nymphaea* species (Nymphaeaceae) as the traditional food item (Vhet-laddu) in Northeast India. Int J Food Agric Veter Sci 3:82–87

Sannigrahi S, Mazumder UK, Pal DK, Mondal A, Roy S (2009) Hepatoprotective potential of flavonoid rich fraction of *Enhydra fluctuans* against CC14—induced oxidative damage in rats. Pharmacology 2:575–586

Sannigrahi S, Mazumder UK, Pal DK, Parida S, Jain S (2010) Antioxidant potential of crude extract and different fractions of *Enhydra fluctuans* Lour. Iran J Pharm Res 9(1):75–82

Sarker MAQ, Mondol PC, Alam MJ, Parvez MS, Alam MF (2011) Comparative study of antitumor activity of three pteridophytes ethanol extracts. J Agric Technol 7(6):1661–1671

Sculthorpe CD (1967) The biology of aquatic vascular plants. Edward Arnold Ltd, London, p 610

Selvakumari E, Shantha S, Purushoth PT, Kumar S (2012) Antiproliferative activity of ethanolic flower extract from *Nymphaea pubescens* Willd. against human cervical and breast carcinoma *in vitro*. Int Res J Pharm 3(1):124–125

Siddique AK, Ahmed S, Iqbal A, Sobhan A, Poddar G, Azim T, Sack DA, Rahman M, Sack RB (2011) Epidemiology of rotavirus and cholera in children aged less than five years in rural Bangladesh. J Health Popul Nutr 29(1):1–8

Sivakumar R, Sunnathi D (2016) Phytochemical screening and antimicrobial activity of ethanolic leaf extract of Alternanthera sessilis (L.) R. Br ex Dc and Alternanthera Philoxeroides (Mart) Griseb. Eur J Pharm Sci 3(3):409–412

Soni P, Singh I (2012) *Marsilea quadrifolia* Linn.: a valuable culinary and remedial fern in Jaduguda, Jharkhand, India. Int J Life Sci Pharm Res 2:L99–L104

Subhashimi T, Krishnaveni V, Reddy CS (2010) Anti-inflammatory activity of leaf extracts of *Alternanthera sessilis*. Hygeia 2:54–56

Sunmathi D, Sivakumar R, Ravikumar K (2016) In vitro anti-inflammatory and antiarthritic activity of ethanolic leaf extract of Alternanthera sessilis (L.) R. BR. Ex DC and Alternanthera philoxeroides (Mart.) Griseb. Int J Adv Pharm Biol Chem 5(2):109–115

Thackeray SJ, Sparks TH, Frederiksen M, Burthe S, Bacon PJ, JR B, Botham MS, Brereton TM, Bright PW, Carvalho L, Clutton-Brock T, Dawson A, Edwards M, Elliott JM, Harrington KR, Johns D, Jones ID, Johns JT, Leech DI, Roy DB, Scott WA, Smith M, Smithers RJ, Winfield IJ, Wanless S (2010) Trophic level asynchrony in rates of phenological change for marine, freshwater and terrestrial environments. Glob Chang Biol 16(12):3304–3331

Tiwari OP, Bhattamisra SK, Singh PN, Kumar V (2009) Adaptogenic anti-stress activity of standardized extract of *Marsilea minuta* L. Pharmacology 1:290–299

Tokoro T, Hosokawa S, Miyoshi E, Tada K, Watanabe K, Montani S, Kayanne H, Kuwae T (2014) Net uptake of atmospheric CO_2 by coastal submerged aquatic vegetation. Glob Chang Biol 20(6):1873–1884

Trenberth KE, Dai A, Van der Schrier G, Jones PD, Barichivich J, Briff KR, Sheffield J (2014) Global warming and changes in drought. Nat Clim Chang 4:17–22

Weng W, Chen J (1996) The eastern perspective on functional food based on traditional Chinese medicine. Nutr Rev 54:S11–S16

Wrona FJ, Prowse TD, Reist JD, Hobbie JE, Levesque LMJ, Vincent WF (2006) Climate change effects on aquatic biota, ecosystem structure and function. AMBIO: J Hum Environ 35:359–369

Macrophytes and Their Role in Wetland Ecosystems

6

Arif J. Siddiqui, Sadaf Jahan, Mohd Adnan, Syed Amir Ashraf, and Ritu Singh

Abstract

Aquatic macrophytes, also known as hydrophytes, hold an important part in wetland ecosystems. This chapter highlights the significance of aquatic macrophytes in wetland ecosystems especially nutrient cycling. The most commonly recognized macrophyte classification includes free-floating and rooted macrophytes, where rooted macrophytes are further categorized into three subclasses, i.e., emergent, floating-leaved, and submerged. These macrophytes have a big impact on biogeochemical processes occurring in the water column as well as sediments. Macrophytes also differ in terms of biomass production, nutrient recycling, and rhizosphere impacts, precisely carbon and oxygen release, along with their ability to act as methane conduit. However, the species of wetland macrophytes are drastically declining due to increasing incidences of cultural eutrophication, leading to their substitution by monoculture species serving as strong competitors. In order to understand the wetland ecosystems in

A. J. Siddiqui (✉) · M. Adnan
Department of Biology, College of Science, University of Hail, Hail, Saudi Arabia
e-mail: Ar.Siddiqui@uoh.edu.sa

S. Jahan
Department of Medical Laboratories, College of Applied Medical Sciences, Majmaah University, Majmaah, Saudi Arabia

S. A. Ashraf
Department of Clinical Nutrition, College of Applied Medical Sciences, University of Hail, Hail, Saudi Arabia

R. Singh
Department of Environmental Science, School of Earth Sciences, Central University of Rajasthan, Ajmer, Rajasthan, India
e-mail: ritu_ens@curaj.ac.in

S. Kumar et al. (eds.), *Aquatic Macrophytes: Ecology, Functions and Services*,
https://doi.org/10.1007/978-981-99-3822-3_6

more depth, sediment microorganisms and wetland macrophytes should be studied together as they play an important role in maintaining the wetland ecosystems.

Keywords

Aquatic system · Macrophyte · Biomass production · Nutrient recycling · Wetland · Ecosystem

6.1 Introduction

Aquatic macrophytes, otherwise known as hydrophytes, form a crucial part of aquatic and wetland ecosystems. These macrophytes are considered as principal components of food chains as they act as a major nutrient source for detritivores and herbivores, i.e., fish, invertebrates, birds, etc., while also serving as an essential carbon source for indigenous microbial communities. Moreover, their leaves, roots, and stems behave as periphyton substrate as well as provide habitat to a variety of amphibians, invertebrates, and fishes (Swe et al. 2021; Germ et al. 2021; Dvořák 1996). Besides, some of the aquatic macrophytes are also critically important to human societies for food, biomass, and building materials (Peters and Lodge 2009; Bornette and Puijalon 2011). Hence, both the presence and absence of aquatic macrophytes have a significant impact on sediment and water dynamics, water column, and biogeochemical processes occurring in the sediments. However, to understand their impact on these underwater mechanisms requires a thorough knowledge of aquatic macrophytes and their associated functions in shallow lakes and wetland ecosystems. Their understanding is also important to deal with a variety of recent practical challenges like wetland restoration, wastewater treatment, invasive species management etc. (Lan et al. 2010). Thus, the major objective of this chapter is to provide a brief overview of several aquatic macrophytes, exploring their various aspects and, more importantly, delving deeper into their significance in wetlands and aquatic ecosystems.

6.2 Role of Aquatic Macrophytes in Wetland Ecosystems

Various aspects of aquatic macrophytes in a wetland ecosystem have been summarized in Fig. 6.1. The plants produce biomass for grazers by utilizing solar energy, CO_2, water, and nutrients. The organic matter produced after senescence provides nutrients and organic carbon to decomposers. Leaves and stems act as gas conduits, carrying oxygen-laden air into the root zones, i.e., rhizosphere, while discharging methane through it. The associated processes are explained in the following sections with a focus on significance of macrophytes in nutrient uptake and cycling.

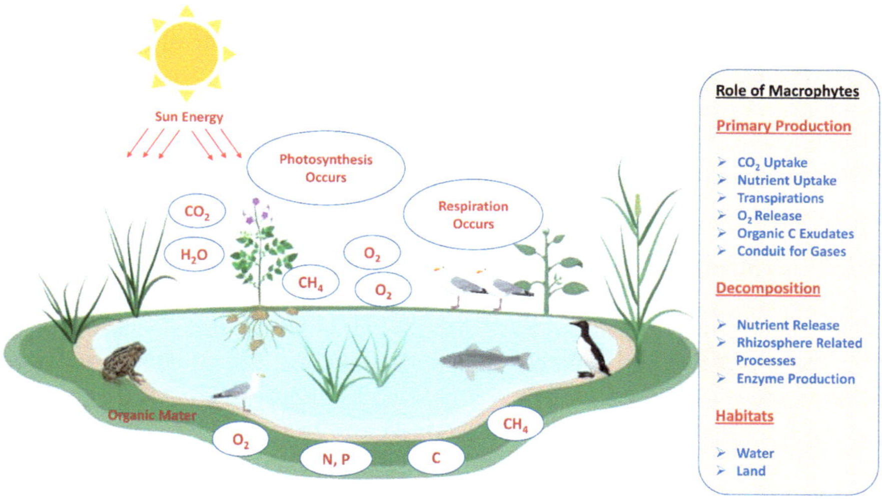

Fig. 6.1 The different functional aspects of macrophytes in wetland ecosystems (Adopted from Rejmankova 2011)

6.2.1 Primary Production

In the hierarchical strata of an ecosystem, primary production holds the very bottom level. It basically represents overall biomass of plants, usually denoted by W, i.e., production per unit area per unit time. The net primary production (NPP) is of particular interest in this context and is basically defined as the total energy produced from biomass minus the energy released through metabolic processes. NPP is measured as dry mass per unit of time, i.e., g/m^2/y, and hence also known as ash-free dry weight or carbon weight, C. NPP is approximately equivalent to the highest seasonal biomass (W_{max}) within a single generation time, primarily in temperate climatic zones. However, in other climatic zones, NPP varies based on different generation times, resulting in NPP being two to three times greater than of W_{max} the tropics. Several environmental conditions like temperature, solar radiation, and available nutrients play a major role in NPP production, for instance, they influence the overall functioning and biomass production of terrestrial ecosystems. However, the primary production in wetland ecosystems is usually restricted by a lack of water and/or nutrients, even when temperature and solar radiation are near optimal (Tanabe et al. 2019). Water scarcity is not an issue in permanently flooded wetlands, hence allowing these wetland ecosystems to become nutrient-rich and most productive on the planet (e.g., papyrus swamp). Similarly, a few more examples of extremely productive macrophytes include typical grasslands that usually form floating meadows on tropical river floodplains. Notable examples can be found in the Magela Creek floodplains in Australia, Amazon floodplains in Brazil, and various other floodplains where the peak biomass, particularly above-ground biomass ones, often exceeds 4000 g/m^2 (Rejmankova 2011). Salinity also has a crucial impact on NPP, and various types of macrophytes have evolved through different mechanisms and adaptations to cope with elevated salinity levels, ensuring a favorable internal osmotic potential (Pezeshki and DeLaune 2012). Some macrophytes routinely employ higher mineral salt uptake to lower their internal osmotic potential (*Typha*, *Eleocharis*), while others utilize organic osmolytes (*Cladium*) (Rejmankova 2011).

6.2.2 Nutrients

While most of the wetlands are experiencing eutrophication at a sluggish or faster rate lately, there are still some wetland areas where nutrient restrictions are being reported, which could be possibly due to lack of nitrogen (N), phosphorus (P), or maybe a combination of both the nutrients (Craft et al. 2007; Berthold and Schumann 2020). The transition from ombrotrophic to minerotrophic nutrient environments leads to a shift from low nutrient concentrations to increased nutrient accessibility. This gradual change along the ombrotrophic to minerotrophic gradient enhances biomass productivity (Keller et al. 2006; Mitsch et al. 2009). The predominant N restriction is common in vascular macrophytic species of swampy and marshy regions, whereas P restriction and co-limitations are related to other infertile bog and fen populations, in the regions dominant with carstic or serpentine minerals (Willby et al. 2001; Rejmankova 2011). The change in composition of species is due to stress or tolerance that increases the nutrient preservation to fast-growing species, helping in their increased growth and height. In addition, monospecific stands are also linked to P to N restriction. However, a shift from N to P limitation might occur as a consequence of high N settings in the atmosphere. These fluctuations have a significant impact on species composition and diversity, as well as on a variety of ecosystem functionalities.

6.2.3 Strategies for Surviving in Nutrient-Limited Environment

The evolution of plants has been classified into two distinct categories for P acquisition in low-nutrient environments, viz., (1) conservation of use and (2) enhanced acquisition or uptake.

Conservation of use: The classic example of conservation approach by plants is the reabsorption of nutrients from old and aging organs into freshly growing plant or new storage organs (Aerts and Chapin 1999). The growth of perennial plant species is not only influenced by the types of nutrients they consume but also by the volume of nutrients they store and further reuse. The storage of different nutrients is carried out through resorption or their transportation from the senescing parts to the storage organs or other growing tissues of the plant. In addition, the conservation helps to determine the association between the live tissues and the dead cells through nutrient content, i.e., N and P (Lambers et al. 1998). There are a few terms that are associated with resorption of nutrients, i.e., (1) resorption efficiency (RE), which primarily indicates the nutrient decline percentage between the green and senescent leaves, and (2) resorption proficiency (RP) signifying the terminal content of nutrients in senescent leaves. These two terms are commonly used to describe the nutrient uptake (Killingbeck 1996). The RE value for normal phosphorus is around 55% (Aerts 1996), while in P-limited environments, it can reach over 80% (Güsewell 2005, Rejmánková 2005). Similarly, nitrogen resorption has slightly lower average values (50%), and it never goes as high as RE for P.

In general, nitrogen fluctuation is minimal in plant tissues when compared to phosphorus (Aerts and Chapin 1999). The terminal N and P concentration in senescent leaves is found to be 3 mg/g and 0.1 mg/g, respectively (Aerts and Chapin 1999). This terminal nutrient concentration of entirely senesced leaves is predominantly observed as a sensitive indicator of nutrient accessibility in wetland ecosystems (Rejmánková 2005). Rejmánková (2005) found that the N/P ratio of healthy and growing tissues is a worthy predictor of P resorption proficiency and efficiency. Likewise, the quality of litter or dead tissues is determined by the amount of nutrients that get resorbed from senescing materials, which have a serious impact on decay rates (Aerts and Chapin 1999, Cai and Bongers 2007).

Enhanced acquisition or uptake: There are two chief ways of enhanced acquisition, i.e., secretion of hydrolytic enzymes and development of symbiotic association of roots with mycorrhizae. Under P-limited conditions, plants release phosphatase, which releases P from organic P-esters. Similarly, peptidase is secreted in plants against N deficiency. In addition, arbuscular mycorrhizae (AM) have also been reported to increase the plant growth under nutrient-limited conditions by facilitating the exchange between phosphorus and carbon sources (Rejmankova 2011).

6.3 Water and Water Quality

Water is the most vital source for the survival of biota. It owns a number of chemical as well as physical properties, which assist the water molecules to act as paramount suited medium for all the life activities. Water covers 71% of the Earth's surface, with nearly 2.5% of that being freshwater (1.2% in wetlands, rivers, streams, lakes, and ponds and 30.1% in groundwater and 68.7% as glaciers (Khatri and Tyagi 2015; USGS 2019). Freshwater ecosystem is one of the scarce natural resources, which have great socioeconomic, aesthetic, ecological, and cultural values, besides playing a significant role in the conservation of genetic resources of both plants and animals. In addition to sources of water and recreation, they also provide food, fish, fodder, wildlife, green manure, medicinal plants, vegetables, timber, and other valuable products, which are important for the sustainable economy of the region. The capabilities of different countries to manage water restoration efficiently decide the future welfare. Very little efforts have been made for the protection and conservation of aquatic ecosystem, particularly in the urban environment despite the existing water bodies providing tremendous aesthetic, ecological, and economic benefits (Alikhani et al. 2021).

Several rural and urban communities depend on lakes, ponds, and reservoirs for surface water supplies and maintenance of livelihood. In recent years, the quality of freshwater in lakes and reservoirs has been deteriorating in developing countries, posing a threat to ecosystem services and all the biotic communities. Eutrophication and nutrient enrichment are two of the most common causes of water quality degradation, resulting in algal blooms and massive macrophyte growth. Water quality basically refers to the state or condition of water resources in relation to biotic species requirements and human needs (Malone and Newton 2020; Ndungu et al. 2014). It is defined as the physical, chemical, and biological characteristics of water, usually in connection to its feasibility to carry out a specific purpose (Stringfellow et al. 2014). Water quality supports the processes of all the ecological systems that upkeep biodiversity in aquatic environments. Regular and periodic monitoring of water quality (physicochemical and biological parameters) provides insights to better understand the lake conditions, which in turn helps in the development of effective management practices (Stringfellow et al. 2014; Park et al. 2014).

Water quality parameters describe the characteristics and suitability of water for specific use and give the exact nature, cause, and level of pollution (Omer 2019). Chemical parameters such as biochemical oxygen demand (BOD), chemical oxygen demand (COD), dissolved oxygen (DO), and oxidation reduction potential (ORP) highlight the nature and measure of organic loads, while total dissolved solids (TDS) and electrical conductivity (EC) provide a measure of inorganic dissolved solids. The concentrations of different forms of nitrogen (nitrate, nitrite, ammonia) and phosphorus (organic and inorganic phosphate) depict the overall nutrient condition. Lake productivity and trophic state depend on light availability, temperature, and nutrient concentration (Dove and Chapra 2015; Peterson and Risberg 2008). Regular and periodic monitoring of physicochemical and biological parameters of lake ecosystems helps in addressing the adverse impacts, thus assisting in the sustainable management of lake ecosystems.

6.4 Wetlands and Lakes

Wetlands are Earth's transition zones, which play an important part in nutrient dynamics and are the world's most productive ecosystems. Wetlands are a diverse group of aquatic habitats that include marsh, fen, peat land/open water, flowing (rivers and streams) or static (lakes and ponds), fresh, brackish, or saltwater with a depth of less than 1–2 m (Chamorro et al. 2015; Balwan and Kour 2021). Some of the important functions of wetlands include maintaining the region's microclimate, flood mitigation, water purification, stream flow maintenance, nutrient recycling, groundwater recharge, drinking water source, fish, fodder, fuel, and societal recreation. Wetland stability is determined by the balance of energy, matter production, and consumption pattern at various trophic levels present in the ecosystem. The interaction of humans with wetlands has been a source of concern in recent decades, owing to unplanned urbanization, which has resulted in massive industrial, commercial, and residential development, as well as increased pollution by domestic and industrial wastewaters, among other things (Kennish 2002; Zhang and Zhang 2021). Wetland loss, degradation, and anthropogenic stress have been affecting the urban wetlands, mainly due to the inflow of sewage, which has altered the chemical integrity of wetlands. The flow of untreated or partially treated wastewater to the water bodies has led to the enrichment of nutrients, resulting in eutrophication. Anoxic conditions due to excess growth, aging, and decay of macrophytes are the major consequences among others that occur due to entry of partially treated and untreated sewage. This has altered the food chain networks, affecting the natural balance or integrity of the ecosystem. Due to overaccumulation of nutrients, these surface water bodies are increasingly becoming eutrophic, resulting in frequent algal blooms and profuse macrophyte growth with periodic successions.

Various physical, chemical, as well as biological parameters govern the quality of water. The physicochemical characteristics determine the kind of organisms present in the water body, and in turn, these organisms alter abiotic factors for their existence (Pal and Chakraborty 2014). The presence of nutrients is responsible for richness in the productivity of aquatic ecosystems. Because primary production is limited or triggered by nutrient availability, nutrients play a critical role in the environmental state of lakes. Increased nutrient levels result in increased lake productivity, which has a cascading effect on the remaining trophic levels. Their availability is influenced by the rates of external and internal inputs, as well as the system's permanent removal by biological, chemical, or physical processes. Carbon (C) and nitrogen (N) are the primary macronutrients. Apart from carbon and nitrogen, phosphorus (P) is also an important nutrient that helps primary producers grow while becoming an essential link in the food chain network. As a result, phosphorus availability is frequently regarded as the single most important factor affecting the overall ecological state of lakes.

Dumping of unprocessed or partially treated waste effluents into surface water bodies has become somewhat rampant in developing countries, which exert nutrient stress in the water body (Khan and Mohammad 2014; Englande Jr. et al. 2015). Disposal of untreated sewage with organic nutrients alters the physicochemical properties of water that induces nutrient stress affecting the ecological health and hence threatening the biotic-abiotic complex interactions (Gavrilescu 2021). Excess nutrients are a major warning to freshwater ecosystems causing a chance of severe pollution problems. Water quality, nutrient percentage, biochemical concentrations, and biotic entities like microbes, planktons, macrophytes and animals in shallow lakes have complex interactions. The study on nutrient status of sediments, water, and plant communities is a prerequisite to understand the functions of wetland ecosystems. The nutrient budget, ecology, and trophic status of lake systems are influenced by plant detritus and sediments. Lake ecosystem plays an important role in trapping and providing a surface for sediment deposition (Mwamburi 2018). Sediments are the primary source of N and P nutrients to the water columns, which give way to benthic-pelagic mixture, impacting the primary productivity while acting as a normal regulator of biological developments in the lakes. The particulate plant detritus are the major cause of nutrient richness and a source of total organic carbon in lake other surface water sediments, while animals and other sources constitute only a minor part (Zimmer 2008).

One of the most important factors determining the dynamics of an aquatic ecosystem is the source of organic matter. Nutrients like C and N keep the various trophic levels in lake ecosystems afloat. The C/N ratio, i.e., the dry weight ratio of the total organic carbon to total nitrogen, has been utilized to evaluate the source of organic matter in sediments. The presence of cellulose is one factor that influences the C/N ratio or the type of organic matter present. Because cellulose is primarily composed of carbon, it has a positive impact on the C/N ratio in plants. The C/N ratio can be employed to estimate the percentage of autochthonous planktonic materials in sediments due to the different protein content of organic matter produced by various organisms (Perdue and Koprivnjak 2007). Many scientists have claimed that C/N ratios in aquatic sediments indicate the overall makeup of organic residues to some extent (Perdue and Koprivnjak 2007; Woodward and Gadd 2019; Achyuthan et al. 2020). The C/N ratio in phytoplankton and zooplankton, which contain proteins as primary nitrogen compounds, is 5 to 6, whereas the C/N ratio in vascular plants of terrestrial regions is 15 or even greater than 20. Since ancient times, the causes of organic matter for lakes have been determined by monitoring the change in C/N ratios within lake sediments. The C/N ratio of algae typically falls within the range of 4 and 10; however, the C/N ratio of terrestrial organic matter remains much higher, greater than 20. Surges in the C/N ratio in sediment profiles have been used to recognize different periods in a lake's history when large amounts of terrestrial organic matter were deposited in the sediments (Mahapatra et al. 2011). Because macrophytes influence nutrient cycling, nutrient dynamics in the vegetated littoral zone of lakes are complicated. Macrophytes absorb nutrients for growth and release them back into the environment, particularly during decaying process (Takamura et al. 2003). Due to huge numbers of the biomass they produce and their ability to accrue huge concentration of nutrients, macrophytes play an important role in nutrient cycling. The primary source of nutrients in macrophytes is sediment, while water is the secondary source (Bowden et al. 2017; Zou et al. 2014).

6.5 Nutrient Dynamics in Wetlands

Wetlands act as nutrient sinks. They sequester bio-available nutrients from incoming waters, store them in sediments, and convert them into organic forms, which may be stored or exported (Pasut et al. 2021). Nutrient sequestration and storage are important in regulating the productivity, water quality, and species diversity of wetlands. Nutrient loading and accumulation are often linked to high primary and secondary productivities in wetlands (Pasut et al. 2021; Shen et al. 2022). Sedimentation, absorption, precipitation, filtration, microbial processes, and plant uptake are the various nutrient reduction mechanisms in wetlands (Bakhshoodeh et al. 2020).

Carbon (C): Biomass and carbon content of macrophytes in tropical wetlands undergo seasonal changes. During summer, areas that are seasonally flooded or dried out become exposed, leading to the germination of buried plants' seeds including the perennial, emergent, and mudflat plants. In monsoon period, mudflat species wiped out, leaving only the emergent species because of increased water level. The submerged species swiftly return as their seeds germinate in the standing water (Thomaz et al. 2009). The emergent species increase in density and become dominant. In winter season, the emergent species bear maximum biomass and nutrients stored belowground in the rhizomes. During summer season, an increase in shoot growth is observed as a result of photosynthesis and translocation of nutrients stored in their rhizomes.

Aquatic macrophytes are known for sequestering large amount of carbon from the atmosphere, which is added to the soil as organic carbon after they die and decompose (Pal et al. 2017). These are the main sources of carbon removal in wetlands while providing steady supply of organic matter. Reducing down the water velocities, providing substrate for particle adsorption, and preventing re-suspension, these all are the ways by which macrophytes stabilize the sediments in the wetlands (Pal et al. 2017; Lin et al. 2002). When macrophytes like *Typha latifolia* and *Scirpus acutus* dominate the wetlands, the rate of carbon sequestration increases. Emergent macrophyte ecosystems are thought to be the world's most productive biomass generators, and after their death, the same biomass is supplied to the wetland floor. While the macrophytic biomass may not fully explain the entire carbon dynamics in wetlands, it is important to document and understand the role played by macrophytes in these ecosystems (Kurniawan et al. 2021). In wetland carbon dynamics, both the aboveground and the belowground productivities are considered as the primary sources of labile carbon (Lolu et al. 2018). Many researchers have been studying the aboveground productivity of macrophytes (Lolu et al. 2018; Dalla Vecchia et al. 2020). For instance, the plant families and the biomass (only aboveground) in boreal continental peatlands were studied by Miller (2011). Similarly, aboveground carbon storage potential of selected macrophytes in newly created wetlands was assessed by Means et al. (2016) however, the research on belowground biomass dynamics is very limited.

Nitrogen (N): The common forms of nitrogen in wastewaters are ammonia, nitrite, nitrate, organic nitrogen, and gaseous nitrogen (e.g., nitrogen dioxide). All these forms are components of N cycle and are biochemically inter-convertible. Unionized ammonia is toxic to life forms and also causes significant dissolved oxygen demand in water system (Min et al. 2020; Lu et al. 2020). The organic form of nitrogen includes soluble (mainly urea and amino acids) and particulate forms. In wetlands, the initial removal of organic nitrogen occurs rapidly in the form of TSS. Besides, microbes in wetlands facilitate conversion of particulate organic nitrogen to ammonia while producing new biomass via decomposition. The removal mechanisms for N in wetlands are as follows:

1. Volatilization: the conversion of ammonium ions in solution to ammonia gas, which is released into the atmosphere. This process occurs only at a pH value above 8 (Tang et al. 2020; Bakhshoodeh et al. 2020).
2. Ammonification: biological transformation of organic N into inorganic N. In oxygenated zone, the process occurs rapidly and decreases as the process

switches from aerobic to anaerobic microflora (Vázquez et al. 2020; Strock 2008).

3. Nitrification: biological oxidation of ammonium ions into nitrite and then to nitrate. This process is supported by chemoautotrophs, which uses CO_2 as carbon source for synthesizing new cells. *Nitrosomonas* and *Nitrobacter* are the two genera of bacteria involved in this process (Norton and Stark 2011; Lancaster et al. 2018).
4. Denitrification: reduction of nitrate to molecular nitrogen. It is a bacterial process in which nitrogen oxides serve as electron acceptor for respiratory electron transport. The optimum pH for denitrification is 7–8 (Skiba 2008; Robertson and Groffman 2015; Tiso and Schechter 2015).

Phosphorus (P): Excessive phosphorus loading in water bodies leads to nutrient enrichment. The major sources of P include agricultural runoff, urban runoff, wastewater effluent, etc. (Min et al. 2020). P is usually present in wastewater as orthophosphate, polyphosphate, and organic phosphorus. Oxidation of P results in conversion of most phosphorus into orthophosphates (PO_4^{3-}, HPO_4^{-2}, etc.). The polyphosphates also undergo hydrolysis in aqueous solution and revert to orthophosphate. The removal mechanisms for phosphorus in a wetland system include various processes such as direct uptake by plants and microorganisms, sedimentation and burial, mineralization, and absorption and precipitation (Vymazal 2007). Pore water has maximum concentrations of nutrient molecules than the lake water. The major components present in tissues of aquatic angiosperms are nitrogen and phosphorus. The nitrogen content varies from 0.035 to 1.492 mg/g, whereas phosphorus ranges from 0.027 to 1.039 mg/g. The N/P ratio varies from 3.46 to 4.42, which indicates nitrogen is the limiting nutrient. Squires and Lesack (2003) assessed macrophyte profusion and dissemination among lakes of the Mackenzie Delta (Squires and Lesack 2003), wherein the organic matter and total nitrogen content of the sediments were observed to increase with the increase in the biomass of macrophytes, while phosphorus showed insignificant variations. *Potamogeton* species were found to be dominant at low and intermediate transparency and moderate sediment organic matter content, while *Chara* and *Ceratophyllum* were dominant at high transparency and high sediment organic matter. A high rate of inorganic and organic sedimentation resulted in the most suitable substrate for the growth of the plants. Besides, the summer biomass was found to be highest, and the winter biomass was almost 40% lower than the annual mean. The content of phosphorus in sediments, water, and tissues of three macrophyte species (*Chara fibrosa*, *Nitella hyalina*, and *Najas marina*) in Myall Lake, Australia, was investigated by Shilla et al. in 2006 (Shilla et al. 2006). Total P in the sediments and water column was found to be significantly higher in the central deep areas of the lake. A significant correlation was observed between the soluble reactive phosphorus and total dissolved phosphorus in sediment pore water, as well as phosphorus content in *N. marina* tissues and total phosphorus in sediments. *Nitella hyalina* species got its most phosphorus from sediment, while the other two species acquired it from water.

During the winter and summer season, Hadad and Maine (2007) measured biomass and phosphorus concentration in the three floating and rooted macrophytes from the Middle Parana River floodplain wetlands. Due to higher biomass, *Pistia stratiotes* showed higher phosphorus concentrations in roots and shoots, whereas *Eichhornia crassipes*, *Pontederia cordata*, and *Typha domingensis* were found to be the most efficient species for P retention. During the exponential growth of *C. papyrus* and *P. mauritianus*, aboveground biomass increased rapidly. The concentration of nitrogen was found highest in young plants and decreased as they grew older. Highest nitrogen levels were found in the young umbels of *C. papyrus* (1.95%) during the first month of growth in the Nakivubo wetland and then dropped to 0.62% by the fifth month. Tubers of *Colocasia esculenta* had the highest nitrogen content in the wetlands of Kirinya and Nakivubo (4.8% and 3.7%, respectively) (Muraza et al. 2013; Aregu et al. 2021). Seasonal dynamics and comparative abundance of the macrophytic species (*Nuphar luteum*, *Najas marina*, *Najas minor*, *Potamogeton lucens*, and *Potamogeton pectinatus*), as well as their nutrient content (total nitrogen and phosphorus), were studied at three locations in Lake Velenjsko jezero (Mazej and Germ 2008). During the early stages of growth, the total phosphorus concentration was observed to be higher, and there was significant seasonal variation. *Nuphar luteum*, a floating-leaved species, displayed more total nitrogen in its aboveground tissues as compared to submerged species. The biomass of *Nuphar luteum*, *Najas minor*, and *Potamogeton lucens* contained fewer nutrients. Seasonal variation in total nitrogen was reported to be less pronounced.

Costa and Henry (2010) investigated the carbon, nitrogen, and phosphorus contents of aquatic macrophyte species in the land-water transition zone of three lakes lateral to the Paranapanema River in Sao Paulo, Brazil. Similarly, Mazej and Germ (2008) looked at the seasonal dynamics and the relationship among the aquatic macrophytes (Mazej and Germ 2008). *Salvinia auriculata* and *Eichhornia azurea* displayed the highest nitrogen and phosphorus content in Camargo Lake during the dry season, while during the rainy season, *Echinochloa polystachya* showed the highest levels of phosphorus and nitrogen. *Cyperus esculentus* had the highest phosphorus and nitrogen concentrations in Coqueiral Lake during the dry season, while in rainy season, *E. polystachya*, *Ludwigia octovalvis*, and *Polygonum spectabile* showed high content of phosphorus and nitrogen. *Myriophyllum aquaticum* showed the highest phosphorus and nitrogen during the dry season in Cavalos Lake, while *C. esculentus* and *E. azurea* displayed highest nitrogen concentrations during the rainy season (Costa and Henry 2010). Esteves and Suzuki (2010) also assessed carbon, nitrogen, and phosphorus content of aquatic macrophytes – *Ceratophyllum demersum*, *Najas marina*, and *Egeria densa* – found in a lagoon of Norte Fluminense (Esteves and Suzuki 2010). Substantial seasonal differences were observed in total P, N, and C content, without a clear pattern between the seasons and macrophytes. Higher concentrations of phosphorus in macrophytes were found in the rainy season. Ławniczak (2011) studied variations of nutrient concentrations in sediment biomass and aboveground biomass production of four evolving macrophytes *Typhetum angustifoliae*, *Glycerietum maximae*, *Phragmitetum australis*, and *Caricetum acutiformis* during the vegetation season in the littoral zone of Lake Niepruszewskie (Ławniczak 2011). N/P ratio was a good indicator of moisture variations than the N/K ratio. Macrophytes were dependent on the N/P ratio in the sediment. *Glyceria maxima* grew in sites with high N and K concentrations, while *Typha angustifolia* was present in sites with the lowest nutrient concentrations. Zhou et al. (2017) investigated the biomass productivity of eight dominating macrophytes and nutrient content in tissues. The biomass of the eight studied macrophyte species declined in the following order: *Typha orientalis* > *Pontederia cordata* > *Pistia stratiotes* > *Hydrilla verticillata* > *Panicum hemitomon* > *Nuphar advena* > *Potamogeton* spp. > *Salvinia* spp. (Zhu et al. 2017). *Typha orientalis* had the maximum amount of total nitrogen (N) (23.4 g·Nm^{-2}) and phosphorus (P) (1.59 g·Pm^{-2}) storage, while *Pistia stratiotes* and *Hydrilla verticillata* exhibited the highest total N (14.6 g·Nm^{-2}) and P (1.04 g Pm^{-2}) net storage capacity among all the studied macrophytes. The study found out that the highest N and P uptake/year was by *Pistia stratiotes* and *Hydrilla verticillata* with 146 kg N/ha/year and 10.4 kg P/ha/year, respectively.

Alonso et al. (2018) assessed the concentrations of nutrients and pollutants in various macrophytes of four peri-urban wetlands of Middle Parana River floodplain and investigated their potential for biomonitoring of the wetland ecosystem (Alonso et al. 2018). The study included *Typha domingensis*, *Eichhornia crassipes*, *Alternanthera philoxeroides*, and *Pistia stratiotes.* The total phosphorus (TP) content was found to be highest in the leaves of *Alternanthera philoxeroides.* In contrast, other species displayed higher TP concentration in their roots. Highest Zn was reported in the roots of *Typha domingensis*, while highest Pb content was found in *Eichhornia crassipes* roots (Alonso et al. 2018). The root length of *Pistia stratiotes* displayed an inverse relationship with the concentration of soluble reactive phosphorus in water. The concentrations of nitrate and ammonium in water were positively correlated with all root anatomical parameters of *Typha domingensis* and *Eichhornia crassipes.* In all these studies, macrophytes showed a high level of lenience, letting them to survive and grow in peri-urban wetlands having contaminated water (Hadad et al. 2021b; Rai 2008; Hadad et al. 2021a).

6.6 Monitoring and Conservation of Wetlands

Monitoring and conservation of wetlands include activities conducted within, in, and around the water body with the goal to protect, restore, or provide their functions and values that were lost, altered, or influenced by human activities (Xu et al. 2019; Reis et al. 2017). Their present situation and valuable services that they provide should be recognized properly in order to manage and protect the remaining wetlands over time while benefiting from their functions.

6.6.1 Biomonitoring

Biological monitoring can be defined as the application or study of specific species or communities (single or multiple groups) that provide information on the physical and/or chemical conditions of their immediate environment based on their presence or absence. Even though the results of water quality analyses evaluate the amount of organic and inorganic pollutants, biological monitoring is widely used lately, due to cost and time constraints. The ecological significance of specific organisms and their use as bio-indicators is explained by their presence or abundance in a particular habitat, as well as their ability to grow and outcompete other organisms under specific water quality conditions (Zaghloul et al. 2020). Phytoplanktons, aquatic macrophytes, macro-invertebrates, and vertebrates have been widely employed to measure the biological integrity of aquatic systems. When biological data is associated with physical and chemical parameters, the effects of contaminants can be better understood. No single species can reveal the entire full range of potential hazards associated with various water uses or deliver all of the information needed to make an adequate assessment (Rocha et al. 2018; Hilt et al. 2021). Biomonitoring has a number of advantages, and it is one of the most important assessments at sub-lethal levels of bioaccumulated contaminants within an organism's tissues, indicating the net amount of pollutants integrated over time (Zhou et al. 2008).

6.6.2 Macrophytes as Bio-indicators

Macrophytes create the world's most productive ecosystems (Hilt et al. 2021). Plants growing in aquatic systems or on a substrate that are deficient in oxygen from time to time as a result of excessive water content are commonly referred to as wetland plants (Tiner 2012; Cronk and Fennessy 2016). They are a diverse group of vascular plants that are commonly referred to as macrophytes. These plants can live in still or moving water, as well as in inundated or non-inundated hydric soils. The diversity and distribution of macrophytes are intimately linked to the system's environmental factors (Dokulil and Teubner 2022; Schneider et al. 2018; La Toya et al. 2013). The importance of vegetation in the optimal functioning of wetlands cannot be overstated. Freshwater macrophytes are important players in biogeochemical cycles and food webs in lentic water bodies (Reitsema et al. 2018). They are major conduits for energy flow in the system as they occupy the base of food chains in wetland ecosystem. They form an essential link between the biotic and inorganic environment through their photosynthetic activity. The chief productivity of wetland plant groups is similar to that of tropical rainforests, but the organic matter produced is passed down the detrital food chain rather than being consumed directly by herbivores (Schlesinger and Bernhardt 2020). Other organisms such as periphyton, macro-invertebrates, fish, and epiphytic bacteria use macrophytes as a home (Schlesinger and Bernhardt 2020). They influence water chemistry by acting as a nutrient sink and pump (by absorbing and transferring nutrients from the sediments to the water column). By absorbing nutrients, metals, and other contaminants, macrophytes purify water. These plants are considered as indicators for water pollution and employed as test system for assessment of pollutant impacts. The macrophytes release oxygen for those plants that grow in and beyond the water surface and help in the water refinement process (by providing habitat for microbial activity and transferring hazardous pollutants into non-toxic or beneficial components) (Hilt et al. 2021; Opitz et al. 2021). The occurrence of different communities of macrophytes is determined by a large number of environmental factors. The macrophyte species diversity changes with respect to the increase in water depth, pollution level, and nutrient concentration associated with an increased internal loading and turbidity of lakes. The macrophyte miscellany and biomass affect the key productivity and complexities of tropic states. An ecologically well-balanced wetland ecosystem sustains many species of macrophytes, but extensive growth of macrophytes obstructs many vital activities. Aquatic macrophytes are commonly distributed in distinct zones from land to deeper water, along with a variation in species composition, zonation, and seasonal succession between tropical and temperate water bodies (Opitz et al. 2021; Hilt et al. 2021).

6.6.3 Phytoremediation Potential of Macrophytes

Phytoremediation is the process of using living green plants to reduce the hazards from contaminated sludge, soil, groundwater, and sediments by removing, degrading, or containing contaminants. It has great potential for the elimination of organic pollutants and toxic metals from wastewater and polluted sites. Some plants are known as hyper-accumulators, capable of accumulating large amounts of metals, exceeding what is required for their normal functions. To recover metals from the wetland ecosystem, it is necessary to harvest and process the plants with suitable methods (Sood et al. 2012; Ahila et al. 2020). Although most phytoremediation activities are carried out in terrestrial ecosystems, macrophytes have been found suitable for removing contamination from shallow water bodies. They play a significant role in the process of restoring and cleaning up polluted aquatic environments. Phytoremediation has been investigated using a variety of wetland plants. The accumulation abilities of the macrophytes for the chemical species determine their potential as phytoremediation agent which depends on plant's biomass and their physiological functions/activities. The advantage of using vascular plants for water treatment includes its ease of sampling and harvesting, rapid growth, high mineral uptake, and tolerance to metal pollution and accumulation (Yan et al. 2020). The accumulation of heavy metals in various plant organs can be element specific and linked to organ's function (Tangahu et al. 2011). Furthermore, environmental factors such as pH, organic matter, and soil texture influence the amount of heavy metals accumulated by plants. The accumulation of heavy metals in various plant organs can be element-specific and linked to the functions of those organs (Tangahu et al. 2011; Yan et al. 2020; Nouri et al. 2009).

6.7 Threats to Macrophytes

Different human actions such as hydrologic alterations, wetland draining/filling, industrial effluents, sewage, intrusion of exotic species, etc. have threaten the macrophytes and the wetland ecosystems (Kennish 2002; Newton et al. 2020). Human-caused hydrologic changes, such as flood control and agriculture, frequently result in reduction in the wetland area or cause alteration in the area's hydrologic regime (Newton et al. 2020). Other activities like groundwater pumping, dams, and irrigation projects have the potential to alter the water level of associated wetlands and disrupt the distribution of the wetland species. Altogether anthropogenic activities have profound impact on the ecological balance and biodiversity of wetland ecosystems.

6.8 Conclusion

Taxonomically, wetland macrophytes are varied and diverse group of plants. Their roles in wetland ecosystems have an influence on a number of processes, including nutrient system cycling and dynamics of food chain and food web. When nutrient availability varies, low productivity and high species miscellany are frequently replaced by highly productive species monocultures. Sediment heterotrophic microbial processes are influenced by the quality and quantity of waste material and root carbon exudates. Changes in nutrient availability affect the stoichiometry of nutrients in plant tissues as well as nutrient resorption. Therefore, macrophytes and sediment-associated microorganisms are much crucial for wetland ecosystems and should be further explored together rather than separately.

References

Achyuthan H, Lone AM, Shah RA, Fousiya A (2020) Climate, C/N ratio and organic matter accumulation: an overview of examples from Kashmir Himalayan Lakes. In: Himalayan weather and climate and their impact on the environment, pp 185–203

Aerts R (1996) Nutrient resorption from senescing leaves of perennials: Are there general patterns? J Ecol 84:597–608

Aerts R, Chapin FS III (1999) The mineral nutrition of wild plants revisited: a re-evaluation of processes and patterns. In Advances in ecological research. Academic Press 30:1–67

Ahila KG, Ravindran B, Muthunarayanan V, Nguyen DD, Nguyen XC, Chang SW, Nguyen VK, Thamaraiselvi C (2020) Phytoremediation potential of freshwater macrophytes for treating dye-containing wastewater. Sustain For 13(1):329

Alikhani S, Nummi P, Ojala A (2021) Urban wetlands: a review on ecological and cultural values. Water 13(22):3301

Alonso X, Hadad HR, Córdoba C, Polla W, Reyes MS, Fernández V, Granados I, Marino L, Villalba A (2018) Macrophytes as potential biomonitors in peri-urban wetlands of the middle Parana River (Argentina). Environ Sci Pollut Res 25(1):312–323

Aregu MB, Asfaw SL, Khan MM (2021) Developing horizontal subsurface flow constructed wetland using pumice and Chrysopogon zizanioides for tannery wastewater treatment. Environ Sys Res 10(1):1–13

Bakhshoodeh R, Alavi N, Oldham C, Santos RM, Babaei AA, Vymazal J, Paydary P (2020) Constructed wetlands for landfill leachate treatment: a review. Ecol Eng 146:105725
Balwan WK, Kour S (2021) Wetland-an ecological boon for the environment. East Afr Sch J Agri Life Sci 4(3):38–48
Berthold M, Schumann R (2020) Phosphorus dynamics in a eutrophic lagoon: uptake and utilization of nutrient pulses by phytoplankton. Front Mar Sci 7:281. https://doi.org/10.3389/fmars.2020.00281
Bornette G, Puijalon S (2011) Response of aquatic plants to abiotic factors: a review. Aquat Sci 73(1):1–14
Bowden WB, Glime JM, Riis T (2017) Chapter 13 - Macrophytes and bryophytes. In: Hauer FR, Lamberti GA (eds) Methods in stream ecology, vol 1, 3rd edn. Academic Press, Boston, pp 243–271. https://doi.org/10.1016/B978-0-12-416558-8.00013-5
Cai ZQ, Bongers F (2007) Contrasting nitrogen and phosphorus resorption efficiencies in trees and lianas from a tropical montane rain forest in Xishuangbanna, south-west China. J Trop Ecol 23 (1):115–118
Chamorro A, Giardino JR, Granados-Aguilar R, Price AE (2015) Chapter 7 - a terrestrial landscape ecology approach to the critical zone. In: Giardino JR, Houser C (eds) Developments in earth surface processes, vol 19. Elsevier, pp 203–238. https://doi.org/10.1016/B978-0-444-63369-9.00007-0
Costa MLR, Henry R (2010) Phosphorus, nitrogen, and carbon contents of macrophytes in lakes lateral to a tropical river (Paranapanema River, São Paulo, Brazil). Acta Limnol Bras 22(2): 122–132
Craft C, Krull K, Graham S (2007) Ecological indicators of nutrient enrichment, freshwater wetlands, Midwestern United States (US). Ecol Indic 7(4):733–750
Cronk JK, Fennessy MS (2016) Wetland plants: biology and ecology. CRC press
Dalla Vecchia A, Villa P, Bolpagni R (2020) Functional traits in macrophyte studies: current trends and future research agenda. Aquat Bot 167:103290
Dokulil MT, Teubner K (2022) Comparative primary production. In: Reference module in earth systems and environmental sciences. Elsevier. https://doi.org/10.1016/B978-0-12-819166-8.00040-2
Dove A, Chapra SC (2015) Long-term trends of nutrients and trophic response variables for the Great Lakes. Limnol Oceanogr 60(2):696–721. https://doi.org/10.1002/lno.10055
Dvořák J (1996) An example of relationships between macrophytes, macroinvertebrates and their food resources in a shallow eutrophic lake. Hydrobiologia 339(1):27–36
Englande AJ, Jr., Krenkel P, Shamas J (2015) Wastewater treatment &water reclamation. Reference module in earth systems and environmental sciences: B978-970-912-409548-409549.409508-409547. doi:https://doi.org/10.1016/B978-0-12-409548-9.09508-7
Esteves BS, Suzuki MS (2010) Limnological variables and nutritional content of submerged aquatic macrophytes in a tropical lagoon. Acta Limnol Bras 22:187–198
Gavrilescu M (2021) Water, soil, and plants interactions in a threatened environment. Water 13(19): 2746
Germ M, Janež V, Gaberščik A, Zelnik I (2021) Diversity of Macrophytes and environmental assessment of the Ljubljanica River (Slovenia). Diversity 13(6):278
Güsewell S (2005) Nutrient resorption of wetland graminoids is related to the type of nutrient limitation. Funct Ecol 19:344–354
Hadad HR, Maine MA (2007) Phosphorous amount in floating and rooted macrophytes growing in wetlands from the Middle Paraná River floodplain (Argentina). Ecol Eng 31(4):251–258
Hadad H, Maine M, Mufarrege M, Di Luca G, Sanchez G, Nocetti E (2021a) Macrophyte importance in contaminant treatment and biomonitoring. In: Removal of emerging contaminants through microbial processes. Springer, pp 435–452
Hadad HR, Maine MA, de las Mercedes Mufarrege M, Di Luca GA, Sanchez GC (2021b) Nutrient concentrations in Macrophytes from lotic and lentic environments of the middle Parana River, Argentina. Ing Univ 25:25

Hilt S, Vermaat JE, van de Weyer K (2021) Macrophytes*. In: Reference Module in Earth Systems and Environmental Sciences. Elsevier. https://doi.org/10.1016/B978-0-12-819166-8.00043-8
Keller JK, Bauers AK, Bridgham SD, Kellogg LE, Iversen CM (2006) Nutrient control of microbial carbon cycling along an ombrotrophic-minerotrophic peatland gradient. J Geophys Res Biogeosci 111(G3)
Kennish MJ (2002) Environmental threats and environmental future of estuaries. Environ Conserv 29(1):78–107
Khan MN, Mohammad F (2014) Eutrophication: challenges and solutions. In: Ansari AA, Gill SS (eds) Eutrophication: Causes, consequences and control, vol 2. Springer, Netherlands, Dordrecht, pp 1–15. https://doi.org/10.1007/978-94-007-7814-6_1
Khatri N, Tyagi S (2015) Influences of natural and anthropogenic factors on surface and groundwater quality in rural and urban areas. Front Life Sci 8(1):23–39. https://doi.org/10.1080/21553769.2014.933716
Killingbeck KT (1996) Nutrients in senesced leaves: Keys to the search for potential resorption and resorption proficiency. Ecology 77:1716–1727
Kurniawan SB, Ahmad A, Said NSM, Imron MF, Abdullah SRS, Othman AR, Purwanti IF, Hasan HA (2021) Macrophytes as wastewater treatment agents: nutrient uptake and potential of produced biomass utilization toward circular economy initiatives. Sci Total Environ 790:148219
La Toya TK, Jacob DL, Hanson MA, Herwig BR, Bowe SE, Otte ML (2013) Macrophytes in shallow lakes: relationships with water, sediment and watershed characteristics. Aquat Bot 109:39–48
Lambers H, Chapin FS III, Pons TL (1998) Plant physiological ecology. Springer, New York. https://doi.org/10.1007/978-1-4757-2855-2
Lan Y, Cui B, Li X, Han Z, Dong W (2010) The determinants and control measures of the expansion of aquatic macrophytes in wetlands. Procedia Environ Sci 2:1643–1651
Lancaster KM, Caranto JD, Majer SH, Smith MA (2018) Alternative bioenergy: updates to and challenges in nitrification metalloenzymology. Joule 2(3):421–441
Ławniczak AE (2011) Nitrogen, phosphorus, and potassium resorption efficiency and proficiency of four emergent Macrophytes from nutrient-rich wetlands. Pol J Environ Stud 20(5):1227–1234
Lin Y-F, Jing S-R, Wang T-W, Lee D-Y (2002) Effects of macrophytes and external carbon sources on nitrate removal from groundwater in constructed wetlands. Environ Pollut 119(3):413–420
Lolu AJ, Ahluwalia AS, Sidhu MC, Reshi ZA, Shah MA, Lolu SJ (2018) Wetlands: an imbroglio of interlinked carbon pools. Trop Ecol 59(3):383–395
Lu J, Zhang Y, Wu J, Wang J (2020) Nitrogen removal in recirculating aquaculture water with high dissolved oxygen conditions using the simultaneous partial nitrification, anammox and denitrification system. Bioresour Technol 305:123037
Mahapatra DM, Chanakya H, Ramachandra T (2011) C: N ratio of sediments in a sewage fed urban Lake. Int J Geol 5(3):86–92
Malone TC, Newton A (2020) The globalization of cultural eutrophication in the Coastal Ocean: Causes and consequences. Front Mar Sci 7:670. https://doi.org/10.3389/fmars.2020.00670
Mazej Z, Germ M (2008) Seasonal changes in the contents of nutrients in five macrophyte species from the lake Velenjsko jezero (Slovenia). Acta Biol Slov 51:3–11
Means MM, Ahn C, Korol AR, Williams LD (2016) Carbon storage potential by four macrophytes as affected by planting diversity in a created wetland. J Environ Manag 165:133–139
Miller CA (2011) The effect of long-term drainage on plant community composition, biomass, and productivity in boreal continental peatlands. University of Guelph
Min L, Zhongsheng Z, Zhe L, Haitao W (2020) Removal of nitrogen and phosphorus pollutants from water by FeCl3-impregnated biochar. Ecol Eng 149:105792
Mitsch WJ, Gosselink JG, Zhang L, Anderson CJ (2009) Wetland ecosystems. John Wiley & Sons
Muraza M, Mayo AW, Norbert J (2013) Wetland plant dominance, density and biomass in Mara River basin wetland upstream of Lake Victoria in Tanzania. Int J Sci Technol Res 2:348–359

Mwamburi J (2018) Lake sedimentary environments and roles of accumulating organic matter in biogeochemical cycling processes and contaminants loading are invasions of water hyacinth in Lake Victoria from 1989 a concern? In: Persistent Organic Pollutants. IntechOpen

Ndungu J, Augustijn DC, Hulscher SJ, Fulanda B, Kitaka N, Mathooko JM (2014) A multivariate analysis of water quality in Lake Naivasha. Kenya Mar Freshw Res 66(2):177–186

Newton A, Icely J, Cristina S, Perillo GME, Turner RE, Ashan D, Cragg S, Luo Y, Tu C, Li Y, Zhang H, Ramesh R, Forbes DL, Solidoro C, Béjaoui B, Gao S, Pastres R, Kelsey H, Taillie D, Nhan N, Brito AC, de Lima R, Kuenzer C (2020) Anthropogenic, direct pressures on coastal wetlands. Front Ecol Evol 8:144. https://doi.org/10.3389/fevo.2020.00144

Norton JM, Stark JM (2011) Chapter fifteen - regulation and measurement of nitrification in terrestrial systems. In: Klotz MG (ed) Methods in enzymology, vol 486. Academic Press, pp 343–368. https://doi.org/10.1016/B978-0-12-381294-0.00015-8

Nouri J, Khorasani N, Lorestani B, Karami M, Hassani A, Yousefi N (2009) Accumulation of heavy metals in soil and uptake by plant species with phytoremediation potential. Environ Earth Sci 59(2):315–323

Omer NH (2019) Water quality parameters. Water quality-science, assessments and policy 16 (18):1–34

Opitz J, Alte M, Bauer M, Peiffer S (2021) The role of Macrophytes in constructed surface-flow wetlands for mine water treatment: a review. Mine Water Environ 40(3):587–605. https://doi.org/10.1007/s10230-021-00779-x

Pal P, Chakraborty K (2014) Importance of some physical and chemical characteristics of water bodies in relation to the incidence of zooplanktons: a review. Indian J Soc Nat Sci 3:102–116

Pal S, Chattopadhyay B, Datta S, Mukhopadhyay SK (2017) Potential of wetland macrophytes to sequester carbon and assessment of seasonal carbon input into the East Kolkata wetland ecosystem. Wetlands 37(3):497–512

Park Y-S, Kwon Y-S, Hwang S-J, Park S (2014) Characterizing effects of landscape and morphometric factors on water quality of reservoirs using a self-organizing map. Environ Model Softw 55:214–221

Pasut C, Tang FH, Hamilton D, Riley WJ, Maggi F (2021) Spatiotemporal assessment of GHG emissions and nutrient sequestration linked to agronutrient runoff in global wetlands. Glob Biogeochem Cycles 35(4):e2020GB006816

Perdue EM, Koprivnjak J-F (2007) Using the C/N ratio to estimate terrigenous inputs of organic matter to aquatic environments. Estuar Coast Shelf Sci 73(1–2):65–72

Peters JA, Lodge DM (2009) Littoral Zone. In: Likens GE (ed) Encyclopedia of inland waters. Academic Press, Oxford, pp 79–87. https://doi.org/10.1016/B978-012370626-3.00218-0

Peterson F, Risberg J (2008) Nutrients: phosphorus, nitrogen sources, impact on water quality-a general overview. Minnesota Pollution Control Agency, Saint Paul, MN, USA

Pezeshki SR, DeLaune RD (2012) Soil oxidation-reduction in wetlands and its impact on plant functioning. Biology (Basel) 1(2):196–221. https://doi.org/10.3390/biology1020196

Rai PK (2008) Heavy metal pollution in aquatic ecosystems and its phytoremediation using wetland plants: an ecosustainable approach. Int J Phytoremediation 10(2):133–160

Reis V, Hermoso V, Hamilton SK, Ward D, Fluet-Chouinard E, Lehner B, Linke S (2017) A global assessment of inland wetland conservation status. Bioscience 67(6):523–533

Reitsema RE, Meire P, Schoelynck J (2018) The future of freshwater Macrophytes in a changing world: dissolved organic carbon quantity and quality and its interactions with Macrophytes. Front Plant Sci 9:629. https://doi.org/10.3389/fpls.2018.00629

Rejmánková E (2005) Nutrient resorption in wetland macrophytes: comparison across several regions of different nutrient status. New Phytol 167(2):471–482

Rejmankova E (2011) The role of macrophytes in wetland ecosystems. J Ecol Environ 34(4): 333–345

Robertson GP, Groffman PM (2015) Chapter 14 - Nitrogen Transformations. In: Paul EA (ed) Soil microbiology, ecology and biochemistry, 4th edn. Academic Press, Boston, pp 421–446. https://doi.org/10.1016/B978-0-12-415955-6.00014-1

Rocha C, Lima D, Cunha M, Almeida J (2018) Aquatic macrophytes and trophic interactions: a scientometric analyses and research perspectives. Braz J Biol 79:617–624

Schlesinger WH, Bernhardt ES (2020) Chapter 7 - wetland ecosystems. In: Schlesinger WH, Bernhardt ES (eds) Biogeochemistry, 4th edn. Academic Press, pp 249–291. https://doi.org/10.1016/B978-0-12-814608-8.00007-4

Schneider B, Cunha ER, Marchese M, Thomaz SM (2018) Associations between Macrophyte life forms and environmental and morphometric factors in a large sub-tropical floodplain. Front Plant Sci 9:195–195. https://doi.org/10.3389/fpls.2018.00195

Shen S, Geng Z, Li X, Lu X (2022) Evaluation of phosphorus removal in floating treatment wetlands: new insights in non-reactive phosphorus. Sci Total Environ 815:152896

Shilla DA, Asaeda T, Kian S, Lalith R, Manatunge J (2006) Phosphorus concentration in sediment, water and tissues of three submerged macrophytes of Myall Lake, Australia. Wetl Ecol Manag 14(6):549–558

Skiba U (2008) Denitrification. In: Jørgensen SE, Fath BD (eds) Encyclopedia of ecology. Academic Press, Oxford, pp 866–871. https://doi.org/10.1016/B978-008045405-4.00264-0

Sood A, Uniyal PL, Prasanna R, Ahluwalia AS (2012) Phytoremediation potential of aquatic macrophyte. Azolla Ambio 41(2):122–137. https://doi.org/10.1007/s13280-011-0159-z

Squires MM, Lesack LF (2003) The relation between sediment nutrient content and macrophyte biomass and community structure along a water transparency gradient among lakes of the Mackenzie Delta. Can J Fish Aquat Sci 60(3):333–343

Stringfellow WT, Domen JK, Camarillo MK, Sandelin WL, Borglin S (2014) Physical, chemical, and biological characteristics of compounds used in hydraulic fracturing. J Hazard Mater 275: 37–54

Strock JS (2008) Ammonification. In: Jørgensen SE, Fath BD (eds) Encyclopedia of ecology. Academic Press, Oxford, pp 162–165. https://doi.org/10.1016/B978-008045405-4.00256-1

Swe T, Lombardo P, Ballot A, Thrane J-E, Sample J, Eriksen TE, Mjelde M (2021) The importance of aquatic macrophytes in a eutrophic tropical shallow lake. Limnologica 90:125910. https://doi.org/10.1016/j.limno.2021.125910

Takamura N, Kadono Y, Fukushima M, Nakagawa M, Kim B-H (2003) Effects of aquatic macrophytes on water quality and phytoplankton communities in shallow lakes. Ecol Res 18(4):381–395

Tanabe Y, Hori M, Mizuno AN, Osono T, Uchida M, Kudoh S, Yamamuro M (2019) Light quality determines primary production in nutrient-poor small lakes. Sci Rep 9(1):4639. https://doi.org/10.1038/s41598-019-41003-9

Tang S, Liao Y, Xu Y, Dang Z, Zhu X, Ji G (2020) Microbial coupling mechanisms of nitrogen removal in constructed wetlands: a review. Bioresour Technol 314:123759

Tangahu BV, Sheikh Abdullah SR, Basri H, Idris M, Anuar N, Mukhlisin M (2011) A review on heavy metals (as, Pb, and hg) uptake by plants through phytoremediation. Int J Chem Eng 2011: 939161. https://doi.org/10.1155/2011/939161

Thomaz S, Esteves F, Murphy K, Dos Santos A, Caliman A, Guariento R (2009) Aquatic macrophytes in the tropics: ecology of populations and communities, impacts of invasions and human use. Trop Bio Conserv Manag 4:27–60

Tiner RW (2012) Defining hydrophytes for wetland identification and delineation

Tiso M, Schechter AN (2015) Nitrate reduction to nitrite, nitric oxide and ammonia by gut bacteria under physiological conditions. PLoS One 10(3):e0119712. https://doi.org/10.1371/journal.pone.0119712

USGS (2019). https://www.usgs.gov/special-topics/water-science-school/science/how-much-water-there-earth

Vázquez E, Teutscherova N, Dannenmann M, Töchterle P, Butterbach-Bahl K, Pulleman M, Arango J (2020) Gross nitrogen transformations in tropical pasture soils as affected by Urochloa genotypes differing in biological nitrification inhibition (BNI) capacity. Soil Biol Biochem 151: 108058

Vymazal J (2007) Removal of nutrients in various types of constructed wetlands. Sci Total Environ 380(1–3):48–65

Willby N, Pulford I, Flowers T (2001) Tissue nutrient signatures predict herbaceous-wetland community responses to nutrient availability. New Phytol 152(3):463–481

Woodward C, Gadd P (2019) The potential power and pitfalls of using the X-ray fluorescence molybdenum incoherent: coherent scattering ratio as a proxy for sediment organic content. Quat Int 514:30–43

Xu T, Weng B, Yan D, Wang K, Li X, Bi W, Li M, Cheng X, Liu Y (2019) Wetlands of international importance: status, threats, and future protection. Int J Environ Res Public Health 16(10):1818. https://doi.org/10.3390/ijerph16101818

Yan A, Wang Y, Tan SN, Mohd Yusof ML, Ghosh S, Chen Z (2020) Phytoremediation: a promising approach for revegetation of heavy metal-polluted land. Front Plant Sci 11:359. https://doi.org/10.3389/fpls.2020.00359

Zaghloul A, Saber M, Gadow S, Awad F (2020) Biological indicators for pollution detection in terrestrial and aquatic ecosystems. Bull Natl Res Cent 44(1):127. https://doi.org/10.1186/s42269-020-00385-x

Zhang X, Zhang Y (2021) Waterscapes for promoting mental health in the general population. Int J Environ Res Public Health 18(22):11792. https://doi.org/10.3390/ijerph182211792

Zhou Q, Zhang J, Fu J, Shi J, Jiang G (2008) Biomonitoring: an appealing tool for assessment of metal pollution in the aquatic ecosystem. Anal Chim Acta 606(2):135–150

Zhou X, He Z, Jones KD, Li L, Stoffella PJ (2017) Dominating aquatic macrophytes for the removal of nutrients from waterways of the Indian River Lagoon basin, South Florida, USA. Ecol Eng 101:107–119

Zhu S, Huang X, Ho S-H, Wang L, Yang J (2017) Effect of plant species compositions on performance of lab-scale constructed wetland through investigating photosynthesis and microbial communities. Bioresour Technol 229:196–203

Zimmer M (2008) Detritus. In: Jørgensen SE, Fath BD (eds) Encyclopedia of ecology. Academic Press, Oxford, pp 903–911. https://doi.org/10.1016/B978-008045405-4.00475-4

Zou R, Li Y, Zhao L, Liu Y (2014) Chapter 15 - exploring the mechanism of catastrophic regime shift in a shallow plateau Lake: a three-dimensional water quality modeling approach. In: Jørgensen SE, Chang N-B, Xu F-L (eds) Developments in environmental modelling, vol 26. Elsevier, pp 411–435. https://doi.org/10.1016/B978-0-444-63249-4.00016-6

Ecotoxicology of REEs in Aquatic Macrophytes and Prospect for Bioremediation of REEs

7

Rajni Choudhary, Sana Parveen, and Shailesh Kumar Patidar

Abstract

Lanthanides (e.g. La, Gd, Nd, Ce) are being used increasingly due to their versatile applications in medical diagnosis, electronics, agriculture, industries, and research; hence, their global utility had been increased drastically in the last few decades. Lanthanide enrichment in agricultural fields due to repetitive use of fertilizer application and REE (rare-earth element)-based fertilizers leads to lanthanide-concentrated agricultural runoff to the water bodies consequently. Through biomagnification in the food chain, there is a risk of enrichment of these REEs in humans. Aquatic macrophytes are one of the important constituents of the food chain along with being a significant primary producer widely studied for metal accumulation and stress physiology. However, very limited knowledge on transfer processes, bioaccumulation, and distribution pattern of REEs is available. Chemical-assisted remediation and omics-guided phytoremediation using hyperaccumulator macrophytes are found to be the most popular methods, but macrophyte-based phytoremediation is preferable as it is eco-friendly and does not create any secondary pollutants. This chapter will cover the significant sources of REE contamination especially lanthanides, ecotoxicological effects on macrophytes, their stress physiology, and bioremediation strategies.

Keywords

Ecotoxicology · Rare-earth elements (RREs) · Stress · Chemical-assisted remediation · Omics-based approaches · Phytoremediation

R. Choudhary · S. Parveen · S. K. Patidar (✉)
Department of Environmental Science, School of Earth Sciences, Ajmer, India
e-mail: shailesh.patidar@curaj.ac.in

S. Kumar et al. (eds.), *Aquatic Macrophytes: Ecology, Functions and Services*,
https://doi.org/10.1007/978-981-99-3822-3_7

7.1 Introduction

IUPAC has defined rare-earth elements (REEs) as a group of 17 elements having similar physical and chemical properties constituting less than 20% of naturally occurring elements (Gwenzi et al. 2018). The average concentration of REEs in Earth's crust is 150 to 220 mg/kg, which occupies nearly $3 \times 10^{-4}\%$ of the total Earth's crust. China leads worldwide for the largest mining and processing of REEs, which accounts for 63% and 90% of total global production, respectively. REEs stand out in the present economy due to their perfect magnetic characteristics and explicit energy levels (Migaszewski and Gałuszka 2015). Therefore, they are proved as an efficient candidate for current electromagnetic-based technologies, superconductors, and high-tech military components (Tesfaye et al. 2021). Lanthanides are a group of 15 elements that mainly belong to those elements having atomic numbers of 58–71 and possess tendency to fill their 4*f* electronic subshell. Elements e.g. scandium and yttrium that were isolated as oxides along with lanthanides are collectively called REEs. REEs can form complexes with anions like phosphate, hydroxide, carbonate, fluoride, and silicates and get occluded to clay and organic matter of soil, thus causing an adverse effect on plants, animals, and human health by entering through the food chain. Further, they also possess good solubility in water which may vary according to the temperature, concentration, pH, and salinity and become a prominent source of water contamination (Trapasso et al. 2021). Furthermore, REE like Gd is stable in dry air, in contrast to other REEs. However, it gets oxidized to Gd_2O_3 in the humid air and slowly absorbs water and often dissolves into acids (Gwenzi et al. 2018). The prevalence of REEs especially lanthanides (Ln) in electronic waste and ineffective management techniques create contamination at each trophic level. Lanthanide pollution requires urgent measures to rehabilitate these precious metals, not only from the economic and technical point of view but also from a public perspective and ecosystem as they are not stable and became emerging environmental contaminants due to their dynamic nature. In developing countries, there are already significant environmental issues and health concerns associated with the illegal trade in e-waste (Dodson et al. 2012). Understanding their possible hazards to biological systems and potential ecotoxicological implications on the biosphere will be crucial. However, compared to other more prevalent metals, knowledge of ecotoxicology for these elements is still inadequate. Bioremediation of REE includes microbial-mediated biosorption, phytoremediation, etc., as corroborated in early studies used for the removal of heavy metals and other contaminants by aquatic plants, which are mostly macrophytes (Igiri et al. 2018). Considering the risk of REEs in aquatic bodies due to surface and agriculture runoff from the e-waste and REE-containing fertilizers, macrophytes (major primary producers after the algae) are significant for phytoremediation as well as ecotoxicological point of view.

7.2 Sources and Pathways of Lanthanide Contamination

REE can be obtained directly from active or inactive mining activities and indirectly from mineral beneficiation along with manufacturing. In terms of aquatic ecosystems, their primary sources are waste and contaminated water generated from medical units, extraction technology, high-tech industries, and recycling plants (Gwenzi et al. 2018). As a result, these are likely to be found in the aquatic environment through domestic sewage, industrial emission, over-land flow, and atmospheric deposition (Trapasso et al. 2021). For instance, wastewater treatment plants are the main source of human-induced Gd in the aquatic environment. Through hydrological processes, REEs in wastewater are ultimately dispersed into aquatic wetlands and croplands. It has been demonstrated that soils in mine-affected areas have REE concentrations 100 times higher than the Earth's crust as a whole (Fig. 7.1).

Ln (especially oxides) is broadly used in modern technology. Ln oxide-based nanoparticle like CeO_2 is widely used in energy and polishing industries and found as one of the largest sources of Ln contamination. Similarly, CeO_2 nanoparticles are one of the oxidized nanoparticles which were manufactured around 9100 t by the year 2016 as estimated by one of the studies. Another current medical technology used Gd_2O_3 nanoparticles as an alternative to Gd chelates in magnetic resonance imaging (MRI) technique to increase its effectiveness. In addition, some Ln oxide-based nanoparticles like La_2O_3, CeO_2, Gd_2O_3, etc. are also proven to be potential antimicrobial agents, which result in their increased market value and manufacturing (Blinova et al. 2020). There have been reports of La and Sm being employed in

Fig. 7.1 Routes of rare-earth element contamination in aquatic system

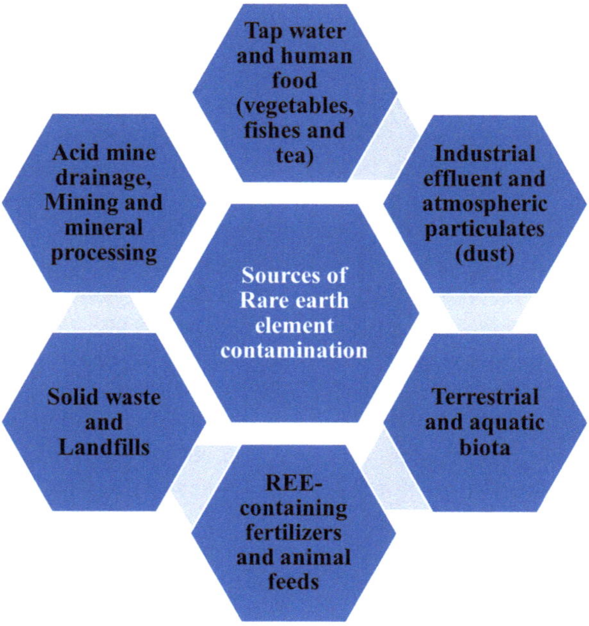

aquatic systems as fluid catalytic cracking catalysts in petroleum refining. Furthermore, lanthanides are extensively used in fertilizers and may persist in the environment as a by-product. REE-containing phosphate fertilizers have increased crop yield in China. For example, Changle is a chiefly used fertilizer in China, which is composed of La_2O_3, CeO_2, Pr_6O_{11}, Nd_2O_3, and other REEs. Therefore, the use of REEs in agriculture may represent a significant pathway of anthropogenic REE's entry into the environment (Blinova et al. 2020).

The widespread usage of REEs in many industrial sectors raises the danger of these elements gaining access to freshwater and the ocean, posing a concern to these systems and, in particular, to the species that live there. Natural REE concentrations in river water can reach up to 200 ng/L, while levels of La and Gd at wastewater outfalls can reach 0.08 and 1.1 mg/L, respectively (Migaszewski and Gałuszka 2015; Trapasso et al. 2021). According to the published data on the background and contamination concentration, the concentration of La ranges from 7.7 to 80.4 μg/L, and for Ce, it ranges from 19.4 to 161 μg/L in contaminated water, while, in river water, it ranges from 19.7 to 74 ng/L, and for Ce, it ranges from 9.67 to 212 ng/L. REE concentrations in natural lakes and rivers rarely surpass 20 g/L, whereas mine drainage can contain up to 24,800 g/L (Cao et al. 2021).

In nations including China, the USA, India, Brazil, and Malaysia, mining operations have been connected to severe environmental and health harm (Adeel et al. 2019; Balaram 2019; Liang et al. 2014; Galhardi et al. 2020). Furthermore, due to inefficient processes, REE recycling is one of the major problems. Extraction of REE-producing minerals has gathered significant consideration in the past few years, probably due to lanthanide augmentation in the environment. Pollution of anthropogenic REEs may be a significant concern in tropical areas (e.g. iron ore residues and discarded mines) (Pagano et al. 2015). Under present national and international law, REEs can be used without restrictions; however, they are emerging contaminants with uncertain impacts (Galdiero et al. 2019).

7.3 Ecotoxicology of Lanthanides in Aquatic Macrophytes

The science of ecotoxicology focuses on how toxic chemicals affect living things, particularly at the population, community, ecosystem, and biosphere levels. REEs build up in soil, biomass, and sediments surrounding mining regions and may contribute to human health risk assessments. Plants require some heavy metals (HMs) for growth but, beyond a certain threshold, become poisonous and interfere with plant metabolism. HM toxicity induces the generation of reactive oxygen species (ROS), which affects the physiological process and even causes plant death (Ansari et al. 2020). The effects on organisms' cellular or biological levels are studied in ecotoxicology, as they offer excellent mechanical and sensitivity values to identify pollutants.

Higher plants (macrophytes) are best living models for examining the toxicity and build-up of different contaminants and are often employed to identify and assess the toxicity (Jiang et al. 2019). So toxicity testing using them will help understand the effect of lanthanides. Furthermore, high-yielding plants are often more environment-friendly than species used in mutagenicity (*Salmonella* sp.) (Corrêa et al. 2016). *Nitellopsis obtusa* internodal cells are susceptible to HMs, and complex combinations at cellular and biochemical levels and their reactions were employed in various investigations (Ahmed et al. 2013).

7.3.1 Bioaccumulation of Lanthanides

According to the findings of several studies, many aquatic macrophytes like *Lemna minor*, *Eichhornia crassipes*, *Azolla caroliniana*, *Wolffia globosa*, *Phragmites communis*, and *Najas marina* are known for their propensity to accumulate HMs (Ansari et al. 2020). Several parameters for determining the accumulation of contaminants are developed. One of the measures is bioaccumulation concentration factor (BCF), which includes different contaminants' level in the organisms with respect to the medium. BCF values are based on the ratio between plant concentrations against surface water, and it was observed that it's similar for most of the REEs throughout the year in all of the studied areas in *Lemna minor*, while BCF values of *Potamogeton pectinatus* were moreover similar for all REEs but differed between study areas, and for some other species, BCF varied in both spatial and temporal scales (Weltje 2003). A reduction in BCF with the atomic number was reported when the concentration ratio in the biota to pore water was measured. Ln levels varied from 39 to 117 mg/kg dry wt. in sediments and 1.7–14 mg/kg dry wt. in macrophytes of Estonia River. It was demonstrated that the use of oil shale fly ash for soil improvement can cause Ln accumulation in plants. Also, continuous mineral fertilizer utilization can raise Ln quantity in plants growing around cultivable land (Blinova et al. 2021). *Pistia stratiotes* roots accumulate more La as compared to the leaf, and it can accumulate La even at very low concentrations (Nazreen et al. 2017).

Low-lying animals (macroalgae and invertebrates) have shown more REE concentration than species in the upper trophic range. It suggests that organisms that feed directly on water and fossils may be more prone to take up REE, and it is less likely to be biomagnified, instead depleted with stratifications (Sun et al. 2020; Ciesielski et al. 2016). The majority of the reported data comes from freshwater habitats (with concentrations of 0.347–80 g/L), while the largest Gd anomalies are observed around heavily industrialized locations. The published investigations found values of 0.36–26.9 ng/L in the marine environment, peaking at a subsurface outfall (409.4 ng/L). The majority of literature on Gd bioaccumulation and consequences focuses on freshwater species, and a wide range of bioaccumulation extents depends on the chemical speciation of Gd complexes. In contrast, there is a lack of field data on the bioaccumulation of Gd in marine species (Trapasso et al. 2021). Recent studies have shown that REEs increase in the hydrosphere and tap water. The bioaccumulation potential of different lanthanide group substances is different. However, current perceptions and reported differences regarding effects remain controversial (Gonzalez et al. 2014; Romero-Freire et al. 2019). Most bioaccumulation studies found in the literature search included only few elements, usually simple lanthanides, which makes it difficult to conclude whether there is a biological tendency for the whole group or not.

7.3.2 Hyperaccumulation of Lanthanides

Hyperaccumulation of metals, metalloids, and other contaminants has been examined in aquatic macrophytes (*Wolffia*, *Azolla*, *Lemna*, *Hydrilla*, *Pistia*, *Eichhornia*, *Typha*, *Crinum*, *Spirodela polyrhiza*, *Alternanthera*, *E. nuttallii*, *Hydrocharis dubia*, *Phragmites*, *Nymphoides peltata*, and *Chrysopogon*); however, there haven't been many studies done on REE hyperaccumulator plants so far. Few studies indicated that a REE hyperaccumulator plant has a bioconcentration factor greater than 1 or has more than 1000 μg/g of total REE in the frond (Liu et al. 2021). REE hyperaccumulators are mostly fern species (Chen et al. 2022) and can absorb REE from the soil even if present in trace amounts (Zhenggui et al. 2001). Five other fern species, of *Asplenium* genus, can hyperaccumulate REE due to their higher BCF for La as shown in Table 7.1.

7.4 Stress Physiology, Avoidance, and Tolerance

7.4.1 Cellular Effects

Studies at the subcellular and plant tissue level have shown that HMs have a deleterious influence on cells and cellular components by causing ROS generation, lipid peroxidation, DNA damage, cell cycle disruption, nutritional deficiency, and

Table 7.1 Rare-earth element hyperaccumulator species

Aquatic macrophytes	Accumulation concentration (μg/g) or BCF	Accumulation site in plant	References
Pronephrium simplex	3000	Lamina	Lai et al. (2005)
Blechnum orientale	1022	Foliage	Chen et al. (2022)
P. triphyllum	1027	Lamina	Avishek and Hazra (2022)
Gleichenia sp. and *Adiantum* sp.	3340.64	Foliage	e Cunha et al. (2012)
Phytolacca americana	1040	Leaf	Yuan et al. (2018)
Dicranopteris pedata	7000	Foliage	Chen et al. (2022)
Dicranopteris strigose	BCF La >1	–	Ozaki et al. (2000)
Dicranopteris linearis	6946.45	Leaf	Shan et al. (2003)

toxicity (Jamla et al. 2021). However, La predominately accumulates in the cell wall of aquatic macrophytes (*Dicranopteris dichotoma*, *Pronephrium simplex*, and *Nymphoides peltata*). It is thought that cell walls act as REE sinks, hence contributing to metal tolerance. Plant cells' wall can alter their metal holding capacity to build up more cations in the apoplast, safeguarding its protoplast. However, La inside the cell wall mostly mediated by cellulose and pectin (79–84%) (Zhang et al. 2015). Moreover, at 1.4–2.8 mg Pr/L, duckweed (*Spirodela polyrhiza*) showed chlorophyll loss and oxidative degradation. The presence of Ce salts enhanced the growth of another pondweed species (*Lemna minor*) up to 139 mg Ce/L and decreases the growth at higher concentrations (Blinova et al. 2020). Lanthanum toxicity was investigated in macrophyte *E. nuttallii* to identify how it affects mineral absorption, malondialdehyde (MDA), chlorophyll, and non-enzymatic antioxidants (Zhang et al. 2015). The findings revealed that La^{3+} arrests the cell at G1/S and S/G2 interphase, which can hinder root growth. It can alter microtubule structure (M. Liu and Hasenstein 2005). High concentrations can inhibit the growth of root cells and, conversely, can stabilize the cytoskeleton at low concentrations. These studies showed that root growth activity usually decreases with mitotic activity in the apical meristem. Hence, disturbed cell growth can determine lanthanide toxicity in plant roots. The lowering of the mitotic index is associated with cell cycle abnormalities, which leads to a decrease in cell activity (d'Aquino et al. 2009; Qin et al. 2015). Only one laboratory study and one monitoring study on La from modified bentonite were found on macrophyte *Elodea nuttallii* with roots embedded inside the sediments. The outcomes were highly variable. The available studies have shown significant differences in certain types of $La(NO_3)_3$ at macrophyte exposure, and already significant cellular effects have occurred in *Hydrocharis dubia* leaves at La concentration (5.56 mg La/L) (Xu et al. 2012). However, studies suggest that when antioxidant stimulation likely controls the high production of ROS to overcome oxidative stress, then unfavourable results are observed. Tissue damage was reported in *L. minor* at high Ln concentrations (Ippolito et al. 2010; Paola et al. 2007). Moreover, beneficial effects are linked with lower concentration and short-term exposure and harmful effects at higher concentration and long-term exposure. La also controls ROS levels in plant cells and functions as a Ca^{2+} channel blocker. Abiotic stress that results in excessive ROS generation damages macromolecules such as DNA and lipids, which ultimately results in cell death (Bailey-Serres and Mittler 2006). The activity of the cell's mitotic machinery and chromosomal structure can both be negatively impacted by an excess of ROS (increased mitotic abnormalities and chromosome abnormalities). However, the mechanism of lanthanides on cellular division and cell structure remains unclear.

7.4.2 Biochemical Effects

Besides these, La causes a variety of physiological and biochemical reactions in the plant tissues that alter nutrient absorption, ROS, decreased glutathione (GSH), non-protein thiols, and phytochelatin (PC) concentrations in *E. nuttallii* that had been exposed to La. According to reports, REEs can harm cytoskeleton and plasma membrane structures, and La^{3+} can diminish ion transport in plants, suppress the activity of the plasma membrane redox system, and inhibit the physiological activity of the proteases (Zhang et al. 2015). The considerable increase in MDA at higher values following La exposure demonstrated the oxidative stress in *E. nuttallii*. It was primarily caused by higher amounts of ROS, such as O_2 and H_2O_2, as well as altered membrane permeability that resulted in increased ion loss (Zhang et al. 2015). Under HM stress, aquatic plants generate enzymatic antioxidants such as superoxide dismutase (SOD), catalase (CAT), ascorbate peroxidase (APX), and dehydroascorbate reductase (DHAR), besides non-enzymatic antioxidants like betaines, proline, and ascorbates (Jamla et al. 2021). Also, MDA (the final product of plasma peroxidation) is considered a sign of lipid peroxidation caused by HM stress (Xu et al. 2016a). If La concentration reached up to 1389 mg, then MDA and H_2O_2 production increases. Many activities of the enzyme within the antioxidant systems and oxido-reduction metabolite glutathione (GSH) were produced significantly at 695 mg La. These can be a typical initial response to stress (Ippolito et al. 2010). *Lemna minor also showed similar responses* (Paola et al. 2007). Praseodymium (Pr) may penetrate *S. polyrhiza*'s cytoplasm and induce oxidative damage, resulting in enhanced lipid peroxidation and lower photosynthetic pigments, protein, and unsaturated fatty acid. On the other hand, Pr stimulates enzymatic (e.g. peroxidase (POD)) and non-enzymatic (e.g. GSH, ascorbate (AsA), and non-protein thiol (NPT)) antioxidants to deal with Pr toxicity. The interaction of Pr with (C(=O) OH), > [C=O], and –OH groups in *S. polyrhiza* may explain the FTIR alterations. When plants were treated with La, it revealed increased K leakage, which is another sign of membrane disturbance. *E. nuttallii* experienced oxidative damage as a result of exposure to La. These findings concur with those found in *Hydrocharis dubia*, *Lemna minor*, and *Hydrilla verticillata*, as well as *Lemna minor* (Zhang et al. 2015).

7.4.3 Molecular Effects

REEs are also known to cause chaperone and protease down-/upregulation, affecting protein metabolism and inhibiting protein synthesis (Xu et al. 2016b). Plants frequently ingest and store even tiny yet harmful quantities of HMs in their edible portions, posing a threat to plant growth, development, and food chains. Increased transcription of GSH-producing genes and PC-biosynthesizing enzymes may be the cause of the declined GSH during La-induced stress. One of the studies also demonstrated that La dramatically boosted PC levels. The activation of PC genes, which are crucial for metal detoxification, may be the cause of the rise in PC levels in *E. nuttallii*. These findings accord with those for *Hydrilla verticillata* exposed to Cu; as a result, La-induced stress in *E. nuttallii* can be neutralized and detoxified by declined GSH, TNP-SH, and PC production (Zhang et al. 2015). Further, in the presence of REEs, the transition of nitrogen structure (inorganic to organic) is enhanced, which helps in the synthesis of proteins and nutrient regulation (Yang et al. 1999). Two cysteine residues in these HM proteins bind to HMs, allowing them to be transported and detoxified. HM ATPase (HMA) is involved in the absorption, translocation, and sequestration of HMs. Several computational approaches (Gene Structure Display Server (GSDS), Multiple Expectation maximizations for Motif Elicitation (MEME), and TBtools) were used to analyse cis regulators, coding sequence region, introns, and 5′- and 3′-untranslated regions (Jamla et al. 2021).

7.4.4 Physiological Effects

To explore the impacts on plant growth/antioxidant systems, *Lemna minor* L. plants were cultivated in the lab with cerium (Ce) ions. Growth increased in plants treated with lower Ce concentrations compared to control plants, but it reduced in plants treated with greater Ce concentrations. In plants treated with higher Ce concentrations, chlorotic symptoms developed along with a decline in carotenoid and chlorophyll. Higher Ce concentrations are harmful to *L. minor*, as evidenced by elevated hydrogen peroxide, antioxidant metabolites, and antioxidant activity (Zicari et al. 2018). The available data indicated a significant impact on the physiology of macrophytes when exposed to certain types of $La(NO_3)_3$ salts. When the leaves of *Hydrocharis dubia* were exposed to 5.56 mg of La, some physiological alterations were observed, and when the concentration enhanced up to 1389 mg of La, total chlorophyll content decreased (Xu et al. 2012).

It has even been discovered that specific components can improve plant development. For instance, *Salvia miltiorrhiza* (bioindicator of REE), when cultivated in the presence of 100 mM Pr, displayed improved plant development and greater secondary metabolite content in plant sections (Fan et al. 2020). Many abiotic stresses cause plants to respond by lowering the number of photosynthetic pigments. This has been seen in various aquatic plants when exposed to REEs, such as *Hydrilla verticillata*, *Hydrocharis dubia*, *Nymphoides peltata*, and *Lemna minor*. This decline in pigment concentration is caused by lipid peroxidation-related disruptions in chlorophyll biosynthesis or degradation, as well as harm to the chloroplast's ultrastructure (Zhang et al. 2015). In ferns, chlorophyll was found with La and Ce in place of magnesium. This modified chlorophyll can replace the original chlorophyll, affecting plants' physiological activity, which seems to be a positive effect (Liu et al. 2018). Studies have shown that REE can enhance the photosynthesis rate. This enhancement in rate is observed due to increased chlorophyll concentration, chloroplast development, and enzyme activity in plants. REEs can change the mineral uptake and control plant growth (Hu et al. 2016). Pr inactivates the photochemical reaction centre (PS II), causing reaction centre functions to be impaired. The decline in plant photosynthetic activity is also due to the significant effects on chloroplast ultrastructure and fluorescence of chlorophyll molecules. In Pr-treated plants, even cell death may occur in response to Pr stress. Further, REEs enhance the transition of nitrogen from inorganic to organic structure, which is helpful for protein synthesis and nutrient balance regulation (Yang et al. 1999).

Plants have a variety of detoxifying strategies, such as compartmentalization inside the cell wall/vacuole and complexation with PCs, to reduce the oxidative damage caused by ROS (Mishra et al. 2006) (Seth et al. 2008). For instance, in *E. nuttallii*, the total NP-SH concentration got increased on La exposure. Generation of PCs and unidentified thiols; stimulation of enzymes (sulphate reduction pathway), including APS reductase and serine acetyltransferase (Zhang et al. 2015); and its participation in the metal-induced oxidative stress response are the possible causes of this. When exposed to La, the plant produced more TNP-SH, demonstrating its capacity to withstand metal stresses (Seth et al. 2008). Furthermore, these findings supported the findings of Seth et al. (2008) and Mishra et al. (2006), demonstrating that plants under Cd stress had lower GSH concentrations than control plants.

7.4.5 Avoidance

Plant cell organelles developed antioxidative responses to fight ROS produced as a result of oxidative stress. A plant's oxidative stress tolerance potential may be determined by the response of antioxidant enzymes. Superoxide dismutase (SOD), glutathione reductase (GR), guaiacol peroxidase (GPX), ascorbate peroxidase (APX), and catalase (CAT) can prevent the effects of ROS. The toxic superoxide ions are sequestered by SOD, which breaks them into H_2O_2 and oxygen. H_2O_2 is then detoxified via CAT, APX, and GPX, which convert H_2O_2 into H_2O and O_2. GR is another important antioxidant enzyme that helps the cell to balance their GSH/GSSG ratio and battle oxidative stress (Sidhu et al. 2016).

Increases in the activity of antioxidant enzymes may be used to define the effects of Ln on seed germination (Hong et al. 2000; Fashui 2002) and improved plant tolerance to Ln stress (Pang et al. 2002). Elevated levels of APX and glutathione (GSH) were found in roots when La and Ln mixture was given, while only La^{3+} increases the APX content in shoots (d'Aquino et al. 2009). Increased antioxidant defences and ROS production that are effectively regulated by antioxidant stimulation will lead to industrial development (Fashui 2002). Increased antioxidant levels may be translated as an indication of the stress caused by Ln (Mittler 2002; Paola et al. 2007). These findings add to our knowledge of the environmental and toxicological mechanisms that underpin REE tolerance in aquatic macrophytes. Further research on the gene expression of aquatic plants is needed to expand our understanding of molecular mechanisms (Xu et al. 2016b).

7.4.6 Tolerance

Low-dose yttrium (Y) accelerated the conversion of putrescine (Put) molecule which helps in the stabilization of DNA into spermidine (Spd) and spermine (Spm) that are involved in cellular metabolism. Simultaneously, it enhances the activity of diamine oxidase (DAO) and S-adenosylmethionine decarboxylase (SAMDC) and prevented oxidative stress. Under HM stress, Y increases photosynthetic pigments and quantum efficiency (PS II) in *P. crispus*, which in turn enhances photosynthesis in the macrophyte. Therefore, the capacity to tolerate HM in *P. crispus* remarkably increased after the exogenous addition of Y. Consequently, it can be inferred from this study that REEs can enhance HM pollution tolerance capacity in the submerged plants (Lyu et al. 2019). The resistivity against HM is mainly due to the presence of some important pigments like anthocyanins and thiols and also due to the presence of antioxidant enzymes (Leão et al. 2014). Duckweeds have a remarkable ability to recover quickly after being exposed to high levels of HMs (Ekperusi et al. 2019). The formation of metal-PC complexes may be implicated in plants' ability to tolerate HM stress; the synthesis of PCs is a critical process that promotes tolerance in plants to heavy metals (Ansari et al. 2020). The *E. nuttallii* can withstand La stress despite its severe impact on various physiological parameters because metal ions are immobilized by the cell wall and sequestered by non-protein thiols as discussed earlier in Sect. 4.3. The plant seems to be an excellent candidate for REE phytoextraction in the aquatic ecosystems because of its rapid development and great accumulation capacity (Zhang et al. 2015). There is debate regarding experimental evidence for interpersonal relationships, physiological impacts, and reactions to stress-inducing chemicals. However, many studies suggested that the induction of antioxidant activity after the treatment of REEs is purely concentration and species dependent. In some plant species like *Oryza sativa*, REEs induce efficient antioxidant activity which tolerates the ROS and thus results in improved aged seed germination of the plant (Ippolito et al. 2010).

7.5 Lanthanide Remediation

Depending on the type of contamination being dealt with, remedial methods might be physical, chemical, thermal, or biological as depicted in Fig. 7.2.

Sorption (chemical) has been offered as a possible technique for long-term water purification and, more recently, rare-earth element (REE) recycling. In a study, six living seaweed species (*Ulva lactuca*, *Osmundea pinnatifida*, *Fucus spiralis*, *Fucus vesiculosus*, *Gracilaria* sp., and *Ulva intestinalis*) were exposed to mono- and multi-element solutions of Y, La, Ce, Pr, Nd, Eu, Gd, Tb, and Dy (1 mol L^{-1}) to examine the impact of REE competition on sorption. Outcomes revealed that mono-element solutions prefer light REE. Heavy REEs have a less strong competitive effect. In contrast to water content (per cent), seaweed surface area is a key determinant of REE sorption, as the surface area is directly proportional to removal and competitive effect (Pinto et al. 2021).

Biosorption is a phytoremediation process comprising physicochemical and biological methods to remove pollutants from wastewater in a profitable manner. Notably, biological materials like seaweeds, biomass, and by-products (e.g. chitosan) have been used for biosorption (Mack et al. 2007). Due to their rapid growth, water ferns like the *Azolla* species (*A. filiculoides* and *A. pinnata*) can create a large amount of biomass in a short period. Moreover, dead *Azolla* recovers more metals (three to seven times) than live organisms from the contaminated

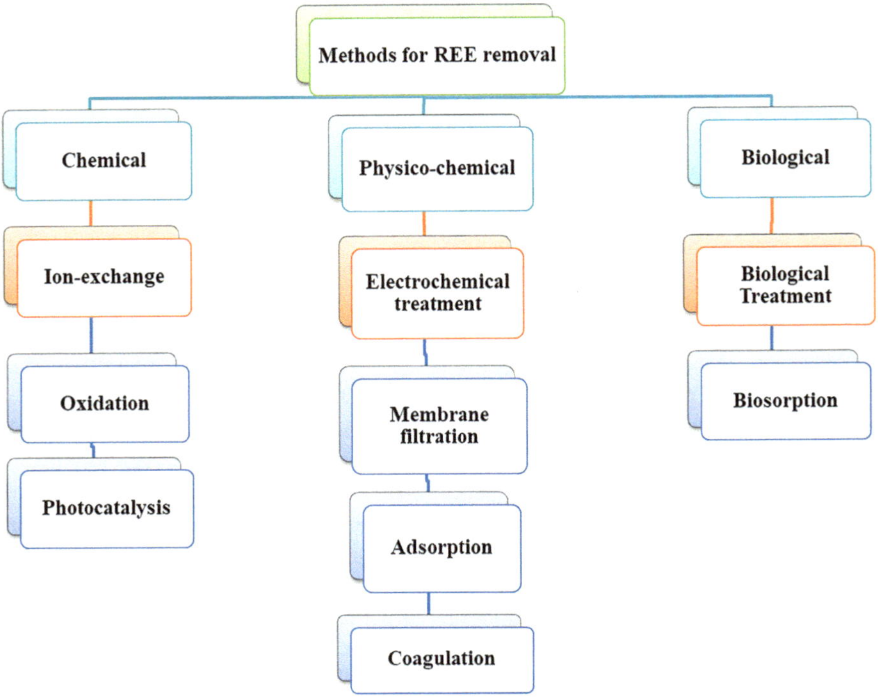

Fig. 7.2 Potential methods for rare-earth element removal

aquatic body (Sood et al. 2012). Thus, dead macrophyte biomass is frequently chosen over living biomass as bio-sorbents to detoxify industrial wastewaters due to economical and other factors such as the absence of growth restrictions, the ease with which the technology can be applied in field settings, and the ability to more easily dispose of (bury) metal-saturated biomass in wastelands. Aquatic plants are being employed to adsorb organic and inorganic contaminants from the aquatic system. Aquatic plants expand quickly and generate enormous amounts of biomass, which can be used for a variety of beneficial activities. Typically, living plants are employed to remove metals because metal ions accumulate in living tissues, whereas adsorption just needs dehydrated plant biomass to eliminate metal toxicity (Priya et al. 2022).

7.5.1 Metal Adsorption/Absorption by Aquatic Macrophytes

For adsorption purposes, a variety of aquatic macrophytes have been employed, including *Eichhornia crassipes*, *Ceratophyllum*, *Lemna*, *Pistia*, *Stratiotes*, *Spirodela polyrhiza*, *Azolla*, and many others. Adsorbents can be altered using different chemical species to boost adsorption effectiveness. Making the method cost-effective and employing nano-adsorbents made from aquatic plants will allow for further research (Singh and Kumar 2022). Aquatic macrophytes are very good at eliminating chemical pollutants. Aquatic macrophytes include watercress, hydrilla, alligator weed, pennywort, duckweed, water hyacinth, etc. By using cutting-edge bioremediation methods, macrophytes' capacity for bioremediation can be improved (Saha et al. 2022). By securing heavy metals, aquatic macrophytes play a crucial part in the operation and maintenance of an aquatic ecosystem's hydrochemistry. Aquatic macrophytes' ability to absorb and excrete HMs depends on whether they are rooted emergent, floating, or immersed. The strong propensity to accumulate HMs in *Potamogeton crispus*, *Phragmites australis*, and *Ceratophyllum demersum* makes them useful for remediating damaged soils, wastewater, and aquatic habitats (Nabi 2021).

Common aquatic weeds like *Pistia stratiotes*, *Lemna*, *Eichhornia crassipes*, etc. showed specific responses to the aquatic contaminants by roots and shoot growth and increase in some of the enzymatic activities commonly. The specificity in the structural variation of leaves and roots in some aquatic plants can improve the bioaccumulation of the contamination. The possible mechanism behind this is due to the tendency of plants to absorb metal in aerial tissues where it is transformed into active form. This ratio of bioaccumulation concentration of metal/heavy metal/ lanthanide in the root to shoot determines the quality of the hyperaccumulator and can be expressed as transfer factor (Ansari et al. 2020). Interestingly, the elimination of REEs was improved by the presence of potentially toxic elements of wastewater like Cd, Pb, Hg, etc. Among the REEs, Gd, La, and Eu were removed in the maximum due to their ability to bind with the sorption site at the macroalgae because of the strong ionic strength. The most possible explanation is the binding of toxic elements with the carbonates preferentially over REEs, which makes it easier for REEs to bind to the sorption sites of macroalgae. For various kinds of free-floating, emergent, and submerged aquatic macrophytes, heavy metal accumulation and removal mechanisms differ significantly. To eliminate these hazardous chemicals from wastewater, highly effective plant species should be selected and used. So there is a wide scope for the bioremediation of REEs along with heavy metals (Costa et al. 2020).

7.5.2 Chemical-Assisted Remediation

The genotype of the macrophyte and the metal to be extracted determine how effective EDTA-assisted phytoextraction will be. However, it was also discovered that the EDTA causes stunted growth and additional disorders. The rising need for green energy and technological revolution necessitates the use of REEs like Nd and Dy. They are vital components, and their recovery for subsequent reuse is critical. To improve the remediation process from the marine macroalgae, it is pertinent to optimize some influential parameters likely to affect the rate of remediation. These parameters included the stock density of the biosorbent, the ionic strength, and the duration of contact. These parameters collectively impact the removal of REEs, but ionic strength became selective in the case of Dy removal which can be justified because of the differences in chemical properties of light and heavy rare-earth elements. The highest removal/recovery for *Gracilaria* sp. was found. *U. lactuca* was the most promising species, removing Nd (98%) and Dy (89%) (Viana et al. 2021). *Azolla* has successfully been used on a large scale to detoxify industrial effluents (Ansari et al. 2020). Similarly, *Eichhornia crassipes* have been used enormously to improve water quality. These species are extremely effective in treating domestic wastewater. Generally, pondweeds are of the *Lemna* genus, which is employed in phytoremediation and ecotoxicology. *Lemna gibba* L. and *Lemna minor* L. have been examined extensively for the removal of REEs and HMs. Also, the growth responses of *Lemna minor* and *Spirodela polyrhiza* were observed to be directly correlated with the aqueous environment. *Wolffia globosa* has a high tolerance for arsenic and efficiently removes it. *U. lactuca* can effectively remove 60–90% of REEs. The high BCF obtained from mass balance analysis implies that REE-enriched algal biomass (up to 1295 g/g) could be a viable and ecologically friendly alternative for ore extraction (Pinto et al. 2020).

The physiochemical environment of the aquatic system is altered by aquatic macrophytes. During strong photosynthetic activities, the existence of another aquatic plant can lower dissolved CO_2 in water. This raises the pH of the water by increasing the amount of dissolved oxygen in the effluent. Moreover, contaminants are absorbed by aquatic macrophytes and stored in their biomass.

7.5.3 Constructed Wetlands

Constructed wetlands (CWs) are a new technology that resembles natural wetlands and helps to get rid of contaminants (organic pollutants, bacteria, antibiotic resistance genes, HMs, metalloids, and REEs). CWs can remove pollutants through complicated processes involving plants, microbes, and substrates/media (Avishek and Hazra 2022). Braun et al. (2018) looked at how Gd affected plant species (*Elodea nuttallii*, *Ceratophyllum demersum*, *E. canadensis*, and *Lemna gibba*). These species can be employed in CWs as biological filters. The bioaccumulation concentration factor (BCF) was 1, implying that the Gd concentration in plant tissues was low compared to external water. It was determined that no Gd complexes were collected in plant tissues, and if found in minute concentration, then it is probably due to the accumulation through nearby waters having comparatively high concentrations of Gd. Approximately 96% of the total REE was removed by 2020 from mining areas in China, according to the Rare Earth Industry Development Plan that recovered REE from mining regions. Moreover, some plants will be cremated to prevent additional pollution (soil/water). There is a paucity of literature confirming single REE uptake, metabolism, and REE destiny and recovery from waste ashes. *P. americana* appears to be capable of accumulating mild concentrations of REEs (Sookrung et al. 2018).

However, only a few attempts were made to use CWs for REE clean-up but proved CW a potential remedy for REE removal. With the right adjustments, CWs can be used to treat REE-containing mine-affected water discharges (pre-treatments). For high-purity metals, specific plant species may be used, followed by metallurgic procedures. REE can also be extracted from the sediments of CWs, where REE concentrates during treatment. When recovered from mineral rocks, REE can also be used as phosphate fertilizers. For REE clean-up, plants like *Phytolacca americana*, *Phragmites australis*, and *Taraxacum officinale* were employed (Wei et al. 2019). REE can be recovered from wetland plants using a variety of extraction processes, including electro-precipitation, leaching, and organic complexations. In the agricultural sector, the extracted REE could be used as a fertilizer and seed/plant growth booster.

7.5.4 Methods of Phytoremediation

Hydrilla verticillata, *Typha angustata*, *Eichhornia crassipes*, *Ipomoea aquatica*, and other aquatic macrophytes can ingest toxic metals from natural water bodies. Based on the method used by various plant species to remove contaminants from soil or water, phytoremediation technology is divided into distinct categories. When pollutants build up in a plant cell, various forms of phytoremediation, including phytoextraction, phyto-sequestration, phyto-degradation, phyto-stabilization, phyto-volatilization, rhizo-remediation, rhizo-filtration, and rhizo-degradation, occur. Aquatic macrophytes can immobilize pollutants from groundwater by absorption and precipitation into roots through phyto-stabilization, sometimes known as phyto-immobilization. Roots of aquatic macrophytes absorb soluble pollutants from water, move them to their leaves, and then use their stomata to volatilize the hazardous material as biomolecules into the environment. This is a promising method of phytoremediation. However, among the several phytoremediation methods potentially used for the REEs and their compounds, phytoextraction is frequently employed to get rid of HMs and REEs from contaminated water and sediments (Das et al. 2022).

7.5.5 Omics Approach

Omics technologies are practical and viable strategies for specifying the role of genomes (genomics), transcriptome (transcriptomics) and non-coding (miRNAomics) RNA transcripts, and metabolites (metabolomics), including metal-related omics (metallomics), used to improve stress tolerance or generate resilient plant systems. Plant stress responses and crop improvement are being elucidated using omics technologies such as genomics, transcriptomics, and metabolomics, along with contemporary techniques such as RNAomics, which are collaboratively considered as multi-omics approaches. HM ATPase (HMA), HM-associated proteins (HMP), multidrug and toxic compound extrusion (MATE), Yellow Stripe-Like (YSL), metal tolerance protein (MTP), and zinc-iron permease (ZIP) were discovered in comparative genomics research focusing on the identification of genes of hyperaccumulator plants (Jamla et al. 2021). These omics-based technologies have provided critical insight and can greatly aid in the determination of the potential of vivid bioremediation approaches. Bio-metallurgy operations depend entirely on the microbial community during the bioleaching of lanthanides from e-waste, and leachate may be processed for recovery using chemical techniques. Omics-based approaches allow the prediction of the activity of microbial consortia during bioleaching and direct towards the bioprospecting of non-cultivable novel strains of bacteria. Combined approaches of bioleaching of the e-waste using microbes and bioaccumulation with bio-extraction using macrophytes are some of the emerging avenues to offer better remediation. Few known plant microbes and metal interactions promote facile uptake, sequestration, and immobilization/detoxification of heavy metals (Shelake et al. 2019).

7.6 Analytical Techniques to Quantify REEs (Lanthanides)

Spectrophotometric methods are among the analytical approaches for quantifying REEs (lanthanides).

- In matrices like plants, synchrotron X-ray absorption spectroscopy is routinely employed to assess REE (lanthanide) utilization, toxicity, and tolerance.
- X-ray absorption spectroscopy yields knowledge about the analyte, including the element's oxidation state, the symmetry of ligand complexes, and the additional atoms near the target analyte.
- Inductively coupled plasma mass spectrometry (ICP-MS), inductively coupled plasma optical emission spectrometry (ICP-OES), laser ablation ICP-MS, and neutron activation analysis are some of the hyphenated techniques employed.
- For the identification of individual elements, high-pressure liquid chromatography-inductively coupled plasma mass spectrometry (HPLC-ICP-MS) and hydride generation-gas chromatography-quartz tube flame atomic absorption spectrometry (HGGC-QFAAS) are used.
- Hydrophilic interaction chromatography with inductively coupled plasma mass spectrometry is another approach for speciation (Gwenzi et al. 2018).

7.7 Conclusion

The current scenario of the accumulation of REEs in plants and their ecotoxicological effects on their physiology is broadly discussed in this chapter. REEs not only impact plant physiology by reducing their growth and enzymatic activity but also cause serious threats to human health. Therefore, remediation of these emerging contaminants is necessary. A bioremediation system based on aquatic macrophytes is a more promising strategy for metal biosorption. These wild aquatic weeds act as a strong barrier which hampers the biomagnification of REEs and HMs into the food nexus. In addition, the effective and sustainable management of these aquatic resources through ecological engineering of suitable and broad array consortium of different kinds of hyperaccumulator macrophytes is still not in the wide practice which is indeed needed. The potential of macrophytes for the bioremediation of REEs may be enhanced through genetic engineering and metabolic engineering.

References

Adeel M, Lee JY, Zain M, Rizwan M, Nawab A, Ahmad M (2019) Cryptic footprints of rare earth elements on natural resources and living organisms. Environ Int 127:785–800

Ahmed A, Darwish E, Hamoda S, Alobaidy M (2013) Effect of putrescine and humic acid on growth, yield and chemical composition of cotton plants grown under saline soil conditions. J Agric Environ Sci 13(4):479–497

Ansari AA, Naeem M, Gill SS, AlZuaibr FM (2020) Phytoremediation of contaminated waters: an eco-friendly technology based on aquatic macrophytes application. Egypt J Aquat Res 46(4): 371–376

Avishek K, Hazra M (2022) Plant based removal and recovery of rare earth elements. In: Environmental technologies to treat rare earth elements pollution: Principles and engineering

Bailey-Serres J, Mittler R (2006) The roles of reactive oxygen species in plant cells. Plant Physiol 141:311

Balaram VJ (2019) Rare earth elements: a review of applications, occurrence, exploration, analysis, recycling, and environmental impact. Geosci Front 10(4):1285–1303

Blinova I, Lukjanova A, Reinik J, Kahru AJ (2021) Concentration of lanthanides in the Estonian environment: a screening study. J Hazard Mater 4:100034

Blinova I, Muna M, Heinlaan M, Lukjanova A, Kahru AJN (2020) Potential hazard of lanthanides and lanthanide-based nanoparticles to aquatic ecosystems: data gaps, challenges and future research needs derived from bibliometric analysis. Nano 10(2):328

Braun M, Zavanyi G, Laczovics A, Berényi E, Szabó S (2018) Can aquatic macrophytes be biofilters for gadolinium based contrasting agents? Water Res 135:104–111

Cao Y, Shao P, Chen Y, Zhou X, Yang L, Shi H (2021) A critical review of the recovery of rare earth elements from wastewater by algae for resources recycling technologies. Resour Conserv Recycl 169:105519

Chen H, Chen H, Chen ZJ (2022) A review of in situ phytoextraction of rare earth elements from contaminated soils. Int J Phytoremediation 24(6):557–566

Ciesielski TM, Pastukhov MV, Leeves SA, Farkas J, Lierhagen S, Poletaeva VI (2016) Differential bioaccumulation of potentially toxic elements in benthic and pelagic food chains in Lake Baikal. Environ Sci Pollut Res Int 23:15593–15604

Corrêa AX, Cotelle S, Millet M, Somensi CA, Wagner TM, Radetski CM (2016) Genotoxicity assessment of particulate matter emitted from heavy-duty diesel-powered vehicles using the in vivo Vicia faba L. micronucleus test. Ecotoxicol Environ Saf 127:199–204

Costa M, Henriques B, Pinto J, Fabre E, Dias M, Soares J (2020) Influence of toxic elements on the simultaneous uptake of rare earth elements from contaminated waters by estuarine macroalgae. Chemosphere 252:126562

d'Aquino L, De Pinto MC, Nardi L, Morgana M, Tommasi F (2009) Effect of some light rare earth elements on seed germination, seedling growth and antioxidant metabolism in Triticum durum. Chemosphere 75(7):900–905

Das S, Das S, Ghangrekar M (2022) Efficacious bioremediation of heavy metals and radionuclides from wastewater employing aquatic macro-and microphytes. J Basic Microbiol 62(3–4): 260–278

Dodson J, Hunt A, Parker H, Yang Y, Clark JJCE (2012) Intensification PP. elemental sustainability: towards the total recovery of scarce metals. Chem Eng Process Process Intensif 51:69–78 e Cunha MCL, Pereira VP, LVS N, ACB N, Vedana LA, Formoso ML (2012) REE distribution pattern in plants and soils from Pitinga mine-Amazon. Brazil Open J Geol 2(04):253

Ekperusi AO, Sikoki FD, Nwachukwu EO (2019) Application of common duckweed (Lemna minor) in phytoremediation of chemicals in the environment: State and future perspective. *Chemosphere*. https://doi.org/10.1016/j.chemosphere.02.025

Fan Z, Zhang K, Wang F, Zhao X, Bai R, Liu B (2020) Effects of rare earth elements on growth and determination of secondary metabolites under in vitro conditions in salvia miltiorrhiza. HortScience 55(3):310–316

Fashui H (2002) Study on the mechanism of cerium nitrate effects on germination of aged rice seed. Biol Trace Elem Res 87:191–200

Hong F, Zhenggui W, Guiwen Z (2000) Effect of lanthanum on aged seed germination of Rice. Biol Trace Elem Res 75(1–3):205–213. https://doi.org/10.1385/BTER:75:1-3:205

Galdiero E, Carotenuto R, Siciliano A, Libralato G, Race M, Lofrano G (2019) Cerium and erbium effects on Daphnia magna generations: a multiple endpoints approach. Environ Pollut 254: 112985

Galhardi JA, Leles BP, de Mello JW, Wilkinson KJ (2020) Bioavailability of trace metals and rare earth elements (REE) from the tropical soils of a coal mining area. Sci Total Environ 781(717): 134484

Gonzalez V, Vignati DA, Leyval C, Giamberini LJ (2014) Environmental fate and ecotoxicity of lanthanides: are they a uniform group beyond chemistry? Environ Int 71:148–157

Gwenzi W, Mangori L, Danha C, Chaukura N, Dunjana N, Sanganyado E (2018) Sources, behaviour, and environmental and human health risks of high-technology rare earth elements as emerging contaminants. Sci Total Environ 636:299–313

Hu H, Wang L, Zhou Q, Huang X (2016) Combined effects of simulated acid rain and lanthanum chloride on chloroplast structure and functional elements in rice. Environ Sci Pollut Res 23:8902–8916

Igiri BE, Okoduwa SI, Idoko GO, Akabuogu EP, Adeyi AO, Ejiogu IK (2018) Toxicity and bioremediation of heavy metals contaminated ecosystem from tannery wastewater: a review. J Toxicol 2018:1. https://doi.org/10.1155/2018/2568038

Ippolito M, Fasciano C, d'Aquino L, Morgana M, Tommasi F (2010) Responses of antioxidant systems after exposition to rare earths and their role in chilling stress in common duckweed (Lemna minor L.): a defensive weapon or a boomerang? Arch Environ Contam Toxicol 58:42–52

Jamla M, Khare T, Joshi S, Patil S, Penna S, Kumar VJ (2021) Omics approaches for understanding heavy metal responses and tolerance in plants. Curr Plant Biol 27:100213

Jiang X, Chen H, Liao Y, Ye Z, Li M, Klobučar GJ (2019) Ecotoxicity and genotoxicity of polystyrene microplastics on higher plant Vicia faba. Environ Pollut 250:831–838

Lai Y, Wang Q, Yan W, Yang L, Huang BJ (2005) Preliminary study of the enrichment and fractionation of REEs in a newly discovered REE hyperaccumulator Pronephrium simplex by SEC-ICP-MS and MALDI-TOF/ESI-MS. J Anal At Spectrom 20(8):751–753

Leão GA, de Oliveira JA, Felipe RTA, Farnese FS, Gusman G (2014) Anthocyanins, thiols, and antioxidant scavenging enzymes are involved in Lemna gibba tolerance to arsenic. J Plant Interact 9(1):143–151

Liang T, Li K, Wang LJ (2014) State of rare earth elements in different environmental components in mining areas of China. Environ. Monit Assess 186:1499–1513

Liu X, Li L, Li M, Su L, Lian S, Zhang B, Li L (2018) AhGLK1 affects chlorophyll biosynthesis and photosynthesis in peanut leaves during recovery from drought. Sci Rep 8(1):2250

Liu C, Yuan M, Liu W-S, Guo M-N, Zheng H-X, Huot H et al (2021) Element case studies: rare earth elements, pp 471–483. https://doi.org/10.1007/978-3-030-58904-2_24

Liu M, Hasenstein K (2005) La3+ uptake and its effect on the cytoskeleton in root protoplasts of Zea mays L. Planta 220:658–666

Lyu K, Wang X, Wang L, Wang G (2019) Rare-earth element yttrium enhances the tolerance of curly-leaf pondweed (Potamogeton crispus) to acute nickel toxicity. Environ Pollut 248:114–120

Mack C, Wilhelmi B, Duncan J, Burgess J (2007) Biosorption of precious metals. Biotechnol Adv 25(3):264–271

Migaszewski ZM, Gałuszka AJ (2015) The characteristics, occurrence, and geochemical behavior of rare earth elements in the environment: a review. Crit Rev Environ Sci Technol 45(5): 429–471

Mishra S, Srivastava S, Tripathi R, Govindarajan R, Kuriakose S, Prasad M (2006) Phytochelatin synthesis and response of antioxidants during cadmium stress in Bacopa monnieri L. Plant Physiol Biochem 44(1):25–37

Mittler R (2002) Oxidative stress, antioxidants and stress tolerance. Trends Plant Sci 7(9):405–410

Nabi M (2021) Monitoring, management. Heavy metals accumulation in aquatic macrophytes from an urban lake in Kashmir Himalaya, India. Environ Nanotechnol Monit Manag 16:100509

Nazreen M, Amalina A, Omar MJ (2017) Bioaccumulation of rare earth element by water lettuce (Pistia stratiotes). Orient J Chem 33(3):1097–1102

Ozaki T, Enomoto S, Minai Y, Ambe S, Makide YJ (2000) A survey of trace elements in pteridophytes Biol. Trace Elem Res 74:259–273

Pagano G, Guida M, Tommasi F, Oral RJ (2015) Health effects and toxicity mechanisms of rare earth elements—knowledge gaps and research prospects. Ecotoxicol Environ Saf 115:40–48

Pang X, Wang D, Xing X, Peng A, Zhang F, Li C (2002) Effect of La3+ on the activities of antioxidant enzymes in wheat seedlings under lead stress in solution culture. Chemosphere 47(10):1033–1039

Paola IM, Paciolla C, d'Aquino L, Morgana M, Tommasi F (2007) Effect of rare earth elements on growth and antioxidant metabolism in Lemna minor L. Firenze University Press 60(1–2) 60(1–2):125–128

Pinto J, Costa M, Henriques B, Soares J, Dias M, Viana T (2021) Competition among rare earth elements on sorption onto six seaweeds. J Rare Earths 39(6):734–741

Pinto J, Henriques B, Soares J, Costa M, Dias M, Fabre E (2020) A green method based on living macroalgae for the removal of rare-earth elements from contaminated waters. J Environ Manag 263:110376

Priya A, Gnanasekaran L, Dutta K, Rajendran S, Balakrishnan D, Soto-Moscoso M (2022) Biosorption of heavy metals by microorganisms: evaluation of different underlying mechanisms. Chemosphere 307:135957

Qin R, Wang C, Chen D, Björn LO, Li S (2015) Copper-induced root growth inhibition of Allium cepa var. agrogarum L. involves disturbances in cell division and DNA damage. Environ Toxicol Chem 34(5):1045–1055

Romero-Freire A, Joonas E, Muna M, Cossu-Leguille C, Vignati D, Giamberini LJ (2019) Assessment of the toxic effects of mixtures of three lanthanides (Ce, Gd, Lu) to aquatic biota. Sci Total Environ 661:276–284

Saha A, Mukherjee P, Roy K, Sen K, Sanyal T (2022) A review on phyto-remediation by aquatic macrophytes: a natural promising tool for sustainable management of ecosystem. Int J Exp 27: 09–31

Seth CS, Misra V, Chauhan L, Singh R (2008) Genotoxicity of cadmium on root meristem cells of Allium cepa: cytogenetic and comet assay approach. Ecotoxicol Environ Saf 71(3):711–716

Shan X, Wang H, Zhang S, Zhou H, Zheng Y, Yu H (2003) Accumulation and uptake of light rare earth elements in a hyperaccumulator Dicropteris dichotoma. Plant Sci 165(6):1343–1353

Shelake R, Waghunde R, Kim J (2019) Plant–microbe–metal (PMM) interactions and strategies for remediating metal ions. In: Plant-Metal Interactions, pp 247–262

Sidhu GPS, Singh HP, Batish DR, Kohli RL (2016) Effect of lead on oxidative status, antioxidative response and metal accumulation in Coronopus didymus. Plant Physiol Biochem 105:290–296

Singh A, Kumar S (2022) Application of aquatic plants dead biomass in remediation of heavy metals pollution by adsorption: a review. Indian. J Sci Technol 15(16):729-35.2

Sood A, Uniyal PL, Prasanna R, Ahluwalia A (2012) Phytoremediation potential of aquatic macrophyte, Azolla. Ambio 41:122–137

Sookrung N, Reamtong O, Poolphol R, Indrawattana N, Seesuay W, Saelim N, Tungtrongchitr A (2018) Glutathione S-transferase (GST) of American cockroach, Periplaneta americana: classes, isoforms, and allergenicity. Sci Rep 8(1):484

Sun T, Wu H, Wang X, Ji C, Shan X, Li F (2020) Evaluation on the biomagnification or biodilution of trace metals in global marine food webs by meta-analysis. Environ Pollut 264:113856

Tesfaye F, Peng H, Zhang MJJ (2021) Advances in the circular economy of lanthanides. JOM 73(1):16–18

Trapasso G, Chiesa S, Freitas R, Pereira E (2021) What do we know about the ecotoxicological implications of the rare earth element gadolinium in aquatic ecosystems? Sci Total Environ 781: 146273

Viana T, Henriques B, Ferreira N, Lopes C, Tavares D, Fabre E (2021) Sustainable recovery of neodymium and dysprosium from waters through seaweeds: influence of operational parameters. Chemosphere 280:130600

Wei X, Zhang S, Shimko J, Dengler R (2019) Mine drainage: treatment technologies and rare earth elements. Water Environ Res 91(10):1061–1068

Weltje L (2003) Bioavailability of lanthanides to freshwater organisms. https://www.elibrary.ru/item.asp?id=5918396

Xu Q, Fu Y, Min H, Cai S, Sha S, Cheng GJ (2012) Laboratory assessment of uptake and toxicity of lanthanum (La) in the leaves of Hydrocharis dubia (Bl.) backer. Environ Sci Pollut Res 19: 3950–3958

Xu Q-M, Wang Y-Z, Liu H, Cheng J (2016a) Soil. Physiological responses and chromosomal aberration in root tip cells of Allium sativum L. to cerium treatments. Plant Soil 409:447–458

Xu T, Su C, Hu D, Li F, Lu Q, Zhang T (2016b) Molecular distribution and toxicity assessment of praseodymium by Spirodela polyrrhiza. J Hazard Mater 312:132–140

Yang X, Yin D, Sun H, Wang X, Dai L, Chen Y (1999) Distribution and bioavailability of rare earth elements in aquatic microcosm. Chemosphere 39(14):2443–2450

Yuan M, Liu C, Liu W-S, Guo M-N, Morel JL, Huot H (2018) Accumulation and fractionation of rare earth elements (REEs) in the naturally grown Phytolacca americana L. in southern China. Int J Phytoremediation 20(5):415–423

Zhang J, Zhang T, Lu Q, Cai S, Chu W, Qiu H (2015) Oxidative effects, nutrients and metabolic changes in aquatic macrophyte, Elodea nuttallii, following exposure to lanthanum. Ecotoxicol Environ Saf 115:159–165

Zhenggui W, Ming Y, Xun Z, Fashui H, Bing L, Ye T (2001) Rare earth elements in naturally grown fern Dicranopteris linearis in relation to their variation in soils in South-Jiangxi region (southern China). Environ Pollut 114(3):345–355

Zicari MA, d'Aquino L, Paradiso A, Mastrolitti S, Tommasi F (2018) Effect of cerium on growth and antioxidant metabolism of Lemna minor L. Ecotoxicol Environ Saf 163:536–543

Efficiency of Aquatic Plants for Remediation of Wastewater

8

Dhananjay Kumar, Urvashi Sharma, Vishal Singh,
Aneet Kumar Yadav, Anita, Sanjeev Kumar, and Narendra Kumar

Abstract

Rampant industrialization, unplanned urbanization, and agricultural activities release an enormous quantity of contaminants in water which adversely affects social and economic development globally. Today, aquatic ecosystems are a major environmental threat due to inorganic (heavy metals and metalloids) and organic contaminants, especially in developing countries. Heavy metals and POP are common environmental pollutants that affect soil, water, and air quality. Heavy metals are of high concern due to their persistent, carcinogenic, the potential of long-distance transport, and bioaccumulation in the food chain. Hence, wastewater must be treated up to adequate level before being discharged into the aquatic system. Traditional treatment approaches are not always very effective in wastewater remediation. Phytoremediation is an eco-friendly and economically sound technique, which has been accepted by several researchers as an alternative to the current high-cost cleanup methods. Aquatic plants are used in this technology to efficiently remove, detoxify, or immobilize heavy metals and persistent organic pollutants. Many aquatic plant species, particularly high growth-rate plants like macrophytes, are currently being studied to determine their potential and effectiveness for phytoremediation applications. Excess

D. Kumar · U. Sharma · A. K. Yadav · Anita · N. Kumar (✉)
Department of Environmental Science, Babasaheb Bhimrao Ambedkar University, Lucknow, Uttar Pradesh, India

V. Singh
Department of Environmental Science, Dr. Rammanohar Lohia Avadh University, Ayodhya, Uttar Pradesh, India

S. Kumar
Department of Geology, School of Earth and Environmental Sciences, Babasaheb Bhimrao Ambedkar University, Lucknow, Uttar Pradesh, India

S. Kumar et al. (eds.), *Aquatic Macrophytes: Ecology, Functions and Services*,
https://doi.org/10.1007/978-981-99-3822-3_8

contaminants in domestic, agricultural, and industrial effluent, such as inorganic and organic pollutants, metals, and pharmaceutical pollutants, can be absorbed by aquatic plants. This chapter deliberates the phytoremediation potential of *Eichhornia crassipes, Azolla*, and duckweeds aquatic macrophytes.

Keywords

Azolla · Duckweeds · *Eichhornia* · Heavy metals · Phytoremediation · Bioremediation

8.1 Introduction

Industries, the major contributors for economic growth and urbanization, along with agriculture have accelerated the development of various sectors associated with the expansion of global population and various luxurious facilities. Over the past few decades, increasing human population and global economy which is highly reliant on the use of natural resources (soil, forest, water, etc.) causes the overexploitation of natural resources and alteration of biogeochemical cycles and other life-supporting natural processes. Water is one of the most essential natural resources that requires for the sustaining life on our planet. Rampant industrialization, unplanned urbanization, and agricultural activities release an enormous quantity of contaminants in water (Adabembe et al. 2022; Kumar et al. 2018). On earth, only 2.5% of the total water budget is freshwater. And out of 2.5% only 0.0072% of the total freshwater is accessible for drinking, agricultural, energy production, and industrial purposes. After agriculture, industrial sector is the second largest consumer of total available freshwater. Globally, industrial sector consumes around 22% of total freshwater, while in high-income countries water consumption percentage may reach up to 60%.

Various industries such as electroplating, battery manufacturing, iron and steel, paper and pulp processing, fertilizers and pesticides, chemicals and drugs manufacturing, mining, textile manufacturing, tannery, and refineries, utilize substantial amounts of freshwater and chemicals. These industries discharge their wastewater, either directly or indirectly, into different water systems. In developing countries, around 70% of the industries are releasing their wastewater without any proper treatment (Okereafor et al. 2020). Untreated industrial and domestic wastewater may contain high level of suspended solids, toxic organic (pesticides, dyes, oils, grease, phenols, etc.) and inorganic (heavy metals, metalloids, radioactive elements, and nutrients like phosphate, nitrate, etc.) contaminants.

8.2 Heavy Metals in the Aquatic Environment

Metals are those elements having properties of high electrical conductivity, malleability, lustrous, and naturally occur in earth's crust in the form of different minerals, but their composition/concentration in the crust may vary among different areas. Heavy metals like arsenic (As), copper (Cu), cadmium (Cd), chromium (Cr), mercury (Hg), lead (Pb), and nickel (Ni) are special class of inorganic pollutants, having high density between 5.306 and 22.00 g/cm^2 and atomic number >20 (Sood et al. 2012; Ali et al. 2013). Heavy metals may exist in more than one valence state, so any metal may be of less or more toxic depending on their valence state. For example, Cr (IV) is more toxic than Cr (III), while As (III) is highly toxic as compared to As (V) and As (VI). Some heavy metals like Mn, Zn, and Fe are quintessential at low concentration for proper biological and physiological function in living organisms. However, at higher concentration these metals become very toxic and cause several adverse health effects. Furthermore, some metals like As, Pb, Cr, Cd, and Hg are very toxic even at very low concentration (Rezania et al. 2016). Chemical composition, level of persistence, and toxicity of contaminants present in industrial wastewater or effluents are very diverse. Among the various water pollutants, heavy metals are major concern due to their persistence and toxic nature. Sources of heavy metals in aquatic system may be natural (mainly soil erosion, weathering, and dissolution of rocks and minerals) and anthropogenic (mining, urban runoff, sewage and industrial discharge, agricultural activities, etc.) (Kumar et al. 2018). Heavy metals enter into the aquatic systems like rivers and ponds either through point source (leakage of septic tanks, industrial discharge, etc.), or through nonpoint source (runoff), or both. Heavy metals can cross the cellular membrane and get easily absorbed in the living cells and cause toxicity.

8.3 Persistent Organic Pollutants (POPs) in Aquatic Environment

Persistent organic pollutants (POPs) are highly toxic, semi-volatile, high fat solubility, low water solubility, anthropogenic organic chemicals that withstand most environmental degradation processes and air-water-soil cycling during long-range transport far from their sources, allowing for bioaccumulation in aquatic organisms, and as a result of biomagnification toward human via food chain (Han and Currell 2017). Because of their bioaccumulation and biomagnification properties, POPs are also known as silent killers (Alharbi et al. 2018). In recent decades, persistent organic pollutants (POPs) have received attention worldwide (Li et al. 2019a, b). They transported regional and global levels via the atmosphere, rivers, estuaries, and oceans (Han and Currell 2017). The international treaty on reduction of POPs signed in Stockholm convention under the United Nation Environmental Programme (UNEP) (Trojanowicz 2020).

Freshwater bodies like river, lake, pond, streams, and wetlands are the sink/reservoirs of the POPs deposited from the atmospheric deposition and other anthropogenic sources (sewage, runoff, industrial effluent, fossil fuel burning, etc.). POPs may enter in the aquatic environment by several routes such as/via effluent runoff, atmospheric deposition, and other anthropogenic activities. Because of their low/poor water solubility, bounded with the particulate matter in surface water sediment. As a result, POPs circulation for long periods sequestrated in the deposits can act as a sink/reservoir of the POPs. However, it can reintroduce disturbance to the aquatic ecosystem food chain, potentially becoming a source of local and global contamination. Their high lipophilicity POPs can bioaccumulate in large amount of the POPs in the tissues of marine organisms (Alharbi et al. 2018).

Polycyclic aromatic hydrocarbon (PAH) is a class of aromatic compounds that are emitted by natural and anthropogenic sources such as pyrolysis, industrial processes, incomplete combustion of organic matter, motor vehicles, incinerators, oil spills, combustion of fossil fuels, and wood. PAHs cause several health disorders (cancer and mutagenic impact) in humans through the aquatic environment's biomagnification process (Jonker and Koelmans 2002). POPs are transported in the aquatic environment through various processes, including adsorption, aggregation, retention, egestion, and direct chemical release (Yu et al. 2019).

POPs are found in almost every surface water bodies due to polluted environment. Because of their low water solubility, they accumulate in fatty tissue of living organisms in the upper trophic level of the food chain posing potential risks to humans such as hormonal disruption, cardiovascular disease, reproductive and neurological disorder, and diabetes and learning disabilities caused by exposure of POPs (Bergman et al. 2014). Polycyclic aromatic hydrocarbons (PAHs) can also be found in estuarine sediments, the environment, water, and soil (Nikitha et al. 2017). Lipophilic compounds are those easily transported across cell membranes by lipoproteins, accumulate in fatty tissues leaves to destruction of living organisms (Alharbi et al. 2018).

POPs have been found in fetuses and embryos, which is surprising in this polluted world. Our children are being born with POPs and have begun to accumulate more pollutants; Furthermore, these pollutants cause defects in female embryos. POP levels are found in all age groups, with the elderly having the highest levels (Li et al. 2021).

8.4 Techniques for Remediation of Wastewater

Several conventional and nonconventional techniques like chemical precipitation, reverse osmosis, oxidation/reduction, ion exchange, electrodialysis, adsorption, bioremediation, filtration, membrane technology, and phytoremediation have been explored for the wastewater remediation. Although, various conventional techniques have been proven efficient in wastewater remediation, but these techniques have also several limitations like high energy requirement, high operation and maintenance cost, and generation of toxic by-products. Among the various remediation techniques, bioremediation is considered as the nature-driven remediation technique, which involves uses of biological process for the removal of heavy metals and POPs from wastewater. It has been proven that bioremediation is an economical remediation technique as compared to other conventional remediation techniques. Phytoremediation, a solar-driven botanical remediation techniques that exploits the potential of plants to reduce, remove, immobilize, or degrade the heavy metals, metalloids, and POPs from the wastewater. Over the past few decades, phytoremediation technique has emerged as one of the most economical, efficient, eco-friendly, and aesthetically acceptable technique for the remediation of wastewater contaminated with heavy metals and POPs (Kumar et al. 2012, 2013) (Table 8.1).

8.5 Phytoremediation as an Emerging Tool for Wastewater Remediation

Phytoremediation is an environmental-friendly approach that exploits plants (terrestrial or aquatic) and potential for in situ and ex situ remediation of toxic metals, metalloids, and POPs from wastewater. This solar-driven botanical remediation has been reported for its removal effectiveness even at shallow and low level of contagion (Rezania et al. 2016; Ali et al. 2013). Application of aquatic macrophytes for wastewater remediation have been well documented by several researchers (Mishra et al. 2008a, b; Bennicelli et al. 2004; Lu et al. 2010; Ramaswami et al. 2001; Sood et al. 2012). Till date, among the different plant species applied for in situ or ex situ wastewater treatment, macrophytes have attracted significant attention among scientific communities working in area of eco-friendly remediation technologies. The aquatic macrophytes used as solar-driven pumps which extract/absorb the heavy metals and POPs from the contaminated water and concentrate/retain them into their root and shoot. In an aquatic ecosystem, macrophytes play a significant role for the functioning and maintenance of hydrochemistry by sequestration of various organic and inorganic toxic chemicals (Eid et al. 2020; Dhir et al. 2009). Some macrophytes have a great capability to accumulate/absorb heavy metals or metalloids up to 100,000 times greater than the concentration present in their respective growing medium (Mishra and Tripathi 2008). Several macrophytes like *Azolla pinnata, Azolla caroliniana, Eichhornia crassipes, Hydrilla verticillata, Lemna minor, Lemna polyrhiza, Pistia stratiotes, Spirodela polyrhiza, Typha latifolia, Typha orientalis*, etc. have been reported for the wastewater remediation (Laet et al. 2019; Pati and Satapathy 2016). Since these macrophytes have remarkable capability to accumulate/transfer heavy metals and POPs present in growing medium (sediments and water) at varying levels, these macrophytes are being used in naturally and constructed wetland for wastewater remediation (Liao and Chang 2004; Mishra et al. 2008a, b; Bhatiya and Goyal 2014). In order to achieve high remediation efficiency, the selection of macrophytes becomes an important factor. Macrophytes must be indigenous and have high tolerance to high concentration and wide range of chemicals and other abiotic stress (Vymazal 2016; Kumar and Dutta 2019; Kumar et al. 2018). Besides several advantages of phytoremediation, this technique has some limitations associated with it (Table 8.2).

8.6 Phytoremediation Potential of Aquatic Macrophytes

Macrophytes are aquatic flora growing in or near water bodies and may be emergent, submerged, or floating. The aquatic macrophytes have unique metabolic and absorption potential and well-developed transportation systems that can take up heavy

Table 8.1 Advantages and disadvantages of different wastewater remediation techniques

Techniques	Advantages	Limitations	References
Chemical precipitation	Simple and economical	Generates sludge and toxic gases, high maintenance cost	Ahmad et al. (2016)
Ion exchange	Effective, efficient in metals recovery	High operational cost, ion exchanger quickly reduces its exchange capacity	Ahmad et al. (2016)
Adsorption	Efficient and simple	Relatively high cost, adsorbent gradually decline its capacity	Fenglian and Wang (2011), Ahmad et al. (2016)
Membrane filtration	Microfiltration and ultrafiltration are highly effective in remediation, no phase change involved, also efficient at low temperature	Complexity, limited flow rates, high cost involved	Sampera et al. (2009), Zhao et al. (2016)
Reverse osmosis	Highly efficient	Membranes sensitive	Fenglian and Wang (2011), Zhao et al. (2016), Ahmad et al. (2016)
Electrodialysis	Less power consumption	Complexity, high operation cost, low permeable flux	Barakat (2011), Ahmad et al. (2016)
Coagulation and flocculation	Simple and cost-effective	Large amount of sludge generation, high chemical consumption, time-taking process	El-Samrani et al. (2008), Ahmad et al. (2016)
Electrochemical treatment	Fast treatment, less maintenance	High chemical consumption, high sludge generation, high power consumption	Sirés et al. (2014), Radjenovic and Sedlak (2015)
Flotation		High cost, efficiency decreases on increasing ion strength	Fenglian and Wang (2011), Ahmad et al. (2016)
Constructed wetland	Eco-friendly, low cost, less maintenance, facilitates reuse and recycle of wastewater	Efficiency depends on retention time and hydraulic load	Gorgoglione and Torretta (2018), Adabembe et al. (2022), Akinbile et al. (2015)
Phytoremediation	Cost-effective, eco-friendly, efficient, aesthetical	Remediation efficiency highly depends on the types of plant species. Efficiency also varies under different growth conditions	Kumar et al. (2012, 2018)

Table 8.2 Advantages and limitations of phytoremediation technique

Advantages	Limitations
Eco-friendly and cost-effective as compared to other conventional methods	Effective up to the depth reached and surface area covered by the root system
Exploit plant's natural potential to accumulate/absorb/degrade the contaminants and conserve of the natural hydrochemistry of aquatic ecosystem	Removal efficiency of plants may vary under different abiotic stresses like high or low temperature, light intensity, etc.
Efficient for the wastewater contaminated with more than one contaminant	Removal efficiency depends on the tolerance potential of plants against abiotic stresses
Hyperaccumulator plants can be used for metal recovery	Possibility of returning of pollutants in the growing environment

metals from wastewater (Rai 2008; Rai 2009; Dixit and Dhote 2010; Fazal et al. 2015). The process removes heavy metals from water through binding or degradation and detoxification and then macrophytes can be subsequently harvested, processed, and disposed. Aquatic macrophytes play a very crucial role in aquatic ecosystem since they offer food and shelter for aquatic animals. Further, the remediation capabilities of macrophytes normally decline from submerged to floating and then to emergent; however, the remediation efficiency may also be affected by the plant species and growing environment (Vymazal 2016; Harguinteguy et al. 2014).

Metal accumulation potential of macrophytes make them unique for study with objectives of testing and modelling of several ecological theories as well as element/nutrient cycles (Rai 2008). Globally, during the last few decades, numerous researches have highlighted the potential of macrophytes for phytoremediation/treatment of wastewater. In several countries, substantial part of GPD is being invested for the restoration and management of water bodies like ponds, lakes, and rivers that were contaminated by heavy metals and POPs. Accordingly, the use of an economically viable and eco-friendly techniques, that is, phytoremediation through aquatic flora for the remediation of wastewater, had achieved a great attention. A number of macrophytes such as *Eichhornia, Azolla, Hydrilla, Pistia, Lemna,* and *Wolffia* have been well documented for their metal and nutrient removal efficiency from the wastewater (Shutes 2001; Liao and Chang 2004; Mishra et al. 2009; Babić et al. 2009; Ali et al. 2020). Till date, more than 150 macrophytes species have been documented for their application in constructed wetlands globally (Ali et al. 2020; Kumar and Dutta 2019). List of some aquatic macrophytes used for phytoremediation are given below (Table 8.3).

Further, phytoremediation potential of some macrophytes is discussed below:

8.6.1 *Eichhornia crassipes*

E. crassipes is an invasive aquatic species belonging to the family Pontederiaceae. It is commonly known as water hyacinth. Over the last few decades, a great interest has been shown by several researchers toward *Eichhornia* because of its high remediation potential along with the higher biomass production rate and tolerance to

Table 8.3 Aquatic macrophytes applied for remediation of heavy metals and metalloids from wastewater

Aquatic macrophytes	Heavy metals/ metalloids	References
Eichhornia crassipes	Eu (III), Zn, As, Ni, Cd, Cr, Cu, Se	Laet et al. (2019), Kumar et al. (2012)
Pistia stratiotes	Hg	Mishra et al. (2009)
Pennisetum purpureum	Cr	Mant et al. (2007)
Brachiaria decumbens	Cr	Mant et al. (2006)
Lemna polyrhiza	Zn, Pb	Sharma and Gaur (1995)
Lemna minor	Pb, Ni, Fe, Cu, As, Hg, Ti	Ali et al. (2020), Boulé et al. (2009), Mishra et al. (2008a, b), Babić et al. (2009)
Lemna gibba	Ur, As	Mkandawire et al. (2004)
Azolla pinnata	Fe, Cu, Hg, Cr, Hg, Cd	Jain et al. (1989), Rai (2008)
Azolla caroliniana	Hg	Bennicelli et al. (2004)
Azolla filiculoides	Pb	Naghipour et al. (2018)
Nasturtium officinale	Cu, Zn, Ni	Kara (2005)
Cyperus alternifolius	Cu, Zn, Pb, Cd	Cheng et al. (2002)
Spirodela polyrhiza	Hg	Mishra et al. (2008a, b)
Cyperus eragrostis *Juncus* spp. *Rumex obtusifolius* *Typha orientalis* *Cyperus ustulatus*	As	Robinson et al. (2006)
Eleocharis acicularis	Fe, Cu, Cr, Ni, Zn, Pb, Mn	Hoang et al. (2009)
Egeria densa	Zn, Cd, Cu	Pietrobelli et al. (2009)
Ceratophyllum demersum	Cd	Aravind et al. (2009)
Potamogeton pusillus	Cu	Monferran et al. (2009)
Vallisneria spiralis	Cr, Cd, Ni	Verma et al. (2008)
	Hg	Rai and Tripathi (2009)
Mentha spp.	Fe	Arora et al. (2008)
Typha latifolia	Zn, Ni, Cu	Sasmaz et al. (2008)
Spirodela polyrhiza	As	Rahman et al. (2008)

fluctuating ecological factors (Laet et al. 2019; Priya and Selvan 2014; Akinbile and Yusof 2012). The potential of *Eichhornia* for the removal of inorganic contaminants like Eu(III), Zn, As, Cd, Cr, Cu, Ni, Se, and Hg in wastewater has been studied extensively (Ali et al. 2020; Liao and Chang 2004; Mishra et al. 2008a, b; Sakakibara et al. 2011). Because of its rapid growth rate and a high accumulation potential, it is most commonly used macrophytes for the wastewater remediation by constructed wetlands (Tiwari et al. 2007; Priya and Selvan 2014). *Eichhornia* has a great potential for accumulation/absorption of heavy metals like Cu, Cd, Pb, Zn, and Ni even if their concentrations in growing medium are very low (Mishra et al. 2008a, b; Sakakibara et al. 2011). Generally, the ratio of accumulated metals in root with shoot has been testified 3 to 15 times higher than shoots. The overall accumulation potential for each metal has been reported as Cu > Pb > Cd > Ni > Zn (Liao and Chang 2004). Further, in an experiment conducted by Jayaweera et al. (2008) it has been testified that *Eichhornia* can also easily grow under oligotrophic environment and simultaneously remove Fe from wastewater (Jayaweera et al. 2008). Kelley et al. (1999) have validated the potential of *Eichhornia* for remediation of lanthanide metal like Eu (III) from contaminated water. Further, they also reported that the maximum absorbed concentration of Eu (III) was found onto the root hairs (Kelley et al. 1999). In an experiment conducted by Mokhtar et al. (2011) founded that *Eichhornia* can efficiently remove approximately 97% of Cu from the growing aquatic medium in 21 days. Similarly, it has also been reported that the water hyacinth can remove 70–90% of Pb, Cu, Fe, and Cr from wastewater of textile industry (Kolawole 2001). Furthermore, Schneider et al. (1995) examined and reported that the biomass of dried root of *Eichhornia* can be used as a biosorbent for the remediation of Pb^{2+}, Cu^{2+}, Cd^{2+}, and Zn^{2+}. They also reported that the efficiency of dried root biosorbent for Pb and Cu was much proficient as compared to the biosorbent of *Mycobacterium phlei* (bacterium), *Candida parapsilosis* (yeast), *Rhizopus oryzae* (fungus), and acacia bark (Schneider et al. 1995).

8.6.2 Azolla

Azolla is a small aquatic fern belonging to the family *Azollaceae* and only pteridophyte which symbiotically associate with diazotrophic, nitrogen-fixing cyanobacteria (Adabembe et al. 2022; Akinbile et al. 2015; Sood et al. 2012). Azolla is fast growing and free floating in nature which helps in easy harvest for recovery of heavy metals from biomass (Sood et al. 2012). The heavy metal remediation efficiency of *Azolla pinnata* has been documented comprehensively (Rai 2008; Sood et al. 2012; Adabembe et al. 2022; Akinbile et al. 2015; Sood and Ahluwalia 2009). In a study, Rai (2008) has examined the metals remediation potential of *A. pinnata* and reported that after 13 days of exposure of Hg and Cd, *A. pinnata* removed 70–94% of Hg and Cd along with the 27–33.9% growth inhibition. Further, the highest growth inhibition was reported in the presence of Hg (II) ions at the concentration on 0.5 mgL^{-1} compared to control (Rai 2008). It has been reported that azolla has efficiency to accumulated Cu and Fe up to 78 times more than that of

Table 8.4 List of some selected species of *Azolla* used for phytoremediation

Species	Heavy metal	Metal accumulation level (μg g^{-1})	References
A. pinnata	Hg	450–940	Rai (2008), Rai and Tripathi (2009)
	Cd	740–2759	Rai (2008), Arora et al. (2004)
	Cr	1095–9125	Arora et al. (2006), Pati and Satapathy (2016)
	Ni	16,252	Arora et al. (2004)
A. imdbricata	Cd	183	Dai et al. (2006)
A. microphylla	Cr (VI)	14,931	Arora et al. (2006)
	Ni	21,785	Arora et al. (2004)
	Cd	1805	
A. caroliniana	As	>120	Zhang et al. (2008)
	Pb	416	Stepniewska et al. (2005)
	Cd	259	
	Cr (VI)	356	Bennicelli et al. (2004)
	Cr (III)	964	
	Hg	578	
A. filiculoides	As	>60	Zhang et al. (2008)
	Cr (VI)	12,383	Arora et al. (2006)
	Cr (III)	1904	Sela et al. (1989)
	Cd	10,441	
	Ni	28,443	Arora et al. (2004)
		43,400	Zhao and Duncan (1997)

the metal concentrations present in growing aquatic medium (Jain et al. 1989). *A caroliniana* has been well documented for its hyperaccumulation potential. In a study, Zhang et al. (2008) examined the variations in As accumulation and tolerance ability among 50 strains of aquatic *Azolla.* They observed that accumulation potential of *A. caroliniana* was twofold more than *A. filiculoides*. Further, it was also reported that both the strains have showed a same level of tolerance to accumulated internal As, while *A. filiculoides* was found more resilient to external As (Zhang et al. 2008) (Table 8.4).

8.6.3 Duckweeds

Duckweeds is a collective name for five genera, namely, *Lemna, Spirodela, Wolffiella, Landoltia, and Wolffia* representing the family *Lemnaceae*. Duckweeds are the smallest and fastest growing angiosperm spread globally in diverse aquatic environment with wide-ranging salinity and nutrient concentration. They can survive in a wide range of temperature (7–35 °C), pH (3.5–10.5), and salinity (154–2276 mg L^{-1}) (Baek et al. 2021). Several species of duckweed, for example, *Lemna trisulca, L. gibba,* and *L. minor,* have been comprehensively reported for heavy metals remediation potential, sewage and secondary effluents treatment, and nutrient removal from wastewater (Sharma and Lenaghan 2022; Para et al. 2012; Ali et al. 2020; Korner et al. 2003). *Lemna* has been investigated for their remediation potential for heavy metals like Cr, Cu, As, Hg, Pb, and Zn and reported that remediation efficiency varied from 3 to 30%. Para et al. (2012). Bokhari et al. 2016 investigated the phytoremediation potential of *L. minor* for heavy metals, viz., Cd, Cu, Pd, and Ni and reported that removal efficiency was higher than 80% for all heavy metals and maximum removal efficiency (99%) was observed for Ni from sewage mixed industrial effluent (Bokhari et al. (2016). They also reported that the uptake and accumulation of Pb in dry biomass of *L. minor* was significantly greater than other metals. Further, they also documented that bio-concentration (BCF) factors were less than 1000 and maximum BCFs were observed for Cu and Pb, that is, 558 and 523.1, respectively (Bokhari et al. (2016). The accumulation, tolerance potential, and its defense mechanisms against several toxic metal ions have also been demonstrated by several authors (Oporta et al. 2006; Malec et al. 2010; Radic et al. 2010). Sekomo et al. (2012) have reported that the duckweed and algal ponds are suitable for remediation of heavy metals like Cr and Zn from textile industry effluents during polishing step. To investigate the metal remediation efficiency, they designed an experimental setup operated at a hydraulic retention time of 7 days under two different metal loading rate and light regimes. They found that the Cr and Zn removal rate by duckweed pond were 94% and 80%, respectively. They also reported that metal removal rate was also affected by light regime and metal loading rate (Sekomo et al. 2012). It has been reported that *Lemna gibba* L. is a promising candidate for the removal of Ur and As contamination (Mkandawire et al. 2004). Mkandawire et al. (2004) reported that *L. gibba* had accumulated 896.9 and 1021 mg kg^{-1} of Ur and As, respectively, exposed for 21 days in laboratory steady-state experiments (Mkandawire et al. 2004).

Phytoremediation efficiency of *Wolffia* has been reported by several authors (Zhang et al. 2009; Garg and Chandra 1994; Boonyapookana et al. 2002). It has been reported that *W. globosa* has 2–10 times higher As accumulation potential than some aquatic species, viz., *S. polyrhiza, W. globosa, L. minor, A. carolina*, and *A. filiculoides* grown in the arsenate contaminated medium (Zhang et al. 2009). Zhang et al. (2009) reported that the *W. globosa* has potential to accumulate >1000 mg kg^{-1} As on dry weight basis and can tolerate As up to 400 mg kg^{-1}. Moreover, at lower concentration, uptake rate was similar for arsenate and arsenite, while at high concentrations arsenite accumulated more rapidly as compared to arsenate. *W. globosa* has been advocated by several authors for being a perfect candidate for the study of As uptake and metabolism as it lacks root to shoot translocation barrier (Zhang et al. 2009; Tel-Or and Forni 2011).

8.7 Conclusion

Several conventional and nonconventional wastewater remediation approaches like ion exchange, coagulation, reverse osmosis, adsorption, membrane filtration, flocculation, nanofiltration, electrodialysis etc. have been applied for wastewater treatment. However, certain limitations like cost involvement, sensitiveness, toxic by-product generation, and extra operating cost involved in sludge disposal make these techniques unsuitable and unsustainable. Phytoremediation is a nature-driven, eco-friendly approach that uses plants for in situ and ex situ remediation of toxic metals and metalloids from wastewater. Phytoremediation with the help of aquatic plants has proven its effectiveness even at shallow and low level of contamination. For an effective phytoremediation, it is recommended that the selected plants species should be perennial, fast growing, tolerant, have high biomass production and survival rate under diverse biotic and abiotic stress environments. Aquatic macrophytes have unique metabolic and absorption capabilities and well-developed transport systems that can take up heavy metals from wastewater. In this chapter, authors have focused on the researches that showed the phytoremediation potential of aquatic macrophytes like *Eichhornia, Azolla, Pistia, Lemna,* and *Wolffia*; however, further intensive research is needed in this concern.

References

Adabembe BA, Fasinmirin JT, Olanrewaju OO, Dada AA, Faloye OT (2022) Phytoremediation of aquaculture wastewater using *Azolla pinnata* and evaluation of its suitability for irrigation purpose. Sustain Water Resour Manag 8:166. https://doi.org/10.1007/s40899-022-00753

Ahmad S, Ali A, Ashfaq A (2016) Heavy metal pollution, sources, toxic effects and techniques adopted for control. Int J Curr Res Aca Rev 4(6):39–58

Akinbile CO, Yusof MS (2012) Water hyacinth (*Eichhornia crassipes*) and lettuce (pistiastratiotes) efectiveness in aquaculture wastewater treatment in Malaysia. Inter J Phytoremed 14(3): 201–211

Akinbile CO, Ogunrinde TA, Man HC, Aziz HA (2015) Phytoremediation of domestic wastewaters in free water surface constructed wetlands using Azolla pinnata. Int J Phytoremediation 18(1): 54–56. https://doi.org/10.1080/15226514.2015.1058330

Alharbi OM, Khattab RA, Ali I (2018) Health and environmental effects of persistent organic pollutants. J Mol Lisq 263:442–453

Ali H, Khan E, Sajad MA (2013) Phytoremediation of heavy metals – concepts and applications. Chemosphere 91:869–881

Ali S, Abbas Z, Rizwan M et al (2020) Application of floating aquatic plants in phytoremediation of heavy metals polluted water: a review. Sustainability 12:1927. https://doi.org/10.3390/su12051927

Aravind P, Prasad MNV, Malec P, Waloszek A, Strzałka K (2009) Zinc protects Ceratophyllum demersum L. (free-floating hydrophyte) against reactive oxygen species induced by cadmium. J Trace Elem Med Biol 23:50–60

Arora A, Sood A, Singh PK (2004) Hyperaccumulation of cadmium and nickel by *Azolla* species. Indian J Plant Physiol 3:302–304

Arora A, Saxena S, Sharma DK (2006) Tolerance and phytoaccumulation of chromium by three *Azolla* species. World J Microbiol Biotechnol 22:97–100

Arora M, Kiran K, Rani S, Rani A, Kaur B, Mittal N (2008) Heavy metal accumulation in vegetables irrigated with water from different sources. Food Chem 111:811–815

Babić M, Radić S, Cvjetko P, Roje V, Pevalek-Kozlina B, Pavlica M (2009) Antioxidative response of Lemna minor plants exposed to thallium (I)-acetate. Aquat Bot 91:166–172

Baek G, Saeed M, Choi H-K (2021) Duckweeds: their utilization, metabolites and cultivation. Appl Biol Chem 64:73. https://doi.org/10.1186/s13765-021-00644-z

Barakat MA (2011) New trends in removing heavy metals from industrial wastewater. Arab J Chem 4:361–377

Bennicelli R, Stepniewska Z, Banach A, Szajnocha K, Ostrowski J (2004) The ability of *Azolla caroliniana* to remove heavy metals Hg(II), Cr(III) and Cr(VI) from municipal wastewater. Chemosphere 55(1):141–146

Bergman Å, Heindel J, Jobling S, Kidd K, Zoeller R (2014) WHO (World Health Organization)/ UNEP (United Nations Environment Programme) Global assessment of the state-of-the-science of endocrine disruptors, vol 121, p A107. https://doi.org/10.1289/ehp.1306695

Bhatiya M, Goyal D (2014) Analyzing remediation potential of wastewater through wetland plants: A review. Environ Prog Sustain Energy 33(1):9–27

Bokhari SH, Ahmad I, Mahmood-Ul-Hassan M, Mohammad A (2016) Phytoremediation potential of *Lemna minor* L. for heavy metals. Int J Phytoremediation 18:25–32

Boonyapookana B, Upatham ES, Kruatrachue M, Pokethitiyook P, Singhakaew S (2002) Phyoaccumulation and phytotoxicity of Cd and Cr in duckweed *Wolffia globosa*. Int J Phytoremediation 4:87

Boulé KM, Vicente JAF, Nabais C, Prasad MNV, Freitas H (2009) Ecophysiological tolerance of duckweeds exposed to copper. Aquat Toxicol 91:1–9

Cheng S, Grosse W, Karrenbrock F, Thoennessen M (2002) Efficiency of constructed wetlands in decontamination of water polluted by heavy metal. Ecol Eng 18:317–325

Dai LP, Xiong ZT, Huang Y, Li MJ (2006) Cadmium induced changes in pigments, total phenolics, and phenylalanine ammonia-lyase activity in fronds of *Azolla imbricata*. Environ Toxicol 21: 505–512

Dhir B, Sharmila P, Pardha Saradhi P, Nasim SA (2009) Physiological and antioxidant responses of *Salvinia natans* exposed to chromium-rich wastewater. Ecotoxicol Environ Saf 72:1790–1797

Dixit S, Dhote S (2010) Evaluation of uptake rate of heavy metals by Eichhornia crassipes and Hydrilla verticillata. Environ Monit Assess 169:367–374

Eid EM, Galal TM, Sewelam NA, Talha NI, Abdallah SM (2020) Phytoremediation of heavy metals by four aquatic macrophytes and their potential use as contamination indicators: A comparative assessment. Environ Sci Pollut Res 27:1–14

El-Samrani AG, Lartiges BS, Villiéras F (2008) Chemical coagulation of combined sewer overflow: heavy metal removal and treatment optimization. Water Res 42:951–960

Fazal S, Zhang B, Mehmood Q (2015) Biological treatment of combined industrial wastewater. Ecol Eng 84:551–558

Fenglian F, Wang Q (2011) Removal of heavy metal ions from wastewaters: A review. J Environ Manag 92:407–418

Garg P, Chandra P (1994) The duckweed *(Wolffia globosa)* as an indicator of heavy metal pollution. Environ Monitor Assess 29:89–95

Gorgoglione A, Torretta V (2018) sustainable management and successful application of constructed wetlands: a critical review. Sustainability 2018(10):3910. https://doi.org/10.3390/su10113910

Han D, Currell MJ (2017) Persistent organic pollutants in China's surface water systems. Sci Total Environ 580:602–625

Harguinteguy CA, Cirelli AF, Pignata ML (2014) Heavy metal accumulation in leaves of aquatic plant Stuckenia filiformis and its relationship with sediment and water in the Suquía river (Argentina). Microchem J 114:111–118. https://doi.org/10.1016/j.microc.2013.12.010

Hoang H, Sakakibara NTHM, Sano S, Hori RS, Sera K (2009) The potential of *Eleocharis acicularis* for phytoremediation: case study at an abandoned mine site. Clean Soil Air Water 37:203–208

Jain SK, Vasudevan P, Jha NK (1989) Removal of some heavy metals from polluted water by aquatic plants: studies on duckweed and water velvet. Biol Wastes 28:115–126

Jayaweera MW, Kasturiachchi JC, Kularatne RKA, Wijeyekoon LJ (2008) Contribution of water hyacinth (Eichhornia crassipes (Mart.) Solms) grown under different nutrients conditions to Fe-removal mechanisms in constructed wetlands. J Environ Manag 87:450–460

Jonker MT, Koelmans AA (2002) Sorption of polycyclic aromatic hydrocarbons and polychlorinated biphenyls to soot and soot-like materials in the aqueous environment: mechanistic considerations. Environ Sci Technol 36(17):3725–3734

Kara Y (2005) Bioaccumulation of Cu, Zn and Ni from the wastewater by treated *Nasturtium* officinal. Int J Environ Sci Technol 2(1):63–67

Kelley C, Mielke R, Dimaquibo D, Curtis AJ, Dewitt JG (1999) Adsorption of Eu(III) onto roots of water hyacinth. Environ Sci Technol 33:1439–1443

Kolawole BS (2001) Cleaning of effluent from textile industry by water hyacinth (B. Sc. Thesis). University of Agriculture, Abeokuta, Ogun state

Korner S, Vermaat JE, Veenstra S (2003) The capacity of duckweed to treat wastewater: ecological considerations for a sound design. J Environ Qual 32:1583–1590

Kumar S, Dutta V (2019) Constructed wetland microcosms as sustainable technology for domestic wastewater treatment: an overview. Environ Sci Pollut Res 26:11662. https://doi.org/10.1007/s11356-019-04816-9

Kumar N, Bauddha K, Dwivedi N, Barman SC, Singh DP (2012) Accumulation of metals in selected macrophytes grown in mixture of drain water and tannery effluent and their phytoremediation potential. J Environ Biol 33:923–927

Kumar N, Bauddha K, Kumar S, Dwivedi N, Singh DC, Barman SC (2013) Accumulation of metals in weed species grown on the soil contaminated with industrial waste and their phytoremediation potential. Ecol Eng 61:491–495

Kumar D, Bharti SK, Anand S, Kumar N (2018) Defluoridation of water with the help of copper phytoremediated Andrographis paniculata plant biomass. J Environ Biol 39:664–670

Laet CD, Matringe T, Petit E, Grison C (2019) Eichhornia crassipes: a powerful bio-indicator for water pollution by emerging pollutants. Sci Rep 9:7326. https://doi.org/10.1038/s41598-019-43769-4

Li C, Yang L, Shi M, Liu G (2019a) Persistent organic pollutants in typical lake ecosystems. Ecotoxicol Environ Saf 180:668–678

Li C, Yang L, Shi M, Liu G (2019b) Persistent organic pollutants in typical lake ecosys-tems. Ecotoxicol Environ Saf 180:668–678

Li Y, Liu M, Hou L, Li X, Yin G, Sun P, Zheng D (2021) Geographical distribution of polycyclic aromatic hydrocarbons in estuarine sediments over China: human impacts and source apportionment. Sci Total Environ 768:145279

Liao S, Chang W (2004) Heavy metal phytoremediation by water hyacinth at constructed wetlands in Taiwan. J Aqua Plant Manage 42:60–68

Lu Q, Zhenli LH, Graetz DA, Stoffella PJ, Yang X (2010) Phytoremediation to remove nutrients and improve eutrophic stormwaters using water lettuce (*Pistia stratiotes* L.). Environ Sci Pollut Resour 17:84–96

Malec P, Maleva MG, Prasad MN, Strzalka K (2010) Response of Lemna trisulca L. (Duckweed) exposed to low doses of Cadmium: thiols, metal binding complexs and photosynthetic pigments as sensitive biomarkers of ecotoxicity. Protoplasma 240(1–4):69–74

Mant C, Costa S, Williams J, Tambourgi E (2006) Phytoremediation of chromium by model constructed wetland. Bioresour Technol 97:1767–1772

Mant C, Costa S, Williams J, Tambourgi E (2007) Phytoremediation of chromium by model constructed wetland. Bioresour Technol 97:767–772

Mishra VK, Tripathi BD (2008) Concurrent removal and accumulation of heavy metals by the three aquatic macrophytes. Bioresour Technol 99(15):7091–7097

Mishra VK, Upadhyay AR, Pandey SK, Tripathi BD (2008a) Concentrations of heavy metals and aquatic macrophytes of Govind Ballabh Pant Sagar an anthropogenic lake affected by coal mining effluent. Environ Monitor Assess 141:49–58

Mishra VK, Upadhyay AR, Pathak V, Tripathi BD (2008b) Phytoremediation of mercury and arsenic from tropical opencast coalmine effluent through naturally occurring aquatic macrophytes. Water Air Soil Pollut 192:303–314

Mishra VK, Tripathi BD, Kim KH (2009) Removal and accumulation of mercury by aquatic macrophytes from an open cast coal mine effluent. J Hazard Mater 172:749–754
Mkandawire M, Taubert B, Dudel EG (2004) Capacity of Lemna gibba L. (Duckweed) for uranium and arsenic phytoremediationin mine tailing waters. Int J Phytoremediation 6:347–362
Mokhtar H, Morad N, Ahmad Fizri FF (2011) Hyperaccumulation of copper by two species of aquatic plants. International Conference on Environmental Science and Engineering IPCBEE 8. IACSIT Press, Singapore
Monferran MV, Sanchez Agudo JA, Pignata ML, Wunderlin DA (2009) Copper-induced response of physiological parameters and antioxidant enzymes in the aquatic macrophytes Potamogeton pusillus. Environ Pollut 157:2570–2576
Naghipour D, Ashrafi SD, Gholamzadeh M, Taghavi K, Naimi-Joubani M (2018) Phytoremediation of heavy metals (Ni, Cd, Pb) by Azolla filiculoides from aqueous solution: a dataset. Data Brief 21:1409–1414. https://doi.org/10.1016/j.dib.2018.10.111
Nikitha T, Satyaprakash M, Vani SS, Sadhana B, Padal SB (2017) A review on pol-ycyclic aromatic hydrocarbons: their transport, fate and biodegradation in the environment. Int J Curr Microbiol Appl Sci 6(4):1627–1639
Okereafor U, Makhatha M, Mekuto L et al (2020) Toxic metal implications on agricultural soils, plants, animals, aquatic life and human health. Int J Environ Res Public Health 17:2204. https://doi.org/10.3390/ijerph17072204
Oporta C, Arce O, Bracck EVD, Bruggen BVD, Vandecasteele C (2006) Experimental study and modelling of Cr (VI) removal from waste water using Lemna minor. Water Res 40:1458–1464
Para LM, Torres G, Arenas AD, Sanchez E and Rodriguez K. 2012. Phytoremediation of low levels of heavy metals using duckweed (Lemna minor). Abiotic stress responses in plants: metabolism, productivity and sustainabilty. Springer New York, NY, 451–463
Pati S, Satapathy KB (2016) Phytoremediation potential of aquatic macrophyte *azolla pinnata* R.Br. and *salvinia molesta* mitchell for removal of chromium from waste water. Int J Sci Environ Technol 5(4):2146–2160
Pietrobelli JMT, de Módenes AN, Fagundes-Klen MRF, Espinoza-Quiñones FR (2009) Cadmium, copper and zinc biosorption study by non-living Egeria densa biomass. Water Air Soil Pollut 202:385–392
Priya ES, Selvan PS (2014) Water hyacinth (Eichhornia crassipes) – an efficient and economic adsorbent for textile effluent treatment – a review. Arab J Chem 10:S3548–S3558
Radic S, Stipanicev D, Cvjetko P, Mikelic IL, Rajcic MM, Sirac S, Kozlina BP, Palvica M (2010) Ecotoxicological assessment of industrial effluent using duckweed (Lemna minor L.) as a test organism. Ecotoxicology 19:216–222
Radjenovic J, Sedlak DL (2015) Challenges and opportunities for electrochemical processes as next generation technologies for the treatment of contaminated water. Environ Sci Technol 49(19): 11292–11302
Rahman MA, Hasegawa H, Ueda K, Maki T, Rahman MM (2008) Influence of EDTA and chemical species on arsenic accumulation in Spirodela polyrhiza L. (duckweed). Ecotoxicol Environ Saf 70:311–318
Rai PK (2008) Phytoremediation of Hg and Cd from industrial effluent using an aquatic free floating macrophyte Azolla pinnata. Int J Phytoremediation 10:430–439
Rai PK (2009) Heavy metal phytoremediation from aquatic ecosystems with special reference to macrophytes. Crit Rev Environ Sci Technol 39:697–753
Rai PK, Tripathi BD (2009) Comparative assessment of Azolla pinnata and Vallisneria spiralis in Hg removal from G.B. Pant Sagar of Singrauli Industrial region, India. Environ Monitor Assess 148:75–84
Ramaswami A, Carr P, Burkhardi M (2001) Plant-uptake of Urenium: hydroponic and soil system studies. Int J Phytoremediation 3(2):189–201
Rezania S, Ponraj M, Md Din MF, Chelliapan S, Md Sairan F (2016) Effectiveness of Eichhornia crassipes in nutrient removal from domestic wastewater based on its optimal growth rate. J Desalinat Water Treat 57(1):360–365

Robinson B, Kim N, Marchetti M, Moni C, Schroeter L, Dijssel CD, Milne G, Clothier B (2006) Arsenic Hyperaccumulation by aquatic macrophytes in the Taupo volcanic zone, New Zealand. Environ Exp Bot 2006(58):206–215

Sakakibara M, Ohmori Y, Ha NTH, Sano S, Sera K (2011) Phytoremediation of heavy metal-contaminated water and sediment by Eleocharis acicularis. Clean Soil Air Water 39:735–741

Sampera E, Rodrígueza M, De la Rubia MA, Prats D (2009) Removal of metal ions at low concentration by micellar-enhanced ultrafiltration (MEUF) using sodium dodecyl sulfate (SDS) and linear alkylbenzene sulfonate (LAS). Sep Purif Technol 65:337–342

Sasmaz A, Obekb E, Hasarb H (2008) The accumulation of heavy metals in Typha lotifolia L. Grown in a stream carrying secondary effluent. Ecol Eng 33:278–284

Schneider IAH, Rubio J, Misra M, Smith RW (1995) Eichhornia crassipes as biosorbent for heavy metal ions. Min Eng 8:979–988

Sekomo CB, Rousseau DPL, Saleha SA, Lensa PNL (2012) Heavy metal removal in duckweed and algae ponds as a polishing step for textile wastewater treatment. 2012. Ecol Eng 44:102–110

Sela M, Garty J, Tel-Or E (1989) The accumulation and effect of heavy metal on the water fern *Azolla filiculoides*. New Phytol 112:7–12

Sharma SS, Gaur JP (1995) Potential of Lemna polyrhiza for removal of heavy metals. Ecol Eng 4: 37–45

Sharma R, Lenaghan SC (2022) Duckweed: a potential phytosensor for heavy metals. Plant Cell Rep 41:2231. https://doi.org/10.1007/s00299-022-02913-7

Shutes RBE (2001) Artificial wetlands and water quality improvement. Environ Int 26:441–447

Sirés I, Brillas E, Oturan MA, Rodrigo MA, Panizza M (2014) Electrochemical advanced oxidation processes: today and tomorrow. A review. Environ Sci Pollut Res 21:8336–8367

Sood A, Ahluwalia AS (2009) Cyanobacterial–plant symbioses with emphasis on Azolla-Anabaena symbiotic system. Indian Fern Journal 26:166–178

Sood A, Perm L, Prasanna UR, Ahluwalia AS (2012) Phytoremediation potential of aquatic macrophyte, Azolla. A Journal of the Human Environment 41:122–137

Stepniewska Z, Bennicelli RP, Balakhnina TI, Szajnocha K, Banach A, Wolin˜ska A (2005) Potential of Azolla caroliniana for the removal of Pb and Cd from wastewaters. Int Agrophys 19:251–255

Tel-Or E, Forni C (2011) Phytoremediation of hazardous toxic metals and organics by photosynthetic aquatic systems. Plant Biosystems 145:224–235

Tiwari S, Dixit S, Verma N (2007) An effective means of biofiltration of heavy metal contaminated water bodies using aquatic weed *Echhornia crassipes*. Environ Monit Assess 129:253–256

Trojanowicz M (2020) Removal of persistent organic pollutants (POPs) from waters and wastewaters by the use of ionizing radiation. Sci Total Environ 718:134425

Verma VK, Tewari S, Rai JPN (2008) Ion exchange during heavy metal bio-sorption from aqueous solution by dried biomass of macrophytes. Bioresour Technol 99(6):1932–1938

Vymazal J (2016) Concentration is not enough to evaluate accumulation of heavy metals and nutrients in plants. Sci Total Environ 544:495–498

Yu F, Yang C, Zhu Z, Bai X, Ma J (2019) Adsorption behavior of organic pollutants and metals on micro/nanoplastics in the aquatic environment. Sci Total Environ 694:133643

Zhang X, Lin A-J, Zhao F-J, Xu G-Z, Duan G-L, Zhua Y-G (2008) Arsenic accumulation by the aquatic fern Azolla: comparison of arsenate uptake, speciation and efflux by A. caroliniana and A. filiculoides¬. Environ Pollut 156(1):1149–1155

Zhang X, Zhao FJ, Huang Q, Williams PN, Sun GX, Zhu YG (2009) Arsenic uptake and speciation in the rootless duckweed *Wolffia globosa*. New Phytol 182:421–428

Zhao M, Duncan JR (1997) Batch removal of sexivalentchromium by *Azolla filiculoides*. Biotechnol Appl Biochem 26:179–182

Zhao X, Liu W, Cai Z, Han B, Qian T, Zhao D (2016) An overview of preparation and applications of stabilized zero valent iron nanoparticles for soil and groundwater remediation. Water Res 100:245–226

Phytoremediation of Organic Contaminants: An Eco-friendly Approach-Based Application of Aquatic Macrophytes

9

Ritu Singh, Avimanu Sharma, Prakriti Goswami, Dhiraj Pradhananga, Deepak Aryal, Soni M. Pradhanang, and Rajesh Kumar

Abstract

Environmental pollution is a serious global issue that threatens public health and the well-being of living beings. Unmatched pollution in aquatic ecosystems necessitates environmentally acceptable, long-term remedial technologies. One of the most effective strategies among the numerous ones implemented so far for environmental remediation is phytoremediation. It is a low-cost, eco-friendly, and long-term method which uses hyperaccumulator plants to convert and/or stabilize the pollutants in soil/water matrices. In recent years, aquatic macrophytes have gained significant attention in the field of environmental cleanup. Aquatic macrophytes are the key component of wetlands that have a substantial impression on ecosystem functions and services. This chapter gives an overview of the aquatic macrophytes and outlines the recent progress made in the use of macrophytes for the remediation of various organic contaminants. In addition, artificial wetlands, their types, and their role in wastewater treatment have been discussed.

R. Singh (✉) · A. Sharma · P. Goswami · R. Kumar
Department of Environmental Science, School of Earth Sciences, Central University of Rajasthan, Ajmer, Rajasthan, India
e-mail: ritu_ens@curaj.ac.in

D. Pradhananga
Department of Meteorology, Tri-Chandra Multiple Campus, Tribhuvan University, Kathmandu, Nepal

The Small Earth Nepal, Kathmandu, Nepal

D. Aryal
Central Department of Hydrology and Meteorology, Tribhuvan University, Kathmandu, Nepal

S. M. Pradhanang
Department of Geosciences, University of Rhode Island, Kingston, RI, USA

S. Kumar et al. (eds.), *Aquatic Macrophytes: Ecology, Functions and Services*,
https://doi.org/10.1007/978-981-99-3822-3_9

Keywords

Environmental pollution · Phytoremediation · Aquatic macrophytes · Organic contaminants · Constructed wetlands · Floating wetlands

9.1 Introduction

Water is considered to be an essential component of life, yet, getting clean as well as safe water has become a significant problem worldwide, notably in countries which are in the developing stage (Tenaw and Assfaw 2022; Zhang et al. 2022; Mueller and Gasteyer 2021). One of the primary reasons is the contamination of water bodies with industrial effluents of varied composition and volume, domestic wastewater/ sewage, surface runoff, accidental spillage, leachates, etc. (Ahmed et al. 2021; Schwarzenbach et al. 2010; Akhtar et al. 2021). These contaminants include various organic and inorganic chemicals, radioactive substances etc. which make the water unusable and harmful to people, animals, and plants owing to their toxicity, carcinogenicity, mutagenicity, teratogenicity, etc. (Jaswal et al. 2022). As per the report of the World Water Development 2017, barely 20% of the wastewater generated around the world is adequately treated, and the rest is dumped into the environment without being cleaned. The presence of organic pollutants in wastewater has become much more common in recent years (Krithiga et al. 2022; Deblonde et al. 2011). This issue is particularly concerning due to the contaminants' toxic nature, their tendency to be semi-volatile, rapid bioaccumulation, and resistance to natural degradation in environmental conditions. Organic pollutants like phenolic compounds, polycyclic aromatic hydrocarbons (PAHs), organic pesticides, and herbicides pose significant challenges as they deteriorate the environmental quality and lead to health problems like cancer, reproductive system disorders, obesity, endocrine disruption, etc., in the exposed individuals (Okoro et al. 2022; Ngo et al. 2015). Phytoremediation is a cost-effective and environment-friendly technology which uses plants to transform, stabilize, or remove a wide range of toxins found in soils, sediments, or water (Etim 2012; Vishnoi and Srivastava 2007; Parveen et al. 2022). Among the plants employed for phytoremediation, aquatic macrophytes have gained worldwide importance due to their exceptional efficiency in removing a wide range of contaminants from wastewater, including heavy metals, radionuclides, explosives, and organic/inorganic pollutants (Kitamura et al. 2022; Justin et al. 2022; Ammeri et al. 2022).

9.2 Organic Contaminants

Organic contaminants are noxious chemicals derived from organic components, which can lead to various ailments in exposed people (Kaur et al. 2020a; Alharbi et al. 2018). Typically, they are originated from the industries or as by-products of manufacturing processes (Jones and De Voogt 1999). These include a wide array of chemicals such as petroleum hydrocarbons, detergents, plastics, organic solvents, insecticides, pesticides, and dyes (Haripriyan et al. 2022). They are cosmopolitan in the environment (Avino and Russo 2018), and due to their long-term negative impacts and chemical complexity, these organic contaminants have serious concerns for both wildlife and humans.

9.2.1 Sources and Effects

Organic contaminants are highly stable in all the matrices of the environment. They are released into the environment via several industrial sources, such as incinerators, heating plants, power plants, as well as furnace of households, transportation, sprays from agriculture, and evaporation from the surfaces of the water, soil, as well as landfills (Haripriyan et al. 2022; Idris et al. 2022). Inadvertent generation can occur during incineration, combustion in chemical plants, forest fires, putrefaction, etc. This form of waste can be seen in various locations and is the result of a type of activities, including the usage of old oil, demolition products from the buildings, and equipment repair. Pollutants, dirt, fuels, oil, liquid, ash, and silt enter the water system through wastewater treatment plants, runoff from agriculture fields, atmospheric deposition, and roads (Du et al. 2022). The main reservoirs for these contaminants are the oceans and seas, where they accumulate from waste disposal facilities, sediments from the river, air deposition, etc. They get a deposited in sediments at the bottom seas, oceans, and large lake, from where they can be released and reenter the environment.

Organic contaminants are particularly hazardous since they are difficult to break down and can accumulate in the body fat of human beings and other animals (Qing Li et al. 2006; Du et al. 2022). Their lipophilic nature makes them a potential candidate for biomagnification along the food chain. Even minute concentrations of these substances could harm the neurological system, create immune system ailments, cause reproductive problems, cancer, etc. (Guo et al. 2019). Besides, biodegradable organic contaminants are also a serious concern as their presence in the aquatic environment leads to a decrease in dissolved oxygen concentration of the exposed water body leading to the deterioration of water quality and posing risks to aquatic communities (Karić et al. 2022). Bacteria and other microorganisms break down biodegradable organic contaminants into simpler organic compounds (Espinosa-Ortiz et al. 2022), and during this process, they consume oxygen, and as their population grows, so does the need for dissolved oxygen (Choi et al. 2022). When potentially harmful wastewater is discharged into a stream, a sequence of events occur over time and distance, leading to the contamination of the water bodies.

9.3 Phytoremediation

Phytoremediation is a method that employs plants to clean up polluted environments. It has been regarded as a natural process, which was first found 300 years ago (Lasat et al. 2000). This technology is a low-cost and environmentally beneficial technique that helps in degrading, stabilizing, immobilizing, removing, transferring, and detoxifying pollutants such as heavy metals, hydrocarbons, pesticides, and many others (Susarla et al. 2002; Fulekar and Jadia 2008; Zhang et al. 2010). Over the last few decades, phytoremediation has become a highly recommended means of detoxifying contaminated soil and water (US EPA 2001). The plants and wild species utilized in this technique help in environmental cleanup via various mechanisms listed below and shown in Fig. 9.1 (Ghosh and Singh 2005; Brunet et al. 2008).

(a) **Phytofiltration or Rhizofiltration**
The process of removal of impurities from wastewater, surface water, or groundwater by plant roots is known as phytofiltration or rhizofiltration (Pivetz

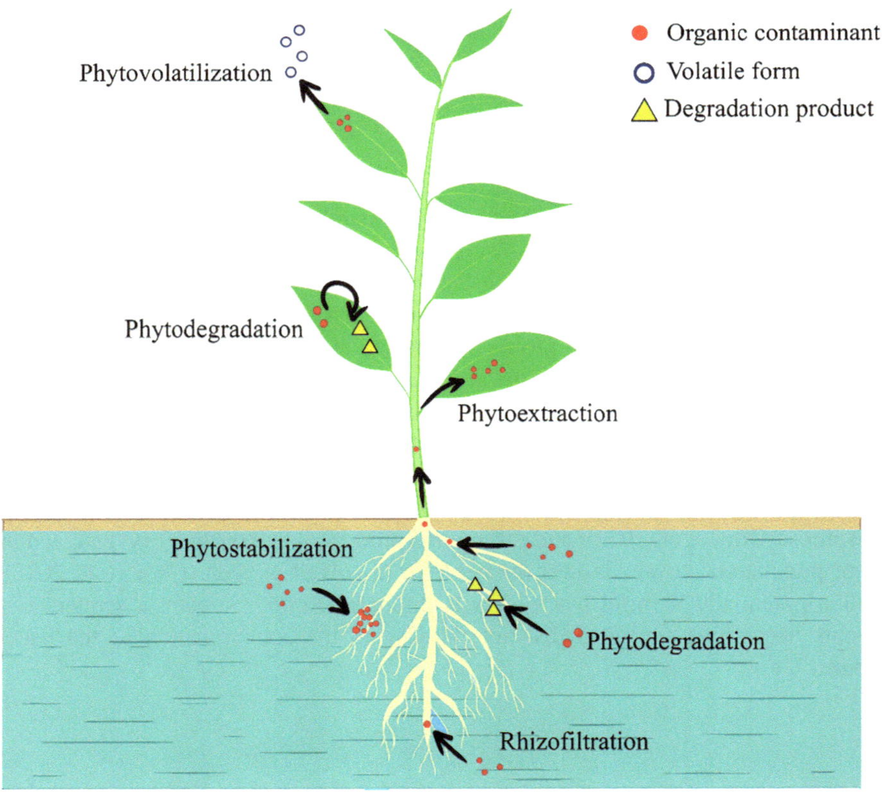

Fig. 9.1 Various mechanisms of phytoremediation in aquatic macrophytes

2001). In greenhouses, plants are cultivated in water rather than soil to develop an extensive root system; following that, the contaminated site's water is used to acclimate the plants and then transferred to the contaminated site, wherein plant roots carry out the filtration of the contaminants.

(b) **Phytostabilization**
Phytostabilization refers to the immobilization of the contaminants belowground, which in turn limits their bioavailability and restricts their movement in the environmental matrix (Jadia and Fulekar 2009). It also reduces the likelihood of biomagnification of the contaminant along the food chain. The stabilization generally involves processes like adsorption on root cell walls, reduction/precipitation in the rhizosphere, absorption/sequestration within root tissues, etc.

(c) **Phytovolatilization**
Phytovolatilization involves the uptake of contaminants from the sediment or water by the plants, which convert them into lesser toxic volatile forms and release them into the atmosphere via leaves during transpiration (Karami and Shamsuddin 2010).

(d) **Phytodegradation**
Phytodegradation, also known as phytotransformation, refers to the utilization of plants to absorb, metabolize, and degrade the organic contaminants within plant tissues. Besides, plant roots are also utilized in conjunction with microorganisms to detoxify the contaminated matrices (Alkorta and Garbisu 2001).

(e) **Phytoextraction**
In this process, plants take up the contaminants from water/sediment/soil and translocate them to aboveground biomass, where they accumulate (Yanai et al. 2006). Unlike phytostabilization, phytoextraction removes the contaminants permanently from the matrix, and thus is more suitable for practical applications.

Previous studies have demonstrated the potential of terrestrial plant species for detoxification, degradation, and removal of a wide array of environmental contaminants from the contaminated matrices (Cunningham and Berti 1993; Chandra et al. 2015). The aquatic macrophytes have also been used in several studies as remediation agents and displayed excellent efficiency for the removal of inorganic contaminants (nitrate, phosphate, radionuclides, heavy metals, etc.) as well as organic contaminants (pharmaceutical drugs, fertilizers, pesticides, etc.) (Dhir 2013; Anand et al. 2017; Mishra and Maiti 2017).

9.4 Aquatic Macrophytes

Aquatic macrophytes play a crucial role in assessing the ecological status of water bodies. Aquatic macrophyte basically refers to those plants which grow in or around water and are visible to the naked eye. It comprises of vascular plants (angiosperms and pteridophytes), bryophytes, and macroalgae flourishing in water bodies (Wersal and Madsen 2012). As primary producers, they form the basis of food webs, share a major part of highly productive aquatic ecosystems, and have a significant impression on ecosystem functions and services. These plants constitute the elementary biotic component of the ecosystem and have an essential role in structuring communities of the aquatic environment and nutrient cycling. As per the biotypes and interactions of the macrophytes with the aquatic environment, they are classified as follows:

(a) **Free-floating macrophytes**
The plant body in free-floating macrophytes remains above water's surface except for the roots. These are generally found in areas with little or no water movement (Rehman et al. 2017). Common examples of free-floating aquatic macrophytes include *Pistia stratiotes* (water lettuce), *Azolla* (water fern), *Ludwigia* sp., *Salvinia* sp., *Eichhornia crassipes* (water hyacinth), etc. These species have demonstrated efficient uptake of variety of heavy metals and inorganic nutrients from polluted waters displaying their remediation potential.
(b) **Submerged macrophytes**
Submerged macrophytes entirely grow under the water body and are most commonly found in shallow, stagnant waters (Schneider and Melzer 2004). These plants may or may not have roots. The plant without roots floats freely underwater, while rooted plants remain attached to any substrate. Examples include *Vallisneria* sp., *Egeria* sp., *Myriophyllum* sp., *Hydrilla* sp., etc.
(c) **Floating-leaved macrophytes**
Floating-leaved macrophytes are submerged macrophytes with their roots attached to the substrate or bottom of the water bodies, and their leaves float on the water's surface (Rejmankova 2011). The plants of the Nymphaeaceae and Potamogetonaceae families are examples of this category.
(d) **Emergent macrophytes**
Emergent macrophytes grew along the banks of rivers, marshes, and lakes. These plants remain rooted in the soil of the water bodies, and their leaves, stems, and flowers rise above the water's surface. These plants grow in shallow water, derive nutrition solely from the soil, and depend on aerial reproduction (Shay and Shay 1986). Examples of this category include *Typha* sp., *Lythrum* sp., and *Phragmites* sp. Figure 9.2 shows the various forms of aquatic macrophytes.

Aquatic macrophytes are known for their strong influence on the microclimate and biogeochemical processes occurring in littoral zones of marine ecosystems and sediment dynamics of freshwater systems (Cott et al. 2008). They also serve as the most effective carbon sinks and play an important role in carbon sequestration (Nag et al. 2023). Apart from these, aquatic macrophytes play an important role in the area of environmental cleanup. Remediation through these plants is considered as an

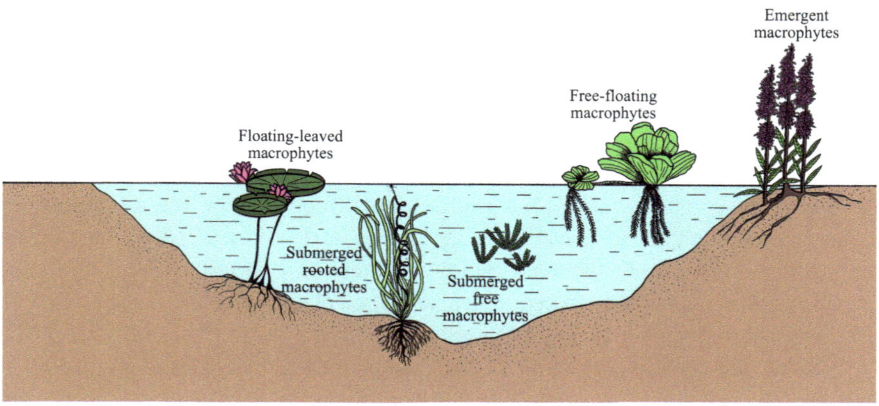

Fig. 9.2 Various types of aquatic macrophytes

easy, cost-effective, eco-friendly, and energy-efficient method of decontamination (Abma et al. 2010). These macrophytes help to remove organic and inorganic impurities as well as engineered nanoparticles from contaminated water. Their self-purification utilities confirm the maintenance of water quality. Aquatic macrophytes are the key component of wetland systems also, contributing much of the total ecosystem biomass (Iamchaturapatr et al. 2007). Understanding the functions of aquatic macrophytes in wetland systems is critical for understanding the elementary processes of the ecosystem and associated issues, such as restoration of ecological integrity in the wetland ecosystem, wastewater treatment, and management of hostile invasive/alien species (Fletcher et al. 2020).

9.5 Utilization of Aquatic Macrophytes for Remediation of Organic Contaminants

Macrophytes have been recognized as ecosystem engineers, which play an important role in the treatment of contaminated water bodies via sediment stabilization, absorption, extraction, sequestration, filtration, degradation, etc. of contaminants (organic/inorganic) (Kumar et al. 2022). In general, macrophytes that could be used for phytoremediation should have several qualities, including (1) the ability to store the contaminant species, (2) the ability to handle high concentrations of the contaminant, (3) fast growth with high biomass, (4) highly branched root system, (5) the ability to be harvested easily, and (6) not being able to be eaten by humans or animals (Arthur et al. 2022; Vymazal 2010, 2013). The following section highlights a few of the species which have been applied successfully for remediation purposes.

9.5.1 *Typha angustifolia* (Narrowleaf Cattail) and *Typha latifolia* (Broaleaf Cattail)

T. angustifolia belongs to the genus *Typha* which is found in a diverse number of wetlands in the Northern Hemisphere. It is an obligate wetland species, meaning they almost always occur in wetlands under natural conditions. Commonly known as narrow-leaf cattail, it is the species utilized most frequently in free-water surface-constructed wetlands. For instance, a mesocosm scale-constructed wetlands were employed to investigate the remediation efficacy of *T. angustifolia* for volatile organic compounds contaminated water, and it displayed a significant decrease in the chemical oxygen demand (COD), total suspended solids (TSS), turbidity, NH_4-N, NO_3-N, and redox potential (Wahab Al-Baldawi et al. 2015). Lee et al. (2018) evaluated the capability of *Typha angustifolia* to uptake norethindrone, a progestin medication. The experiment was carried out in hydroponic solutions containing 0.5 and 2.0 mg/L of norethindrone and results revealed 90% removal within 21 days. The relative growth rates of *T. angustifolia* ranged from 1.821– to 2.589. The findings of this study indicated that *T. angustifolia* has a great capacity to adapt and mitigate the toxicity of norethindrone. Recently, Lei et al. (2022) investigated the removal of 11 micropollutants, including personal care products and pharmaceutical compounds by *T. angustifolia* along with two other aquatic macrophytes (*Phragmites australis and Juncus effuses*) at two different temperatures simulating low- and high-temperature conditions. All the three species successfully removed all the target contaminants within 21 days. In particular, the uptake of propranolol was found to be highest for *Phragmites australis*. Besides, the removal efficacy of the *T. angustifolia and J. Effuses* were observed to be significant at both the temperature conditions signifying their suitability for practical applications.

Another species, *T. latifolia*, commonly known as broadleaf cattail, has been used effectively to remove organic contaminants from the environment. For instance, Moore et al. (2013) reported significant removal of atrazine, diazinon, and permethrin from simulated agricultural runoff water. Similarly, excellent attenuation of pesticides, namely imazalil, and tebuconazole, has been reported in the study conducted by Lv et al. (2016a), wherein the plant species were observed to not only uptake the pesticides from the solution but also metabolize them. The removal of antimicrobials, that is, ciprofloxacin and sulfamethoxazole were investigated with *T. latifolia L.* and *Panicum virgatum L.* (Adesanya et al. 2021). The plant uptake of both contaminants was studied at 5 and 10 μg/L, and it was noticed that the accumulation of ciprofloxacin and sulfamethoxazole is significantly higher in *T. latifolia L.* The high values of bioconcentration factor in *T. latifolia L.* further confirmed that *T. latifolia L.* could be utilized as a phytoremediation agent for the treatment of water matrices contaminated with antimicrobials. In a recent study, Pérez et al. (2022) tested the accumulation of three model contaminants, namely carbamazepine, fluoxetine, and gemfibrozil (representing pharmaceutical and personal care products) in *T. latifolia* after chronic exposure and established its relationship with transpired water and physicochemical properties. All the contaminants showed different rates of accumulation in the tissues of the plant wherein the contaminants mass in roots and rhizome showed a positive correlation with lipophilicity and distribution coefficient, while in leaves, it was correlated with the water transpirated and ionization behavior.

9.5.2 *Eichhornia crassipes* (Water Hyacinth)

Eichhornia crassipes is a fast growing photosynthetic macrophyte, which has the ability to grow rapidly under stressful conditions even (Auchterlonie et al. 2021). This aquatic weed can double itself in 5–15 days, classifying this otherwise useful species as "invasive" all across the United States, Europe, and Asia (Dersseh et al. 2019a, b). While high tolerance and growth are the desired factors in phytoremediation, uncontrolled growth can result in more of a bane than a boon. Various studies have shown that the plant is capable of extracting contaminants from water, thereby providing a possibility for phytoremediation. For instance, *E. crassipes* harvested from surface water was found to contain few pharmaceuticals (nonsteroidal anti-inflammatory drugs and antiretroviral medicines), which were thought to be extracted from contaminated water by plants' roots (Mlunguza et al. 2020a, b; Amos Sibeko et al. 2019). Similarly, Mlunguza et al. (2020a) noticed high amount of antiretroviral drugs (ARD) ranging from 7.4 to 29.0 g kg^{-1} in plant and roots of *E. crassipes*. The uptake of drugs (nonsteroidal anti-inflammatory) by *E. crassipes* roots, followed by their translocation into different plant parts such as leaves and stems, have also been reported in studies of Mlunguza et al. (2020b) and Amos Sibeko et al. (2019). Recently, Deng et al. (2022) employed *E. crassipes* for the removal of ciprofloxacin antibiotic and reported ~84% removal with major accumulation in the root tissues. Besides, plant absorption, hydrolysis, and microbial degradation were identified as secondary removal mechanisms. *E. crassipes* has been employed not only for pharmaceuticals, but extensive research has been conducted on the removal of dyes also. For instance, Sharma et al. (2021) observed color removal rates ranging from 79 to 90.8% for cationic dyes and 33.3 to 62.8% for anionic dyes. In another study by Victor et al. (2016), 82.5% of chemical oxygen demand (COD) removal was noticed in textile wastewater.

9.5.3 *Pistia stratiotes* (Water Lettuce)

It is a hyperaccumulator plant species that has a significantly high potential to remove nutrients (organic and inorganic), suspended particles, heavy metals, etc. The phytoremediation ability of this plant can decrease toxic levels of wastewater and contribute to water quality improvement. The efficacy of *P. stratiotes* in eliminating contaminants, including organic pollutants, total phosphorus (TP), and nitrogen (NH_3-N), from livestock wastewater was investigated for the first time by Chen et al. (2014). In low quantities of livestock wastewater, the plant displayed better removal efficiency. The remediation efficiency of plant for various initial concentrations of livestock wastewater within 8 days was COD (68–82%) > NH_3-N

(57–69%) > TP (27–45%). Water lettuce has also been used to remove pesticides. For instance, Prasertsup and Ariyakanon (2011) investigated chlorpyrifos removal abilities of *P. stratiotes* L. and reported maximum chlorpyrifos elimination (100%) within 4–5 days. From the above examples, it is evident that water lettuce not only removes COD, NH_3, and TP but can also remove pesticides such as chlorpyrifos from the wastewater.

9.5.4 *Azolla* (Water Fern)

Azolla is the only pteridophyte known for its symbiotic association with a diazotrophic cyanobacterium called *Anabaena azollae* which helps in fixing atmospheric nitrogen. This ability makes it easier for *Azolla* to grow in N-deficient sites. Along with rapid growth, high tolerance, and high biomass makes it a more promising candidate for phytoremediation compared to other macrophytes. *Azolla rongpong* was investigated for its efficacy to remove dyes such as, acid green 3, acid red 88, acid blue 15, and acid orange 7 from an aqueous solution (Padmesh et al. 2006). The maximum removal of 83.33 mg/g was noticed for acid green 3 (AG3) at optimum conditions of temperature (30 °C) and pH (2.5). Similarly, Zazouli et al. (2014b) attempted to remove acid blue 15 dye (AB15) from an aqueous solution using *Azolla,* and they observed 98% AB15 removal, at a pH of 3, contact time of 90 min, an absorbent dose of 10 g/L, and AB15 concentration of 10 ppm.

9.5.5 *Vallisneria spiralis* (Eelgrass)

Vallisneria spiralis is a submerged macrophyte that grows in almost all light and water conditions. Commonly considered as an aquarium plant, it entered other countries due to the aquarium trade. Because of its ability to form runners, *V. spiralis* can propagate easily, labeling itself as invasive in some countries. However, it has been known to show phytoremediation abilities in some studies. For instance, Yuan et al. (2014) tested the removal of polycyclic aromatic hydrocarbons (PAHs) using *V. spiralis L.* PAHs are known for their toxic and carcinogenic properties, which are often released into the atmosphere due to human activities like automobile emissions and can end up in water bodies via atmospheric deposition. The study concluded that oxygen released by the roots of *V. spiralis* caused an increase in the growth of PAHs-degrading bacteria which in turn assisted microbial degradation of PAHs. The synergistic relationship between the roots of *V. spiralis* and microbes present in the rhizosphere was also established in another study wherein the remediation potential of *V. spiralis* was investigated in sediments spiked with 80–800 mg/kg of phenanthrene (Yan et al. 2011). Recently, Yan et al. (2022) investigated the microbial communities along rhizosphere and non-rhizosphere sediments of *V. natans* in the sites contaminated with high concentrations of PAHs and its subsequent rhizodegradation. The study revealed that *V. natans* significantly influence the sediment microbial communities that might result in various kinds of degradation mechanisms between rhizosphere and non-rhizosphere. In a mesocosm study, *V. spiralis* displayed great sorption capacities for fluoroquinolone antibiotics such as ciprofloxacin, though it triggered an extended effect on the epiphytic biofilm microbial communities (Ohore et al. 2021).

9.5.6 *Potamogeton crispus* (Curlyleaf Pondweed)

Commonly known as the curly pondweed or curled pondweed, *P. crispus* is a widespread, lowland aquatic macrophyte with turions and creeping root stalks as its significant physical features. A case study in the Anzali wetland by Norouznia and Hamidian (2014) described the phytoremediation potential of *P. crispus*, which is a common aquatic macrophyte native to the Anzali wetland. Meng and Chi (2015) conducted an experiment for 54-days using four plant densities to determine their influence on the remediation of soil sediments contaminated with PAHs. The results demonstrated an increase in plant density and a decrease in plant growth rate. At the end of the 54-day trial, the plant increased the dissipation ratios of pyrene and phenanthrene in soil sediments by 6.5% to 26.2% and 0.5% to 13.4%, respectively. Sediment organic matter has been reported to have a strong influence on PAHs dissipation enhancements during the remediation of PAHs with *P. crispus*. In a 36-day experiment, plant enhanced the dissipation ratios of PAHs model compounds, while sediment organic matter displayed an inverse relationship. The results suggested that the increase in the bioavailability by *P. crispus* is the main factor responsible for enhanced PAHs dissipation (Liu et al. 2018). Trueman and Erber (2013) compared *P. illinoensis*, an invasive species with *P. crispus*, native to Illinois, for the removal of estrogenic compounds and Bisphenol A (BPA) and observed that *P. illinoensis* accumulates higher levels of estrogens and BPA as compared to *P. crispus*. The authors suggested that though invasive species are undesirable, they could be considered for improving the health of contaminated ecosystems. Atrazine, a chlorinated triazine systemic herbicide, has also been reported to be degraded to diaminochlorotriazine and hydroxy atrazine by *P. crispus* in a 60-day experiment suggesting the remediation potential of *P. crispus* for organochlorine pesticides (Li et al. 2019). Table 9.1 lists a few other aquatic macrophytes which have been applied for the removal of organic contaminants.

9.6 Aquatic Macrophytes and Artificial Wetlands

In recent years, artificial wetland systems have emerged as an economical, effective, and environment-friendly solution for aquatic bodies having heavy loads of toxic, hazardous contaminants. The most important component of an artificial wetland is the selection of plant species that could effectively remove the mixture of contaminants and, at the same time, could tolerate heavy loads of toxic contaminants and environmental pressures (Zhang et al. 2010; Vymazal 2010). These wetlands

Table 9.1 List of aquatic macrophytes used for remediation of organic contaminants

Image	Common name(s)	Scientific name	Macrophyte type	Contaminants removed/ wastewater treated	Reference
	Water lettuce, water cabbage, Nile cabbage	*Pistia stratiotes*	Free-floating aquatic macrophyte	Organic pollutants, total phosphorus (TP), and nitrogen (NH_3-N), BOD, and COD	Chen et al. (2014), Prasertsup and Ariyakanon (2011)
	Water Fern, feathered mosquito fern, water velvet	*Azolla pinnata*	Floating aquatic fern	Dyes, paper mill effluent, poultry farm wastewater, etc.	Kooh et al. (2019), Maldonado et al. (2022), Sivakumar et al. (2017), Jangwattana and Iwai (2010)
	Eelweed, eelgrass, tape grass	*Vallisneria spiralis*	Perennial submerged macrophyte	Polycyclic aromatic hydrocarbons (PAHs), fluoroquinolone antibiotics, sulfonamides	Qin et al. (2019), Zhang et al. (2019), Zhu et al. (2020)
	Curly pondweed, crisp-leaved pondweed, curled pondweed	*Potamogeton crispus*	Submerged macrophyte	Polycyclic aromatic hydrocarbons (PAHs), phenanthrene, pyrene, dibutyl phthalate, phthalic acid esters	Liu et al. (2018), Chi and Cai (2012), Chi and Yang (2012), Meng and Chi (2015, 2017)
	Broadleaf cattail, common cattail, bulrush, broadleaf reedmace	*Typha latifolia*	Perennial herbaceous emergent plant	Norethindrone, brewery wastewater, pesticide residues	Lee et al. (2018), Gebeyehu et al. (2018), Tolcha et al. (2020)
	Purple loosestrife, spiked loosestrife, purple Lythrum	*Lythrum salicaria*	Perennial, emergent, flowering plant	Pharmaceuticals, pesticides, herbicides, fungicides, total phosphorus (TP), total nitrogen (TN), chemical oxygen demand, and ammonia nitrogen (NH_4^+–N)	Xiao et al. (2021), Brunhoferova et al. (2021)

	Common water hyacinth, floating water hyacinth	*Pontederia crassipes* (formerly *Eichhornia crassipes*)	Free-floating aquatic macrophyte	Phenols, sulfachloropyridazine, fluoroquinolones, tetracycline, organic dyes, chlorpyrifos	Anudechakul et al. (2015), Madikizela (2021), Lima and Asencios (2021), Kamilya et al. (2022)
	Mosquito fern, duckweed fern, fairy moss, water fern	*Azolla filiculoides*	Free-floating freshwater fern	Azo dye, bisphenol A, pyrocatechol, sewage water, dairy wastewater	Khataee et al. (2013), Zazouli et al. (2014a, b), Rezooqi et al. (2021), Sundararaman et al. (2021)
	Hornwort, rigid hornwort, coontail, or coon's tail	*Ceratophyllum demersum*	Submerged, free-floating macrophyte	Perfluoroalkyl acids (PFASs), dyes, endocrine disruptors, BOD, COD and sewage/municipal wastewater	Foroughi et al. (2010), Kulasekaran et al. (2014), Ewadh (2020), Li et al. (2021), Al-Nabhan and Al-Abbawy (2021), Zhao et al. (2021)
	Common duckmeat, greater duckweed, great duckmeat	*Spirodela polyrhiza*	Free-floating macrophyte	Chlorobenzenes, organochlorine compounds, 4-tert-butylphenol, pharmaceuticals, endocrine disrupting chemicals, pesticides	Dosnon-Olette et al. (2011), Ogata et al. (2013), Reis et al. (2014), Li et al. (2017), Dhir (2020)
	American elodea, common elodea, Anacharis, Canada waterweed	*Elodea densa*	Submerged freshwater perennial macrophyte	DDT, hexachloroethane, phenanthracene, chloropyrifos, municipal wastewater (BOD, COD, TOC)	Dhir et al. (2009), Białowiec et al. (2019), Kumar et al. (2022)
	Eurasian watermilfoil or spiked watermilfoil	*Myriophyllum spicatum*	Submerged macrophyte	Perchlorate/simazine, DDT, dyes, endocrine disruptors	Keskinkan and Lugal Göksu (2007), Dhir et al. (2009), Zhao et al. (2021)

(continued)

Table 9.1 (continued)

Image	Common name(s)	Scientific name	Macrophyte type	Contaminants removed/ wastewater treated	Reference
	Duckweed	*Lemna Minor*	Free-floating macrophyte	Chlorinated phenols, fluorinated phenols, herbicides (isoproturon and glyphosate), pharmaceutical compounds	Day and Saunders (2004), Reinhold and Saunders (2006a, b), Tront et al. (2007), Dosnon-Olette et al. (2011), Radulović et al. (2020), Maldonado et al. (2022)
	Common reed	*Phragmites australis*	Emergent macrophyte	Chiral pesticides, organophosphate pesticides, herbicides, dyes, veterinary pharmaceuticals	Carvalho et al. (2012), Kankılıç et al. (2016), Lv et al. (2017), Echeverri González et al. (2019), Olisah et al. (2021)
	Soft rush, common rush, bog rush, or mat rush	*Juncus effuses*	Emergent macrophyte	Chiral pesticides, herbicide, triazole pesticide, personal care products, pharmaceutical compounds, dyes, wastewater (BOD, COD)	Bouldin et al. (2006), Liu et al. (2013), Lv et al. (2016a, b), Lyu et al. (2018), Vo et al. (2019), Bebba et al. (2019)

have specific dimensions, locations, substrates, hydraulic conditions, and retention times. In comparison to conventional treatment methods, artificial wetlands have several advantages, which include (a) the use of natural remedial agents (plants and microorganisms), (b) less energy intensive, (c) low operation and maintenance costs, (d) use of renewable energy, (e) self-sustenance, etc. (Shutes 2001; Vymazal 2010, 2013; Parde et al. 2021).

9.6.1 Types of Artificial Wetlands

9.6.1.1 Constructed Wetlands (CWs)

A constructed Wetland (CW) is a treatment system that mimics and improves the performance of naturally existing wetland's purification processes (Kaur et al. 2020b). Aquatic plants (such as reeds and duckweed), naturally occurring microorganisms, and a filter bed (often made of soil, sand, and/or gravel) are used in this system (Vymazal 2010). They have been widely employed for pollutant removal due to various advantages that CWs have over conventional wastewater (WW) treatment methods, including simple, eco-friendly, cost-effective, and robust processes with minimal operational costs (Vymazal 2013). CWs have been employed on a modest scale for urban WW treatment as well as to reduce wastewater contamination caused by runoff from the agricultural fields. It is characterized as a land-based wastewater treatment system that improves water quality by utilizing natural treatment processes like soil, plants, water, and microbes (Brix 1994). In CWs, physical, chemical, and biological processes, including absorption, volatilization, sedimentation, filtration, oxidation-reduction, precipitation, chelation, microbial degradation, photodegradation, etc., can all occur concurrently, resulting in the efficient removal of various pollutants from wastewater (Kochi et al. 2020). CWs have shown significant results for the removal of contaminants such as hormones, pesticides, veterinary medicines, plasticizers, surfactants, flame retardants, PPCPs, pharmaceuticals, and various industrial compounds, in addition to the removal of organic matter, nutrients, suspended solids, and metals (Gorito et al. 2018). The plants absorb and translocate various organic contaminants from wastewater generally via diffusion process, as there are no specialized transporters found in plant roots (Zhang et al. 2023).

9.6.1.2 Types of Constructed Wetlands (CWs)

Based on the direction of the flow of the water to be treated, CWs have been categorized as follows:

(a) **Superficial/surface flow (SF) CWs**
Superficial/surface flow CWs appear much similar to a natural wetland wherein varied types of aquatic macrophytes (floating, emergent, or submerged) are planted which remain rooted in a shallow layer of the submerged substrate. Herein, the surface of wastewater remains above the substrate. In contrast to the surface layer, the bottom layers of water and the substrate remains anaerobic.

The system remains insulated at the bottom to avoid any infiltration of water. Its low capital and operational cost are some of the benefits of SF CWs over other types of CWs (Choudhary et al. 2011). These types of CWs are generally employed when the flow of wastewater is highly unpredictable (Kadlec and Wallace 2008). Besides, it has been applied for the remediation of acid mine drainage and runoff from the agriculture (Choudhary et al. 2011).

(b) **Subsurface flow (SSF) CWs**

Subsurface flow CWs have been further classified as horizontal and vertical subsurface flow CWs. In horizontal subsurface flow (HSSF) CWs, wastewater flows below the substrate, while in vertical subsurface flow (VSSF) CWs, it remains above the substrate. Both systems comprise of a porous substrate layer (soil, gravels, and rocks) with different types of aquatic macrophytes (Swarnakar et al. 2022). The wastewater flows across the substrate, where it encounters the communities of microbes present in the rhizosphere, which further assists in carrying out the degradation of the contaminants (Luederitz et al. 2001). Horizontal subsurface flow CWs have been applied successfully in many countries such as Denmark, Australia, the United Kingdom, North America, Asia, and Africa for treating effluents of varied industries such as textile, paper and pulp, leather, food, tanneries, and distilleries (Vymazal 2010). HSSF CWs have also been applied for removing contaminants from landfill leachates, fish farms, airport and highway runoffs, etc. (Comeau et al. 2001; Karrh et al. 1999; Schulz et al. 2003; Revitt et al. 2004; Wojciechowska and Obarska-Pempkowiak 2008). VSSF CWs are considered as more efficient than horizontal subsurface flow CWs and also require less land space; however, their operation and maintenance cost is comparatively higher (Cooper 1999; Weedon 2001). These systems have been used for the remediation of domestic sewage and municipal wastewater (Brix and Arias 2005; Molle et al. 2005) and industrial effluents (Veenstra 1998; Kern and Idler 1999; Aslam et al. 2007).

(c) **Hybrid flow CWs**

Hybrid flow CWs are a combination of horizontal and vertical subsurface flow CWs. These systems are meant to significantly improve the remediation efficacy of wastewater in comparison to single CWs (Vymazal 2010). This system overcomes the setbacks of individual CWs systems improving the overall efficacy of CWs for the removal of a wide array of contaminants (Swarnakar et al. 2022). Generally, these systems require large land areas for construction and skilled manpower for maintenance and operation. Hybrid flow CWs are operational in many countries for the effective treatment of sewage, leachates, and industrial wastewater (Lin et al. 2003; Kinsley et al. 2007; Haydar et al. 2020; Fernandez-Fernandez et al. 2020). Figure 9.3 shows all the three types of CWs.

9.6.1.3 Performance of CWs for Organic Contaminants

In a pilot-scale horizontal subsurface flow (HSSF) CWs planted with *P. australis*, a continuous injection experiment was carried out to assess the removal of three pharmaceuticals (ibuprofen (IB), diclofenac (DCF), naproxen (NPX), and two

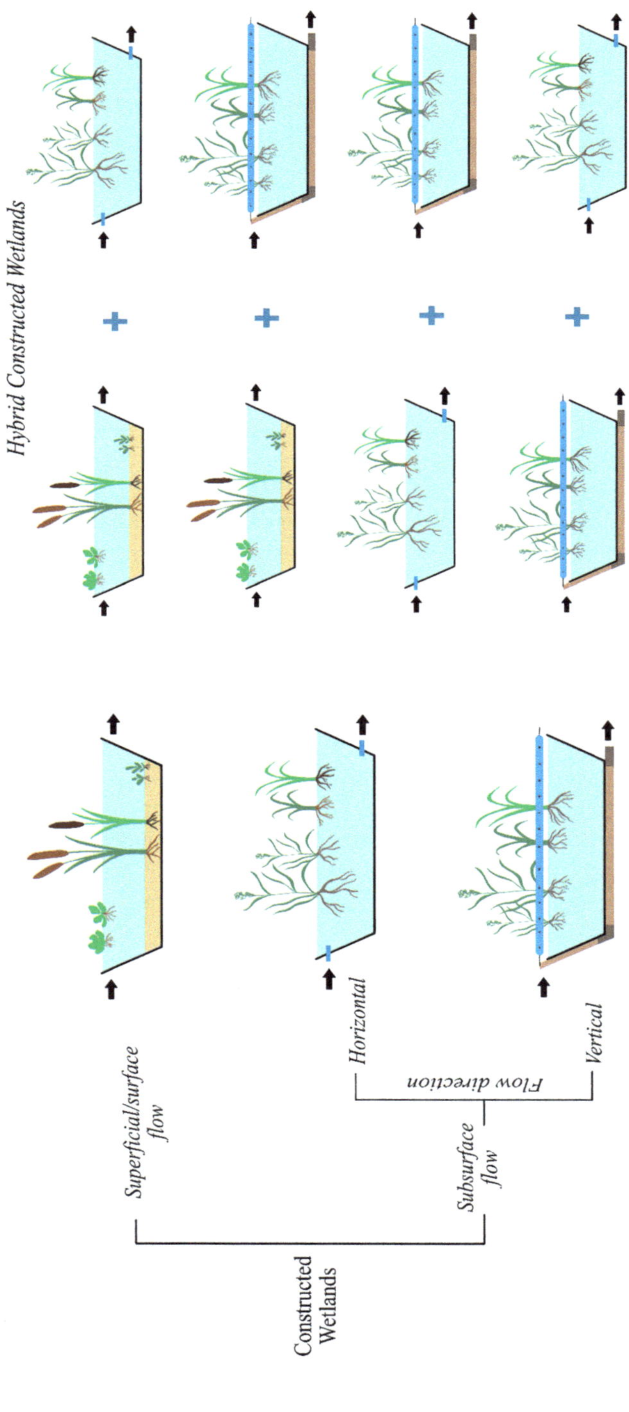

Fig. 9.3 Types of constructed wetlands

PCPs, that is, tonalide and bisphenol A (BPA) (Ávila et al. 2010) from municipal wastewater. As an initial treatment, an anaerobic reactor was used, followed by two small size wetlands (B1 and B2) working parallelly and were connected to another bigger wetland (B3). The removal percentage of the system for the target contaminants ranged from 97% to 99%, and sorption and aerobic/anaerobic biodegradation was identified as the major mechanism behind the removal. In the case of ibuprofen, aerobic conditions favored the removal (50% in smaller HSSF CWs and 99% in B3), while naproxen and diclofenac showed greater removal in anaerobic conditions (93% in smaller HSSF CWs). In the case of tonalide and BPA removal (94% and 83%, respectively), their sorption over suspended particles was recognized as the primary mechanism. Besides, aerobic biodegradation was also suggested as an important mechanism in BPA removal based on the intermediate degradation products identified during the study. Chen et al. (2016a, b) investigated the ability of *Thalia dealbata* and *Iris tectorum* to remove 8 antibiotics (erythromycin, monensin, sulfamethoxazole (SMX), leucomycin, azithromycin, trimethoprim, clarithromycin, sulfamethazine, and sulphapyridine), and 12 antibiotic resistance genes (ARGs) from domestic wastewater in CWs with three types of flow, that is, SF, HSSF, and VSSF. The concentrations of the antibiotics and ARGs were reported to be reduced by 75.8–98.6% and 63.9–84%, respectively. Among varied flow types CWs, SSF displayed better removal efficiency, and biodegradation was revealed as the major removal mechanism. Other removal mechanisms included adsorption over the surface of the substrate and plant uptake.

Salcedo et al. (2018) investigated the removal of alkylphenols (AFs), atrazine (ATZ), methyldihydrojasmonate (MDHJ), galaxolide (GAL), and caffeine (CAF) in horizontal subsurface flow CWs having three kinds of macrophytes, viz., *Typha latifolia*, *Cyperus papyrus*, and *Phragmites australis*, two types of substrates (volcanic gravel and river gravel), and varied hydraulic retention times (HRT: 1, 3, and 5 days). The results revealed that the employed CWs could remove AFs, ATZ, MDHJ, GAL, and CAF, with removal efficiencies of 79%, 56%, 70%, 95%, and 79%, respectively. At 5 days HRT, ATZ showed the highest removal with 98.35% in CWs with tezontle which is a volcanic gravel and *Typha latifolia*, while GAL displayed the minimum removal efficiency of around 21.62% in a river gravel unplanted system. In another study, Gorito et al. (2018) investigated the removal of micropollutants (MPs) such as 2-ethylhexyl-4-methoxycinnamate (EHMC), isoproturon, ATZ, erythromycin, clarithromycin, perfluorooctanesulfonic acid (PFOS), fluoxetine, and norfluoxetine from a freshwater aquaculture effluent in a VSSF CWs having *Phragmites australis*. By spiking MPs @100 ng/L to the same matrix, a broader multicomponent set of 36 MPs were reported to be removed from VSSF CWs was also studied simultaneously and in both cases, all MPs except EHMC showed remediation efficiency greater than 87%.

Xiao et al. (2021) studied different combinations of floating-leaved, submerged, and emergent aquatic macrophytes, namely *Nymphaea L*, *Vallisneria natans* (Lour.) *Hara*, *Lythrum salicaria L*, *Hydrilla verticillata*, and *Myriophyllum spicatum* to simulate structural plant communities in CWs and compared their phytoremediation abilities. Four pollutants, viz., TP, TN, COD, and ammonia nitrogen (NH_4^+–N), were studied under different concentration levels. The combinations of the macrophytes displayed varied abilities toward contaminant removal. For instance, the LNH combination showed the highest removal of COD, while LNV displayed the highest TP and TN removal. The study concluded that the combination of *Nymphaea L, Lythrum salicaria L., and Vallisneria natans* (Lour.) *Hara* was the best among other combinations tried and could be potentially applied for the restoration of the river ecosystem. The effectiveness of horizontal (HSF) and vertical subsurface flow (VSF) CWs was investigated for the removal of six emerging pollutants from the wastewater, which included diethyl phthalate (DEP), di-n-octyl phthalate (DNOP), bis(2-ethylxexyl) phthalate (DEHP), di-isobutyl phthalate (DIBP), tris(1-chloro-2-propyl) phosphate (TCPP), and caffeine (CAF) (Gikas et al. 2021). Three HSF and three VFS CWs were employed with two types of plants (*P. australis* and *T. latifolia*), two HRT, and two wastewater feeding techniques. The removal rate of phthalate esters was found to be relatively higher in VSF CWs, though HSF CWs performance was quite satisfactory. Furthermore, HRT influenced the removal in HSF CWs, but no effect was noticed in VSF CWs, while plant types made no significant differences in removal rates in both cases. It was further revealed that adsorption of the target contaminants onto the substrate and biodegradation was the major mechanism behind the remediation of the phthalates in CWs. Recently, SMX and dimethyl phthalate (DMP) removal was studied in vertical flow CWs with biochar, zeolite, vermiculite, peat, and sand as substrate material (Li et al. 2023). The system showed synergistic removal of both the contaminants along with total nitrogen with physicochemical adsorption as the major mechanism, and intraparticle diffusion was noticed as the controlling factor.

9.6.1.4 Floating Wetlands

Floating treatment wetlands (FTWs) represent another variant of CWs that has gained significant attention in recent years. They have been employed for the treatment of stormwater and wastewater, eliminating varied organic and inorganic pollutants (Li and Katul 2020; Colares et al. 2020). These systems consist of artificial floating platforms and aquatic macrophytes. The platforms are generally made up of buoyant materials such as polyethylene, polypropylene, polyurethane or polyvinyl alcohol foam, coconut coir bed, bamboo, etc. The macrophytes are cultivated hydroponically on these buoyant platforms, which suspend the plant shoots above the water's surface and roots below it. The aquatic macrophytes which are employed have a dense network of roots and rhizomes that hangs down in the water, and the plants get the required nutrients from the water (Nichols et al. 2016). The root network of the plants provides a large surface area for the growth of microorganisms, forming a slimy layer of biofilms that serves as a biologically active surface supporting different physical and biochemical processes responsible for contaminant removal (Walker et al. 2017). The biofilm provides a habitat for a number of bacterial communities and is essential for removing nutrients like nitrogen and phosphorus through nitrification, denitrification, adsorption, etc. (Lucke et al. 2019). Besides, it also impedes the turbulence, water flow, increases sedimentation, and traps and filters suspended particles (Shahid et al. 2018). The macrophytes not only provide the biofilm more space to grow upon, but they also release oxygen through their roots during daytime which in turn affects the redox potential in the water column, determines how nitrogen changes, how certain phytotoxins are broken down, and how microorganisms break down organic matter in an aerobic environment. Root exudates are the organic compounds that plants release in addition to oxygen. These compounds influence biological processes like denitrification (Shahid et al. 2018). Roots extend down into the water column as the plants grow, where they can intercept and remediate toxins or act as a substrate for microorganisms that can remediate them (or a combination of both). Plant selection for FTWs is heavily influenced by two factors: how well the roots disseminate throughout the water column and how actively the leaves dissipate the polluted water. Some advantages of FTWs over subsurface CWs are that the plant roots stay in constant contact with the wastewater and that there is no need for media (like sand or gravel) throughout the system. No requirement of substrate eliminates the risk of substrate clogging, which is a frequent problem in CWs, and also makes FTWs more economical.

Several studies investigated the removal of organic contaminants from wastewater in FTWs using a variety of aquatic macrophytes. For instance, a large-scale FTWs was installed in the city of Faisalabad, Pakistan, that treated a mix of industrial effluents and urban wastewater (40% and 60%, respectively) (Afzal et al. 2019). The macrophytes which were used included *Phragmites australis*, *Leptochloa fusca*, *Brachiaria mutica*, *Canna indica*, *Typha domingensis*, and *Rosa indica*. The study was conducted over 3 years, and results revealed a significant reduction in BOD (88%), COD (79%), TDS (65%), and heavy metals. The cost for treating 60 million m^3/year was estimated to be US$0.00026/$m^3$. In another study of Chakwal, Pakistan, FTWs were employed in the field to clean up crude oil-contaminated water in an oil exploration pit using *Phragmites australis, Typha domingensis, Leptochloa fusca, and Brachiaria mutica* along with hydrocarbon-degrading bacteria (Afzal et al. 2019). After 18 months of the experiment, >97% reduction was seen in BOD, COD, and hydrocarbons, and TDS and heavy metals showed >80% reduction. In another study, the effect of bacterial augmentation in FTWs planted with *P. australis* was investigated for remediation of diesel-contaminated water. The experiments having both plant and bacteria displayed a highest reduction in hydrocarbons (95.8%), BOD (97.7%), COD (98.6%), TOC (95.2%), phenol (98.9%), and toxicity along with an increase in plant growth. The study suggested that the augmentation of hydrocarbon-degrading bacteria could be a promising solution for the treatment of diesel-contaminated water matrices. Hwang et al. (2020) investigated the feasibility of using FTWs seeded with *Canna flaccida* to remove two pharmaceuticals (carbamazepine CBZ and acetaminophen APAP) and one herbicide (atrazine, ATZ) from contaminated water. In 378 L mesocosms, FTWs with varied plant densities were prepared and studied for the removal of pharmaceuticals over a 12-week period. The ability of the planted FTWs to remove the contaminants varied. The plant density showed no effect on APAP and ATZ dissipation; however, a significant influence was observed in the case of CBZ. All the residues of APAP were removed in 2 weeks, while the residues of ATZ were removed within 12 weeks. In the case of

CBZ, only 29–36.7% removal was observed due to FTWs. Similarly, Hwang et al. (2021) investigated the efficiency of *Acorus gramineus* and *Canna hybrida* for the removal of APAP, ATZ, perfluorooctanoic acid (PFOA), CBZ, 17β-estradiol (E2), and SMX, in FTW mesocosm for 17 weeks. Among all the contaminants, APAP and E2 showed the fastest removal in both the plants, while ATZ and SMX showed complete removal in *Canna hybrida*-treated FTWs, and in *Acorus gramineus* treated FTWs, removal up to 87.6% and 97.1% was noticed, respectively. CBZ, on the contrary, showed ~82% removal in both the plants' treatment and PFOA displayed the lowest removal. The study concluded that *Canna* sp. could be a potential candidate for remediating organic contaminants in FTWs.

9.7 Conclusion

This chapter illustrates that aquatic macrophytes are efficient remediation agents which could provide a sustainable solution for water quality improvement and ecosystem restoration. All the forms of aquatic macrophytes, that is, emergent free-floating and submerged types, have shown great phytoremediation abilities for dyes, pharmaceutical compounds, personal care products, etc. Aquatic macrophytes have been employed efficiently in artificial wetland systems and used for the remediation of sewage, industrial effluents, domestic wastewater, agricultural runoff, urban and leachates, etc., reducing heavy loads of biochemical oxygen demand, chemical oxygen demand, suspended solids, total nitrogen, total phosphorus, and toxic organic compounds. The remediation efficiency of both types of artificial wetlands, that is, constructed wetlands and floating wetlands, have been demonstrated in numerous studies. Several physicochemical and biological processes act simultaneously in these wetland structures, assisting the removal of hazardous contaminants. Besides the plant species, the type of substrate/buoyant material applied and hydraulic retention time are the major factors controlling the contaminant removal rates. Furthermore, the interaction between plant and microbe plays a crucial role in the removal of target contaminants in artificial wetlands; hence, in-depth studies should be conducted further to improve their potential for the remediation of recalcitrant emerging contaminants.

References

Abbas N, Butt MT, Ahmad MM, Deeba F, Hussain N (2021) Phytoremediation potential of Typha latifolia and water hyacinth for removal of heavy metals from industrial wastewater. Chem Int 7(2):103–111

Abma WR, Driessen W, Haarhuis R, Van Loosdrecht MCM (2010) Upgrading of sewage treatment plant by sustainable and cost-effective separate treatment of industrial wastewater. Water Sci Technol 61(7):1715–1722

Adesanya T, Zvomuya F, Farenhorst A (2021) Phytoextraction of ciprofloxacin and sulfamethoxaxole by cattail and switchgrass. Chemosphere 279:130534

Afzal M, Arslan M, Müller JA, Shabir G, Islam E, Tahseen R, Khan QM (2019) Floating treatment wetlands as a suitable option for large-scale wastewater treatment. Nat Sustain 2(9):863–871
Ahmed J, Thakur A, Goyal A (2021) Industrial wastewater and its toxic effects. In Biological treatment of industrial wastewater, 2021, pp. 1–14
Akhtar N, Syakir Ishak MI, Bhawani SA, Umar K (2021) Various natural and anthropogenic factors responsible for water quality degradation: a review. Water 13(19):2660
Al-Nabhan EAM, Al-Abbawy DA (2021) Improving wastewater quality by using Ceratophyllum demersum L. IOP Conf Ser Earth Environ Sci 910(1):012086
Alharbi OM, Khattab RA, Ali I (2018) Health and environmental effects of persistent organic pollutants. J Mol Liq 263:442–453
Alkorta I, Garbisu C (2001) Phytoremediation of organic contaminants in soils. Bioresour Technol 79(3):273–276
Ammeri RW, Hassen W, Hidri Y, Di Rauso Simeone G, Hassen A (2022) Macrophyte and indigenous bacterial co-remediation process for pentachlorophenol removal from wastewater. Int J Phytoremediation 24(3):271–282
Amos Sibeko P, Naicker D, Mdluli PS, Madikizela LM (2019) Naproxen, ibuprofen, and diclofenac residues in river water, sediments and Eichhornia crassipes of Mbokodweni river in South Africa: an initial screening. Environ Forensic 20(2):129–138
Anand S, Bharti SK, Dviwedi N, Barman SC, Kumar N (2017) Macrophytes for the reclamation of degraded waterbodies with potential for bioenergy production. Phytoremediation potential of bioenergy plants, pp. 333–351
Anning AK, Korsah PE, Addo-Fordjour P (2013) Phytoremediation of wastewater with Limnocharis flava, Thalia geniculata and Typha latifolia in constructed wetlands. Int J Phytoremediation 15(5):452–464
Anudechakul C, Vangnai AS, Ariyakanon N (2015) Removal of chlorpyrifos by water hyacinth (Eichhornia crassipes) and the role of a plant-associated bacterium. Int J Phytoremediation 17(7):678–685
Arthur GD, Aremu AO, Kulkarni MG, Stirk WA, Davies TC, Penaye J, Van Staden J (2022) Phytoremediation potential of Beta vulgaris L.(Swiss chard) using soil from the vicinity of Kette-Batouri Goldmine (Eastern). S Afr J Bot 151:713–719
Aslam MM, Malik M, Baig MA, Qazi IA, Iqbal J (2007) Treatment performances of compost-based and gravel-based vertical flow wetlands operated identically for refinery wastewater treatment in Pakistan. Ecol Eng 30(1):34–42
Auchterlonie J, Eden CL, Sheridan C (2021) The phytoremediation potential of water hyacinth: a case study from Hartbeespoort dam, South Africa. South Afr J Chem Eng 37:31–36
Ávila C, Pedescoll A, Matamoros V, Bayona JM, García J (2010) Capacity of a horizontal subsurface flow constructed wetland system for the removal of emerging pollutants: an injection experiment. Chemosphere 81(9):1137–1142
Avino P, Russo MV (2018) A comprehensive review of analytical methods for determining persistent organic pollutants in air, soil, water and waste. Curr Org Chem 22(10):939–953
Bebba AA, Labed I, Zeghdi S, Messaitfa A (2019) Purification performance of Typha latifolia, Juncus effusus and Papyrus cyperus in arid climate: influence of seasonal variation. J Water Chem Technol 41(6):396–401
Bertrin V, Boutry S, Jan G, Ducasse G, Grigoletto F, Ribaudo C (2016) Effects of wind-induced sediment resuspension on distribution and morphological traits of aquatic weeds in shallow lakes. J Limnol 76:84–96
Białowiec A, Sobieraj K, Pilarski G, Manczarski P (2019) The oxygen transfer capacity of submerged plant Elodea densa in wastewater constructed wetlands. Water 11(3):575
Bonanno G, Cirelli GL (2017) Comparative analysis of element concentrations and translocation in three wetland congener plants: Typha domingensis, Typha latifolia and Typha angustifolia. Ecotoxicol Environ Saf 143:92–101

Bouldin JL, Farris JL, Moore MT, Smith S Jr, Cooper CM (2006) Hydroponic uptake of atrazine and lambdacyhalothrin in Juncus effusus and Ludwigia peploides. Chemosphere 65(6):1049–1057
Brix H (1994) Functions of macrophytes in constructed wetlands. Water Sci Technol 29(4):71–78
Brix H, Arias CA (2005) The use of vertical flow constructed wetlands for on-site treatment of domestic wastewater: New Danish guidelines. Ecol Eng 25(5):491–500
Brunet J, Repellin A, Varrault G, Terryn N, Zuily-Fodil Y (2008) Lead accumulation in the roots of grass pea (Lathyrus sativus L.): a novel plant for phytoremediation systems? C R Biol 331(11): 859–864
Brunhoferova H, Venditti S, Schlienz M, Hansen J (2021) Removal of 27 micropollutants by selected wetland macrophytes in hydroponic conditions. Chemosphere 281:130980
Carvalho PN, Basto MCP, Almeida CMR (2012) Potential of Phragmites australis for the removal of veterinary pharmaceuticals from aquatic media. Bioresour Technol 116:497–501
Centers for Disease Control and Prevention (2009) Arsenic factsheet [Fact sheet]. https://www.cdc.gov/biomonitoring/pdf/arsenic_factsheet.pdf
Chandra R, Saxena G, Kumar V (2015) Phytoremediation of environmental pollutants: an eco-sustainable green technology to environmental management. Advances in biodegradation and bioremediation of industrial waste, 1–29
Chen J, Nie Q, Zhang Y, Hu J, Qing L (2014) Eco-physiological characteristics of Pistia stratiotes and its removal of pollutants from livestock wastewater. Water Sci Technol 69(12):2510–2518
Chen J, Ying GG, Wei XD, Liu YS, Liu SS, Hu LX et al (2016a) Removal of antibiotics and antibiotic resistance genes from domestic sewage by constructed wetlands: effect of flow configuration and plant species. Sci Total Environ 571:974–982
Chen Y, Vymazal J, Březinová T, Koželuh M, Kule L, Huang J, Chen Z (2016b) Occurrence, removal and environmental risk assessment of pharmaceuticals and personal care products in rural wastewater treatment wetlands. Sci Total Environ 566:1660–1669
Chi J, Cai X (2012) Effects of nitrogen on the removal of dibutyl phthalate from surface water in the presence of Potamogeton crispus L. Ecol Eng 41:70–73
Chi J, Yang Q (2012) Effects of Potamogeton crispus L. on the fate of phthalic acid esters in an aquatic microcosm. Water Res 46(8):2570–2578
Choi KR, Ahn YJ, Lee SY (2022) Bacterial conversion of CO2 to organic compounds. J CO2 Utilizat 58:101929
Choudhary AK, Kumar S, Sharma C (2011) Constructed wetlands: an approach for wastewater treatment. Elixir Pollut 37(8):3666–3672
Colares GS, Dell'Osbel N, Wiesel PG, Oliveira GA, Lemos PHZ, da Silva FP et al (2020) Floating treatment wetlands: a review and bibliometric analysis. Sci Total Environ 714:136776
Comeau Y, Brisson J, Réville JP, Forget C, Drizo A (2001) Phosphorus removal from trout farm effluents by constructed wetlands. Water Sci Technol 44(11–12):55–60
Connor R, Renata A, Ortigara C, Koncagül E, Uhlenbrook S, Lamizana-Diallo BM, et al. (2017) The united nations world water development report 2017. Wastewater: the untapped resource. The United Nations World Water Development Report
Cooper P (2005) The performance of vertical flow constructed wetland systems with special reference to the significance of oxygen transfer and hydraulic loading rates. Water Sci Technol 51(9):81–90
Cott PA, Sibley PK, Somers WM, Lilly MR, Gordon AM (2008) A review of water level fluctuations on aquatic biota with an emphasis on fishes in ice-Covered Lakes 1. J Am Water Resour Assoc 44(2):343–359
Cunningham SD, Berti WR (1993) Remediation of contaminated soils with green plants: an overview. In Vitro Cell Dev Biol Plant 29:207–212
Day JA, Saunders FM (2004) Glycosidation of chlorophenols by Lemna minor. Environ Toxicol Chem Int J 23(3):613–620
Deblonde T, Cossu-Leguille C, Hartemann P (2011) Emerging pollutants in wastewater: a review of the literature. Int J Hyg Environ Health 214(6):442–448

Deng Y, Qian X, Wu Y, Ma T, Xu X, Li J, Yan Y (2022) Effects of ciprofloxacin on Eichhornia crassipes phytoremediation performance and physiology under hydroponic conditions. Environ Sci Pollut Res 29(31):47363–47372

Dersseh MG, Kibret AA, Tilahun SA, Worqlul AW, Moges MA, Dagnew DC, Melesse AM (2019a) Potential of water hyacinth infestation on lake tana, Ethiopia: a prediction using a GIS-based multi-criteria technique. Water 11(9):1921

Dersseh MG, Melesse AM, Tilahun SA, Abate M, Dagnew DC (2019b) Water hyacinth: review of its impacts on hydrology and ecosystem services—lessons for management of Lake Tana. Extreme hydrology and climate variability, 237–251

Dhir B (2013) Phytoremediation: role of aquatic plants in environmental clean-up, vol 14. Springer, New Delhi, pp 1–111

Dhir B (2020) Green technologies for the removal of agrochemicals by aquatic plants. In: Agrochemicals detection, treatment and remediation. Butterworth-Heinemann, pp 569–591

Dhir B, Sharmila P, Saradhi PP (2009) Potential of aquatic macrophytes for removing contaminants from the environment. Crit Rev Environ Sci Technol 39(9):754–781

Dodds WK, Whiles MR (2010) Microbes and plants. Freshwater ecology, pp. 185–219. https://doi.org/10.1016/B978-0-12-374724-2.00009-X

Dos Santos Utmazian MN, Wenzel WW (2007) Cadmium and zinc accumulation in willow and poplar species grown on polluted soils. J Plant Nutr Soil Sci 170(2):265–272

Dosnon-Olette R, Couderchet M, Oturan MA, Oturan N, Eullaffroy P (2011) Potential use of Lemna minor for the phytoremediation of isoproturon and glyphosate. Int J Phytoremediation 13(6):601–612

Du Y, Xu X, Liu Q, Bai L, Hang K, Wang D (2022) Identification of organic pollutants with potential ecological and health risks in aquatic environments: Progress and challenges. Sci Total Environ 806:150691

Dushenkov S (2003) Trends in phytoremediation of radionuclides. Plant Soil 249(1):167–175

Espinosa-Ortiz EJ, Rene ER, Gerlach R (2022) Potential use of fungal-bacterial co-cultures for the removal of organic pollutants. Crit Rev Biotechnol 42(3):361–383

Etim EE (2012) Phytoremediation and its mechanisms: a review. Int J Environ Bioenergy 2(3): 120–136

Ewadh HM (2020) Removal of methylene blue by coontail (Ceratophyllum demersum) using phytoremediation concept. Plant Arch 20(1):2677–2681

Fahid M, Arslan M, Shabir G, Younus S, Yasmeen T, Rizwan M et al (2020) Phragmites australis in combination with hydrocarbons degrading bacteria is a suitable option for remediation of diesel-contaminated water in floating wetlands. Chemosphere 240:124890

Fernandez-Fernandez MI, Vega PMDL, Jaramillo-Morán MA, Garrido M (2020) Hybrid constructed wetland to improve organic matter and nutrient removal. Water 12(7):2023

Fletcher J, Willby N, Oliver D., Quilliam RS (2020) Phytoremediation using aquatic plants. Phytoremediation: In-situ Applications, 205–260

Foroughi M, Najafi P, Toghiani A, Honarjoo N (2010) Analysis of pollution removal from wastewater by Ceratophyllum demersum. Afr J Biotech 9(14):2125–2128

Fulekar MH, Jadia CD (2008) Phytoremediation: the application of vermicompost to remove zinc, cadmium, copper, nickel and lead by sunflower plant. Environ Eng Manag J 7(5):547

Gebeyehu A, Shebeshe N, Kloos H, Belay S (2018) Suitability of nutrients removal from brewery wastewater using a hydroponic technology with Typha latifolia. BMC Biotechnol 18:1–13

Ghosh M, Singh SP (2005) A review on phytoremediation of heavy metals and utilization of it's by products. Appl Ecol Environ Res 3(1):1–18

Gikas GD, Papaevangelou VA, Tsihrintzis VA, Antonopoulou M, Konstantinou IK (2021) Removal of emerging pollutants in horizontal subsurface flow and vertical flow pilot-scale constructed wetlands. PRO 9(12):2200

González AFE, Zúñiga-Benítez H, Peñuela GA (2019) Removal of herbicide 2, 4-D using constructed wetlands at pilot scale. Emerg Contamin 5:303–307

Gorito AM, Ribeiro AR, Gomes CR, Almeida CMR, Silva AM (2018) Constructed wetland microcosms for the removal of organic micropollutants from freshwater aquaculture effluents. Sci Total Environ 644:1171–1180

Guo W, Pan B, Sakkiah S, Yavas G, Ge W, Zou W et al (2019) Persistent organic pollutants in food: contamination sources, health effects and detection methods. Int J Environ Res Public Health 16(22):4361

Haripriyan U, Gopinath KP, Arun J, Govarthanan M (2022) Bioremediation of organic pollutants: a mini review on current and critical strategies for wastewater treatment. Arch Microbiol 204(5): 1–9

Haydar S, Anis M, Afaq M (2020) Performance evaluation of hybrid constructed wetlands for the treatment of municipal wastewater in developing countries. Chin J Chem Eng 28(6):1717–1724

Hwang JI, Li Z, Andreacchio N, Ordonez Hinz F, Wilson PC (2020) Potential use of floating treatment wetlands established with Canna flaccida for removing organic contaminants from surface water. Int J Phytoremediation 22(12):1304–1312

Hwang JI, Hinz FO, Albano JP, Wilson PC (2021) Enhanced dissipation of trace level organic contaminants by floating treatment wetlands established with two macrophyte species: a mesocosm study. Chemosphere 267:129159

Iamchaturapatr J, Yi SW, Rhee JS (2007) Nutrient removals by 21 aquatic plants for vertical free surface-flow (VFS) constructed wetland. Ecol Eng 29(3):287–293

Idris MO, Kim HC, Yaqoob AA, Ibrahim MNM (2022) Exploring the effectiveness of microbial fuel cell for the degradation of organic pollutants coupled with bio-energy generation. Sustain Energy Technol and Assess 52:102183

Jadia CD, Fulekar MH (2009) Phytoremediation of heavy metals: recent techniques. Afr J Biotechnol 8(6)

Jangwattana R, Iwai CB (2010) Using Azolla pinnata for wastewater treatment from poultry farm. IJERD 1(2):23

Jaswal R, Bedi A, Bedi I, Jaiswar A, Jasrotia RS (2022) Phytoremediation of soil and water. In Phytoremediation. Academic Press, Boston, MA, pp 239–262

Ji G, Sun T, Zhou Q, Sui X, Chang S, Li P (2002) Constructed subsurface flow wetland for treating heavy oil-produced water of the Liaohe oilfield in China. Ecol Eng 18(4):459–465

Jones KC, De Voogt P (1999) Persistent organic pollutants (POPs): state of the science. Environ Pollut 100(1–3):209–221

Justin LD, Olukanni DO, Babaremu KO (2022) Performance assessment of local aquatic macrophytes for domestic wastewater treatment in Nigerian communities: a review. Heliyon: e10093

Kadlec RH, Wallace S (2008) Treatment wetlands. CRC Press, Boca Raton, FL

Kamilya T, Yadav MK, Ayoob S, Tripathy S, Bhatnagar A, Gupta AK (2022) Emerging impacts of steroids and antibiotics on the environment and their remediation using constructed wetlands: a critical review. Chem Eng J 451:138759

Kankılıç GB, Metin AÜ, Tüzün İ (2016) Phragmites australis: an alternative biosorbent for basic dye removal. Ecol Eng 86:85–94

Karami A, Shamsuddin ZH (2010) Phytoremediation of heavy metals with several efficiency enhancer methods. Afr J Biotechnol 9(25):3689–3698

Karen DJ, Joab BM, Wallin JM, Johnson KA (1998) Partitioning of chlorpyrifos between water and an aquatic macrophyte (Elodea densa). Chemosphere 37(8):1579–1586

Karić N, Maia AS, Teodorović A, Atanasova N, Langergraber G, Crini G et al (2022) Bio-waste valorisation: agricultural wastes as biosorbents for removal of (in) organic pollutants in wastewater treatment. Chem Eng J Adv 9:100239

Karrh JD, Moriarty J, Kornuc JJ, Knight RL (1999) Sustainable management of aircraft anti/de-icing process effluents using a subsurface-flow treatment wetland. In: Wetlands & Remediation II: Second International Conference on Wetlands & Remediation, pp 187–195

Kaur K, Badru R, Singh PP, Kaushal S (2020a) Photodegradation of organic pollutants using heterojunctions: a review. J Environ Chem Eng 8(2):103666

Kaur R, Talan A, Tiwari B, Pilli S, Sellamuthu B, Tyagi RD (2020b) Constructed wetlands for the removal of organic micro-pollutants. In Current developments in biotechnology and bioengineering. Elsevier, Amsterdam, pp 87–140

Kern J, Idler C (1999) Treatment of domestic and agricultural wastewater by reed bed systems. Ecol Eng 12(1–2):13–25

Keskinkan OLCAYTO, Lugal Göksu MZ (2007) Assessment of the dye removal capability of submersed aquatic plants in a laboratory-scale wetland system using ANOVA. Braz J Chem Eng 24:193–202

Khataee AR, Movafeghi A, Vafaei F, Salehi Lisar SY, Zarei M (2013) Potential of the aquatic fern Azolla filiculoides in biodegradation of an azo dye: modeling of experimental results by artificial neural networks. Int J Phytoremediation 15(8):729–742

Khodadad Hosseini E, Rabbani M, Mooraki N, Derakhshi P (2022) Evaluation of living Azolla filiculoides performance for the removal of total nitrogen, phosphorus, sodium, potassium, COD, BOD, and TDS from dairy waste: full factorial design. Iran J Chem Chem Eng 41(4): 1151–1161

Kinsley CB, Crolla AM, Kuyucak N, Zimmer M, Lafleche A (2007) Nitrogen dynamics in a constructed wetland system treating landfill leachate. Water Sci Technol 56(3):151–158

Kitamura RSA, Brito JCM, Silva de Assis HC, Gomes MP (2022) Physiological responses and phytoremediation capacity of floating and submerged aquatic macrophytes exposed to ciprofloxacin. Environ Sci Pollut Res 30(1):622–639

Klink A (2017) A comparison of trace metal bioaccumulation and distribution in Typha latifolia and Phragmites australis: implication for phytoremediation. Environ Sci Pollut Res 24(4): 3843–3852

Kochi LY, Freitas PL, Maranho LT, Juneau P, Gomes MP (2020) Aquatic macrophytes in constructed wetlands: a fight against water pollution. Sustainability 12(21):9202

Kooh MRR, Dahri MK, Lim LB, Lim LH, Lee SL (2019) Phytoextraction capability of Azolla pinnata in the removal of rhodamine B from aqueous solution: artificial neural network and random forests approaches. Appl Water Sci 9:1–9

Krithiga T, Sathish S, Renita AA, Prabu D, Lokesh S, Geetha R, Sillanpaa M (2022) Persistent organic pollutants in water resources: fate, occurrence, characterization and risk analysis. Sci Total Environ 831:154808

Kulasekaran A, Gopal A, John Alexander J (2014) A study on the removal efficiency of organic load and some nutrients from sewage by Ceratophyllum demersum. L J Mater Environ Sci 5(3): 859–864

Kumar S, Thakur N, Singh AK, Gudade BA, Ghimire D, Das S (2022) Aquatic macrophytes for environmental pollution control. In Phytoremediation Technology for the Removal of Heavy Metals and Other Contaminants from Soil and Water, pp. 291–308

Lasat MM, Pence NS, Garvin DF, Ebbs SD, Kochian LV (2000) Molecular physiology of zinc transport in the Zn hyperaccumulator Thlaspi caerulescens. J Exp Bot 51(342):71–79

Lee JY, Vairappan CS, Saibeh K (2018) Potential of Typha angustifolia L. in removing norethindrone from water. Transac Sci Technol 5(2):58–67

Lei Y, Rijnaarts H, Langenhoff A (2022) Mesocosm constructed wetlands to remove micropollutants from wastewater treatment plant effluent: effect of matrices and pre-treatments. Chemosphere 305:135306

Lei Y, Carlucci L, Rijnaarts H, Langenhoff A (2023) Phytoremediation of micropollutants by Phragmites australis, Typha angustifolia, and Juncus effuses. Int J Phytoremediation 25(1): 82–88

Li S, Katul G (2020) Contaminant removal efficiency of floating treatment wetlands. Environ Res Lett 15(10):1040b7

Li J, Zhou Q, Campos LC (2017) Removal of selected emerging PPCP compounds using greater duckweed (Spirodela polyrhiza) based lab-scale free water constructed wetland. Water Res 126:252–261

Li H, Qu M, Lu X, Chen Z, Guan S, Du H, Zhu D (2019) Evaluation of the potential of Potamogeton crispus and Potamogeton crispus on phytoremediation of atrazine. Int J Environ Anal Chem 99(3):243–257

Li XQ, Hua ZL, Wu JY, Gu L (2021) Removal of perfluoroalkyl acids (PFAAs) in constructed wetlands: considerable contributions of submerged macrophytes and the microbial community. Water Res 197:117080

Li G, Huang Z, Li H, Zhang Z, Cui L (2023) Synergistic removal of sulfamethoxazole and dimethyl phthalate by five constructed wetland substrates. Chemosphere 318:137936

Lima HDP, Asencios YJ (2021) Eichhornia crassipes (Mart.) Solms (natural or carbonized) as biosorbent to remove pollutants in water. SN Appl Sci 3:1–18

Lin YF, Jing SR, Lee DY (2003) The potential use of constructed wetlands in a recirculating aquaculture system for shrimp culture. Environ Pollut 123(1):107–113

Liu F, Xiao R, Wang Y, Li Y, Zhang S, Luo Q, Wu J (2013) Effect of a novel constructed drainage ditch on the phosphorus sorption capacity of ditch soils in an agricultural headwater catchment in subtropical central China. Ecol Eng 58:69–76

Liu S, Meng F, Ding Z, Chi J (2018) Phytoremediation of PAH-contaminated sediments with different organic matter contents by Potamogeton crispus L. Int J Phytoremediation 20(13): 1317–1323

Lucke T, Walker C, Beecham S (2019) Experimental designs of field-based constructed floating wetland studies: a review. Sci Total Environ 660:199–208

Luederitz V, Eckert E, Lange-Weber M, Lange A, Gersberg RM (2001) Nutrient removal efficiency and resource economics of vertical flow and horizontal flow constructed wetlands. Ecol Eng 18(2):157–171

Lv T, Zhang Y, Casas ME, Carvalho PN, Arias CA, Bester K, Brix H (2016a) Phytoremediation of imazalil and tebuconazole by four emergent wetland plant species in hydroponic medium. Chemosphere 148:459–466

Lv T, Zhang Y, Zhang L, Carvalho PN, Arias CA, Brix H (2016b) Removal of the pesticides imazalil and tebuconazole in saturated constructed wetland mesocosms. Water Res 91:126–136

Lv T, Carvalho PN, Casas ME, Bollmann UE, Arias CA, Brix H, Bester K (2017) Enantioselective uptake, translocation and degradation of the chiral pesticides tebuconazole and imazalil by Phragmites australis. Environ Pollut 229:362–370

Lyu T, Zhang L, Xu X, Arias CA, Brix H, Carvalho PN (2018) Removal of the pesticide tebuconazole in constructed wetlands: design comparison, influencing factors and modelling. Environ Pollut 233:71–80

Madikizela LM (2021) Removal of organic pollutants in water using water hyacinth (Eichhornia crassipes). J Environ Manag 295:113153

Maldonado I, Moreno EG, Vilca FZ (2022) Application of duckweed (Lemna sp.) and water fern (Azolla sp.) in the removal of pharmaceutical residues in water: state of art focus on antibiotics. Sci Total Environ 838:156565

Meng F, Chi J (2015) Phytoremediation of PAH-contaminated sediments by Potamogeton crispus L. with four plant densities. Transac Tianjin Univ 21(5):440–445

Meng F, Chi J (2017) Effect of Potamogeton crispus L. on bioavailability and biodegradation activity of pyrene in aged and unaged sediments. J Hazard Mater 324:391–397

Mishra S, Maiti A (2017) The efficiency of Eichhornia crassipes in the removal of organic and inorganic pollutants from wastewater: a review. Environ Sci Pollut Res 24:7921–7937

Mlunguza NY, Ncube S, Mahlambi NP, Chimuka L, Madikizela LM (2020a) Determination of selected antiretroviral drugs in wastewater, surface water and aquatic plants using hollow fibre liquid phase microextraction and liquid chromatography - tandem mass spectrometry. J Mater 382:121067

Mlunguza NY, Ncube S, Mahlambi PN, Chimuka L, Madikizela LM (2020b) Optimization and application of hollow fiber liquid-phase microextraction and microwave-assisted extraction for

the analysis of non-steroidal anti-inflammatory drugs in aqueous and plant samples. Environ Monit Assess 192(8):557. https://doi.org/10.1007/s10661-020-08527-4
Molle P, Liénard A, Boutin C, Merlin G, Iwema A (2005) How to treat raw sewage with constructed wetlands: an overview of the French systems. Water Sci Technol 51(9):11–21
Moore MT, Tyler HL, Locke MA (2013) Aqueous pesticide mitigation efficiency of Typha latifolia (L.), Leersia oryzoides (L.) Sw., and Sparganium americanum Nutt. Chemosphere 92(10): 1307–1313
Moore MT, Locke MA, Kröger R (2017) Mitigation of atrazine, S-metolachlor, and diazinon using common emergent aquatic vegetation. J Environ Sci 56:114–121
Mueller JT, Gasteyer S (2021) The widespread and unjust drinking water and clean water crisis in the United States. Nat Commun 12(1):1–8
Nag SK, Ghosh BD, Sarkar UK, Das BK (2023) An appraisal of carbon capture and sequestration in few selected wetlands of West Bengal. Environ Dev Sustain:1–16
Ngo HH, Guo W, Zhang J, Liang S, Ton-That C, Zhang X (2015) Typical low cost biosorbents for adsorptive removal of specific organic pollutants from water. Bioresour Technol 182:353–363
Nichols P, Lucke T, Drapper D, Walker C (2016) Performance evaluation of a floating treatment wetland in an urban catchment. Water 8(6):244
Norouznia H, Hamidian AH (2014) Phytoremediation efficiency of pondweed (Potamogeton crispus) in removing heavy metals (Cu, Cr, Pb, As and Cd) from water of Anzali wetland. Int J Aquat Biol 2(4):206–214
Ogata Y, Toyama T, Yu N, Wang X, Sei K, Ike M (2013) Occurrence of 4-tert-butylphenol (4-t-BP) biodegradation in an aquatic sample caused by the presence of Spirodela polyrrhiza and isolation of a 4-t-BP-utilizing bacterium. Biodegradation 24:191–202
Ohore OE, Zhang S, Guo S, Addo FG, Manirakiza B, Zhang W (2021) Ciprofloxacin increased abundance of antibiotic resistance genes and shaped microbial community in epiphytic biofilm on Vallisneria spiralis in mesocosmic wetland. Bioresour Technol 323:124574
Okoro HK, Pandey S, Ogunkunle CO, Ngila CJ, Zvinowanda C, Jimoh I, Adeniyi AG (2022) Nanomaterial-based biosorbents: adsorbent for efficient removal of selected organic pollutants from industrial wastewater. Emerg Contamin 8:46–58
Okunowo W, Ogunkanmi LA (2010) Phytoremediation potential of some heavy metals by water hyacinth. Int J Biol Chem Sci 4(2)
Olisah C, Rubidge G, Human LR, Adams JB (2021) A translocation analysis of organophosphate pesticides between surface water, sediments and tissues of common reed Phragmites australis. Chemosphere 284:131380
Padmesh TVN, Vijayaraghavan K, Sekaran G, Velan M (2006) Application of Azolla rongpong on biosorption of acid red 88, acid green 3, acid orange 7 and acid blue 15 from synthetic solutions. Chem Eng J 122(1–2):55–63
Parde D, Patwa A, Shukla A, Vijay R, Killedar DJ, Kumar R (2021) A review of constructed wetland on type, treatment and technology of wastewater. Environ Technol Innovat 21:101261
Parveen S, Bhat IU, Khanam Z, Rak AE, Yusoff HM, Akhter MS (2022) Phytoremediation: in situ alternative for pollutant removal from contaminated natural media: a brief review. Biointerf Res Appl Chem
Pérez DJ, Lombardero LR, Franco MDR, Doucette WJ (2022) Differential bioaccumulation patterns of pharmaceutical and personal care products in Typha Latifolia (cattail) as function of physicochemical properties and transpiration rate. SSRN 4112742
Pivetz BE (2001) Ground water issue: phytoremediation of contaminated soil and ground water at hazardous waste sites. Environ Protect Act 2001:1–36
Pollard SJ, Hough RL, Kim KH, Bellarby J, Paton G, Semple KT, Coulon F (2008) Fugacity modelling to predict the distribution of organic contaminants in the soil, oil matrix of constructed biopiles. Chemosphere 71(8):1432–1439
Prasertsup P, Ariyakanon N (2011) Removal of chlorpyrifos by water lettuce (Pistia stratiotes L.) and duckweed (Lemna minor L.). Int J Phytoremediation 13(4):383–395

Qin Z, Zhao Z, Adam A, Li Y, Chen D, Mela SM, Li H (2019) The dissipation and risk alleviation mechanism of PAHs and nitrogen in constructed wetlands: the role of submerged macrophytes and their biofilms-leaves. Environ Int 131:104940

Qing Li Q, Loganath A, Seng Chong Y, Tan J, Philip Obbard J (2006) Persistent organic pollutants and adverse health effects in humans. J Toxic Environ Health A 69(21):1987–2005

Radulović O, Stanković S, Uzelac B, Tadić V, Trifunović-Momčilov M, Lozo J, Marković M (2020) Phenol removal capacity of the common duckweed (Lemna minor L.) and six phenol-resistant bacterial strains from its rhizosphere: in vitro evaluation at high phenol concentrations. Plants 9(5):599

Rahman MA, Hasegawa H (2011) Aquatic arsenic: phytoremediation using floating macrophytes. Chemosphere 83(5):633–646

Rana V, Maiti SK (2018) Municipal wastewater treatment potential and metal accumulation strategies of Colocasia esculenta (L.) Schott and Typha latifolia L. in a constructed wetland. Environ Monit Assess 190(6):1–15

Rehman F, Pervez A, Khattak BN, Ahmad R (2017) Constructed wetlands: perspectives of the oxygen released in the rhizosphere of macrophytes. Clean Soil Air Water 45(1)

Reinhold DM, Saunders FM (2006a) Development and application of oxygen production rate assessment to uptake of fluorinated organics by Lemna minor. Environ Toxicol Chem Int J 25(11):3039–3045

Reinhold DM, Saunders FM (2006b) Phytoremediation of fluorinated agrochemicals by duckweed. Trans ASABE 49(6):2077–2083

Reis ARD, Tabei K, Sakakibara Y (2014) Oxidation mechanism and overall removal rates of endocrine disrupting chemicals by aquatic plants. J Haz Mat 265:79–88

Rejmankova E (2011) The role of macrophytes in wetland ecosystems. J Ecol Environ 34(4): 333–345

Revitt DM, Shutes RBE, Jones RH, Forshaw M, Winter B (2004) The performances of vegetative treatment systems for highway runoff during dry and wet conditions. Sci Total Environ 334: 261–270

Rezooqi AM, Mouhamad RS, Jasim KA (2021) The potential of Azolla filiculoides for in vitro phytoremediation of wastewater. J Phys Conf Ser 1853(1):012014

Rizwana M, Darshan M, Nilesh D (2014) Phytoremediation of textile waste water using potential wetland plant: eco sustainable approach. Int J Interdiscipl Multidiscipl Stud 1(4):130–138

Salcedo JJP, Montes GHE, Frómeta AEN, Osorio AC, Negrete JM (2018) Removal of organic micropollutants from riverine waters using constructed wetlands: a mesocosms experiment. Int J Appl Eng Res 13:15740–15748

Salt DE, Smith RD, Raskin I (1998) Phytoremediation. Annu Rev Plant Biol 49(1):643–668

Sand-Jensen K, Prahl C, Stokholm H (1982) Oxygen release from roots of submerged aquatic macrophytes. Oikos 38:349–354

Schneider S, Melzer A (2004) Sediment and water nutrient characteristics in patches of submerged macrophytes in running waters. Hydrobiologia 527(1):195–207

Schulz C, Gelbrecht J, Rennert B (2003) Treatment of rainbow trout farm effluents in constructed wetland with emergent plants and subsurface horizontal water flow. Aquaculture 217(1–4): 207–221

Schwarzenbach RP, Egli T, Hofstetter TB, Von Gunten U, Wehrli B (2010) Global water pollution and human health. Annu Rev Environ Resour 35(1):109–136

Shahid MJ, Arslan M, Ali S, Siddique M, Afzal M (2018) Floating wetlands: a sustainable tool for wastewater treatment. Clean Soil Air Water 46(10):1800120

Sharma R, Saini H, Paul DR, Chaudhary S, Nehra SP (2021) Removal of organic dyes from wastewater using Eichhornia crassipes: a potential phytoremediation option. Environ Sci Pollut Res 28:7116–7122

Shay JM, Shay CT (1986) Prairie marshes in western Canada, with specific reference to the ecology of five emergent macrophytes. Can J Bot 64(2):443–454

Shutes RBE (2001) Artificial wetlands and water quality improvement. Environ Int 26(5–6): 441–447
Singh N, Balomajumder C (2021) Phytoremediation potential of water hyacinth (Eichhornia crassipes) for phenol and cyanide elimination from synthetic/simulated wastewater. Appl Water Sci 11(8):1–15
Sivakumar D, Shankar D, Durai Arasu T, Ragothaman VM (2017) Paper mill pollutants reduction-Azolla pinnata. Environ Sci Indian J 13(3):138
Sood A, Uniyal PL, Prasanna R, Ahluwalia AS (2012) Phytoremediation potential of aquatic macrophyte, Azolla. Ambio 41(2):122–137
Sudarsan J, Annadurai R, Kumar KS, Nithiyanantham S (2018) Heavy metal removal using different parts of Typha latifolia. J Bionanosci 12(1):87–91
Suman J, Uhlik O, Viktorova J, Macek T (2018) Phytoextraction of heavy metals: a promising tool for clean-up of polluted environment? Front Plant Sci 9:1476
Sundararaman S, Senthil Kumar P, Deivasigamani P, Jagadeesan AK, Devaerakkam M, Al-Hashimi A, Choi D (2021) Assessing the plant phytoremediation efficacy for Azolla filiculoides in the treatment of textile effluent and redemption of Congo red dye onto Azolla biomass. Sustainability 13(17):9588
Susarla S, Medina VF, McCutcheon SC (2002) Phytoremediation: an ecological solution to organic chemical contamination. Ecol Eng 18(5):647–658
Swarnakar AK, Bajpai S, Ahmad I (2022) Various types of constructed wetland for wastewater treatment-a review. In IOP Conference Series: Earth and Environmental Science (Vol. 1032, No. 1, 012026). IOP Publishing
Tenaw D, Assfaw A (2022) Households willingness to pay for improved urban water supply in Dire Dawa city administration: the role of socio-economic factors and water supply-related perceptions. Sustain Water Resour Manag 8(1):1–12
Tolcha T, Gemechu T, Megersa N (2020) Flower of Typha latifolia as a low-cost adsorbent for quantitative uptake of multiclass pesticide residues from contaminated waters. S Afr J Chem 73: 22–29
Tront JM, Reinhold DM, Bragg AW, Saunders FM (2007) Uptake of halogenated phenols by aquatic plants. J Environ Eng 133(10):955–961
Trueman RJ, Erber L (2013) Invasive species may offer advanced phytoremediation of endocrine disrupting chemicals in aquatic ecosystems. Emir J Food Agric:648–656
United States Environmental Protection Agency (1998) Introduction to phytoremediation, EPA 600/R-99/107
United States Environmental Protection Agency (2001) Citizen's Guide to Phytoremediation. Retrieved 4 January, 2011, from http://www.clu-in.org/download/citizens/citphyto.pdf
Veenstra S (1998) The Netherlands. In: Vymazal J, Brix H, Cooper PF, Green MB, Haberl R (eds) Constructed wetlands for wastewater treatment in Europe. Backhuys Publishers, Leiden, pp 289–314
Victor KK, Séka Y, Norbert KK, Sanogo TA, Celestin AB (2016) Phytoremediation of wastewater toxicity using water hyacinth (Eichhornia crassipes) and water lettuce (Pistia stratiotes). Int J Phytoremediation 18(10):949–955
Vishnoi SR, Srivastava PN (2007) Phytoremediation–green for environmental clean. In Proceedings of Taal 2007: the 12th World lake conference, Vol. 1016, p. 1021
Vo HNP, Koottatep T, Chapagain SK, Panuvatvanich A, Polprasert C, TMH N, Nguyen NL (2019) Removal and monitoring acetaminophen-contaminated hospital wastewater by vertical flow constructed wetland and peroxidase enzymes. J Environ Manag 250:109526
Vymazal J (2010) Constructed wetlands for wastewater treatment. Water 2(3):530–549
Vymazal J (2013) Emergent plants used in free water surface constructed wetlands: a review. Ecol Eng 61:582–592
Wahab Al-Baldawi IA, Abdullah SRS, Suja F, Anuar N, Idris M (2015) Phytoremediation of contaminated ground water using Typha angustifolia. Water Pract Technol 10(3):616–624

Walker C, Tondera K, Lucke T (2017) Stormwater treatment evaluation of a constructed floating wetland after two years operation in an urban catchment. Sustainability 9(10):1687
Weedon CM (2001) Compact vertical flow reed beds: design rationale and early performance. IWA Macrophyt Newslett 23:12–20
Wersal RM, Madsen JD (2012) Aquatic plants their uses and risks. A review of the global status of aquatic plants. FAO, Rome
Wojciechowska E, Obarska-Pempkowiak H (2008) Performance of reed beds supplied with municipal landfill leachate. Wastewater treatment, plant dynamics and management in constructed and natural wetlands, 251–265
Xiao H, Peng S, Liu X, Jia J, Wang H (2021) Phytoremediation of nutrients and organic carbon from contaminated water by aquatic macrophytes and the physiological response. Environ Technol Innovat 21:101295
Yan ZS, Guo HY, Song TS, Hu Y, Jiang HL (2011) Tolerance and remedial function of rooted submersed macrophyte Vallisneria spiralis to phenanthrene in freshwater sediments. Ecol Eng 37(2):123–127
Yan H, Yan Z, Wang L, Hao Z, Huang J (2022) Toward understanding submersed macrophyte Vallisneria natans-microbe partnerships to improve remediation potential for PAH-contaminated sediment. J Hazard Mater 425:127767
Yanai J, Zhao FJ, McGrath SP, Kosaki T (2006) Effect of soil characteristics on Cd uptake by the hyperaccumulator Thlaspi caerulescens. Environ Pollut 139(1):167–175
Yang XE, Long X, Ni W, Fu C (2002) Sedum alfredii H: a new Zn hyperaccumulating plant first found in China. Chin Sci Bull 47(19):1634–1637
Yang C, Wang B, Wang H, He Z, Pi Y, Zhou J, Fu T (2022) Removal of organochlorine pesticides and metagenomic analysis by multi-stage constructed wetland treating landfill leachate. Chemosphere 301:134761
Yuan Y, Zeng G, Liang J, Li X, Li Z, Zhang C, Yu X (2014) Effects of landscape structure, habitat and human disturbance on birds: a case study in East Dongting Lake wetland. Ecol Eng 67:67–75
Zazouli MA, Yousefi Z, Yazdani-Charati J, Mahdavi Y (2014b) Application of Azolla Filiculoides biomass in acid black 1 dye adsorption from aqueous solution. Iran J Health Sci 2(3):24–32
Zazouli MA, Mahdavi Y, Bazrafshan E, Balarak D (2014a) Phytodegradation potential of bisphenolA from aqueous solution by Azolla Filiculoides. J Environ Health Sci Eng 12:1–5
Zhang Z, Gao X, Qiu B (2008) Detection of phytochelatins in the hyperaccumulator Sedum alfredii exposed to cadmium and lead. Phytochemistry 69(4):911–918
Zhang BY, Zheng JS, Sharp RG (2010) Phytoremediation in engineered wetlands: mechanisms and applications. Procedia Environ Sci 2:1315–1325
Zhang Y, Rajkumar MC, Freitas H (2016) Beneficial role of bacterial endophytes in heavy metal phytoremediation. J Environ Manag 174:14–25
Zhang L, Zhang Y, Liu L (2019) Effect of submerged macrophytes Vallisneria spiralis L. on restoring the sediment contaminated by enrofloxacin in aquaculture ponds. Ecol Eng 140: 105596
Zhang Y, Deng J, Qin B, Zhu G, Zhang Y, Jeppesen E, Tong Y (2022) Importance and vulnerability of lakes and reservoirs supporting drinking water in China. Fundament Res 3:265
Zhang H, Wang XC, Zheng Y, Dzakpasu M (2023) Removal of pharmaceutical active compounds in wastewater by constructed wetlands: performance and mechanisms. J Environ Manag 325: 116478
Zhao C, Zhang G, Jiang J (2021) Enhanced phytoremediation of bisphenol a in polluted lake water by seedlings of Ceratophyllum demersum and Myriophyllum spicatum from in vitro culture. Int J Environ Res Public Health 18(2):810
Zhu L, Xu H, Xiao W, Lu J, Lu D, Chen X et al (2020) Ecotoxicological effects of sulfonamide on and its removal by the submerged plant Vallisneria natans (Lour.) Hara. Water Res 170:115354

Remediation of Heavy Metals by Different Aquatic Macrophytes

10

Monalisa Mohanty

Abstract

Our environment is deteriorating day by day with the progress in human civilization and growing demands. Increased industrialization, urbanization, opencast mining activities, and various anthropogenic activities generate and discharge a huge quantity of toxic contaminants that lead to serious pollution problems in the environment through contaminated soil, water, and air. To contend with these problems, we need to develop innovative phytotechnological approaches to the possible uses of aquatic macrophytes for the effective removal of many toxic elements especially toxic heavy metals from contaminated water bodies and industrial effluents. It is a trending research nowadays to effectively use these aquatic plants for natural in situ phytoremediation of polluted water bodies. Employing various floating aquatic macrophytes, with high bioaccumulative potential and adsorption of contaminants through rhizofiltration technology, can be exploited, and used in biotechnological and bioengineering applications to remove metals. The current synthetic review elaborates on the role of various macrophytes in remediation of heavy metal toxicity from water. Aquatic macrophytes, viz., *Eichhornia crassipes, Hydrilla verticillata, Jussiaea repens, Lemna minor, Pistia stratiotes* and *Trapa natans*, *Azolla microphylla*, and *Salvinia molesta* show immense potentiality in phytoextraction and rhizofiltration of heavy metals, and render the wastewater bodies with lesser contaminants. Being a diverse group of plants both from a morphological and anatomical point of view, macrophytes are used effectively in removing and reducing the heavy metal contaminants like Cr, Cd, Hg, Zn, Ni, and Pb from polluted water bodies.

M. Mohanty (✉)
Laboratory of Environmental Biotechnology, Department of Biotechnology, Rama Devi Women's University, Bhubaneswar, Odisha, India
e-mail: monalisamohanty@rdwu.ac.in

S. Kumar et al. (eds.), *Aquatic Macrophytes: Ecology, Functions and Services*,
https://doi.org/10.1007/978-981-99-3822-3_10

Keywords

Aquatic macrophytes · Heavy metal · Phytoremediation · Wastewater

10.1 Introduction

A number of challenges have been elevated in the field of environmental pollution with the introduction of vast scientific, industrial, and technological progress along with rapid urbanization, industrialization, agricultural activities, and discharge of effluent wastewater particularly from opencast mining activities and mineral-processing industries (Mishra et al. 2013). Water pollution and deterioration of aquatic ecosystems due to heavy metal stress is one focused issue in today's global change (Ali et al. 2020; Mishra et al. 2013). Heavy metal pollution in water bodies is considered a principal source of environmental contamination which poses a severe threat to aquatic flora, fauna, and, human health. Application of conventional treatment technologies which include different physical and chemical methods such as reverse osmosis, ion exchange, chemical precipitation, adsorption, and solvent extraction can remove the heavy metal pollutants, viz., Pb, Cu, Cd, Zn, Cr, Ni, and As from wastewater. But these are found expensive, time consuming, environment damaging, and ineffective as reported in several studies by scientists (Ali et al. 2020; Bai et al. 2018; Dhir 2013; Dhir et al. 2009; Manorama Thampatti et al. 2020; Mishra et al. 2013; Pajevic et al. 2003; Ugya et al. 2019). Due to differential chemical speciation of heavy metals in the wastewater, it becomes too difficult to remove heavy metals from the wastewater. Its persistent accumulation in the biota is due to the nonbiodegradable property, and easy mobility from one to other trophic levels (Ali et al. 2020; Kamel 2013; Mohanty and Patra 2010).

Phytoremediation is a cost-effective emerging green clean technology which uses different plant species for the removal of various toxic contaminants. Aquatic macrophytes have proved with competence for aquaremediation of heavy metal contaminants from wastewater. Several potent aquatic macrophytes are identified for the removal of excess heavy metal content from water bodies among those water hyacinth (*Eichhornia crassipes*), water lettuce (*Pistia stratiotes*), and duckweed (*Lemna minor*) and some other aquatic plants are prominent phytoaccumulator of metals as found from different research reports time to time (Ali et al. 2020; Kamel 2013; Mohanty and Patra 2010). Innovative approaches in phytoremediation could be a potential solution for this challenge (Dada et al. 2021; Jain et al. 2019). A brief note on use of aquatic plants in phytoremediation is collected and presented here to enumerate the broad applicability of phytoremediation (Mohanty 2015; Biswasi et al. 2014; Home and Muthigo 2017; Kamel 2013; Manorama Thampatti et al. 2020; Mishra et al. 2013; Newete and Byrne 2016). Evaluation of heavy metal-induced toxic impacts on aquatic flora and fauna is an alarming feature for sustainable development of the environment (Ali et al. 2020; Home and Muthigo 2017). Plants and animals showed adverse effects such as altered metabolism, growth diminution, decreased biomass, and increased metal accumulation in tissues due to heavy metal stress. Many researchers investigated on native free-floating, submerged, and emergent aquatic plants grown in contaminated water bodies such as *Azolla microphylla* Kaulf & *Salvinia molesta* Mitchell, water hyacinth (*Eichhornia crassipes*), water ferns (*Salvinia minima*) duckweeds (*Lemna minor*, *Spirodela intermedia*), water lettuce (*Pistia stratiotes*), water cress (*Nasturtium officinale*), parrot feathers (*Myriophyllum spicatum*), hornwort (*Ceratophyllum demersum*), pondweed (*Potamogeton crispus*), *Mentha aquatica*, *Vallisneria spiralis*, smooth cordgrass (*Spartina alterniflora*), common reed (*Phragmites australis*), cattail (*Typha latifolia* plants rush (*Scirpus* spp.), common reed (*Phragmites*), and smartweed (*Polygonum hydropiperoides*) for assessing their heavy metal phytoremediation potential. They exhibit tremendous tolerance potentiality against various toxic heavy metals like cadmium, lead, iron, zinc, chromium, nickel, and copper, as reported by researchers (Ali et al. 2020; Bai et al. 2018; Home and Muthigo 2017; Ingole and Bhole 2003; Pajevic et al. 2003; Rai 2009; Wani et al. 2017).

The aquaremediation technique describes the advantage of aquatic macrophytes over other plants for alleviation of heavy metal toxicity stress. Its potential phytoremediation efficiency is owing to its fast, prevalent and easy growth, increased growth rate, enhanced biomass, common availability, low cost maintenance of plants, and wide range tolerance to toxic pollutants. Different effluent treatment plants in industries releasing heavy metals can use these green tools in purification systems. These have gained more attention worldwide (Dhir 2013; Krems et al. 2013; Kumar and Deswal 2020; Mohanty and Patra 2010; Rai 2009). The aquaremediation depends on two noteworthy factors such as selection of appropriate phytoremediation technology and biomass to biofuel and biofeed (animal feed, biogas, and compost generation). Biosorption and bioaccumulation are two important aspects of this remediation technology. Aquatic plants like *Lemna* possess starch, cellulose, and hemicelluloses which upon hydrolysis produces economically important products like lactic acid, ethanol, and others (Ali et al. 2020). Therefore, it becomes a new promising feature to use these aquatic weeds or macrophytes for an eco-sustainable environment as reported by Ali et al. (2020). Besides aquaremediation potential, these plants are also in use for the production of sugar through enzymatic hydrolysis (*Pistia stratiotes* and *Eichhornia crassipes*), used as a food source for waterbirds, source of shelter for insect larvae, and small mollusks (*Azolla* spp., *Wolffia* spp., *Spirodela* sp., and duckweeds). Nowadays, fishes have also gained more avenue for remediation purposes.

10.2 Source of Heavy Metals

Heavy metals occur naturally as heterogenous group of elements in the earth's crust. They show varied chemical properties and interfere with biological functions (Briffa et al. 2020; Rajeswari 2014). Heavy metals are generally defined as metals with relatively high densities (>5 g/cm^3), high atomic weights, and more atomic number (>20 gm). Briffa and coworkers (Briffa et al. 2020) reported that out of 116 elements, around 19 are heavy metals which remarkably vary from the other 97 known elements and are with many similar physical and chemical properties. Among the designated 19 heavy metals, lead, cadmium, and mercury do not have any biological connotation or usefulness and pose toxicity stress to different living organisms. Other heavy metals like chromium, copper, manganese, nickel, tin, and zinc become nondegradable and unrecovered when dispersed in the biosphere. As a result, heavy metal pollution leads to permanent environmental effects (Briffa et al. 2020; Rajeswari 2014; Tchounwou et al. 2012).

With their multifaceted applications in industrial, domestic, agricultural, medical, textile, aerocraft engineering, and other technologies along with wide distribution in the environment, their cycling in terrestrial and aquatic environment pose a great risk on human health and the environment (Gautam et al. 2016; Briffa et al. 2020; Masindi and Muedi 2018; Tchounwou et al. 2012).

Heavy metal contaminants like arsenic, cadmium, cobalt, nickel, lead, antimony, vanadium, zinc, platinum, palladium, and rhodium are mainly released into the air through motor vehicle emissions. Industrial leachates, sewage sludge, and acid rain are the major causes of pollution of groundwater, lakes, streams, and rivers with heavy metals (Rajeswari 2014; Tchounwou et al. 2012). Through different food chains these heavy metals enter into plants and other living forms. Through the uptake of contaminated water, and due to its persistent nature which prohibits them to metabolize through plant systems, it gets accumulated in different parts/organelles of plants especially in vacuoles. Heavy metals can accumulate in organisms as they are hard to metabolize (Briffa et al. 2020; Rajeswari 2014; Tchounwou et al. 2012). Industrial production from foundries, smelters, oil refineries, petrochemical plants, pesticide application, chemical manufacturers, opencast mining activities, electronic waste (e-waste), discharge of untreated effluents industries, sewage sludge, and various other diffuse sources like metal piping, traffic and combustion by-products from coal-burning power stations are the primary causes of heavy metal contamination in the environment (Briffa et al. 2020; Masindi and Muedi 2018; Rajeswari 2014; Tchounwou et al. 2012). The secondary source of heavy metals are agricultural runoffs that contain pesticides, insecticides, fertilizers, and more. Other natural sources include volcanic explosion, metal evaporation, corrosion of metals, and sediment resuspension, erosion of soil, and geological weathering for generation of toxic heavy metal (Briffa et al. 2020). Management of heavy metals released from electronic waste (e-waste), particularly through the disposal of used computers and mobiles, contains over 1000 different toxic materials which is a matter of great concern UNEP/GPA (2006) and many of which impairs human health.

10.3 Impact of Heavy Metals on Environmental Pollution

Anthropogenic activity like metal mining, smelting of ores, release from metal-based foundries, and industries, leachates from landfills, waste dumps, excretion, livestock and chicken manure, runoffs, automobiles, and roadworks are the chief causes of heavy metal pollution in the environment (Pachana et al. 2010). Localization of a high amount of the metal is the main cause of the increased toxicity of heavy metals.

Different impacts of heavy metals causing environmental pollution are listed in Table 10.1. Chromium pollution in water and soil is a significant environmental threat that has high accumulation in human and animal tissues, leading to toxic and detrimental health effects (Chung et al. 2014).

Two major origins of water pollution with heavy metals are urbanization and industrialization. Runoffs from urban, rural, and industrial areas containing heavy metals will get deposited in the sediments. Trace amounts of heavy metals can induce toxicity in plants, human diseases and toxic to other animals. Class, physico-chemical nature, and physiological role of the metal present in the environment, type of organism exposed, and the period of exposure of the organism's life to the metals are the different parameters that decide the toxicity of heavy metals, as this will affect the entire food chain with biomagnification.

10.4 Water Pollution by Heavy Metals

The release of mine effluents from mining activities into the surrounding water bodies affects aquatic flora and fauna. It destroys the nutrients and thus causes habitat destruction. This also leads to deterioration of water quality and reduction in the biodiversity.

Frequency of entry, amount, and chemical speciation of metal in drainage, and the buffering capacity of sink are different factors which determine the severity and extent of heavy metal damage (Masindi and Muedi 2018). A list of metals causing toxicity above the prescribed limit is given in Fig. 10.1.

Acid mine drainage from different mining companies are the chief sources of metal pollution in freshwater which has high levels of toxic heavy metals. As a result of which it enters the food chain and leads to serious health problems in humans as evidenced from different investigations carried out by researchers (Masindi and Muedi 2018; Tchounwou et al. 2012; Wani et al. 2017). This figure briefly explains the relative toxicity of various metals citing the required level of mercury, cadmium, and lead in small amounts.

10.5 Heavy Metal-Induced Toxicity in Plants with Special Reference to Aquatic Plants

Arsenic, cadmium, chromium, lead, and mercury show very high toxicity even at low quantities. For this reason they are considered as the priority metals in regard to public health significance. Due to high toxicity they induce multiple organ damage, even at lower levels of exposure (Briffa et al. 2020). Some of them are classified as known and some are probable human carcinogens (USEPA and the International Agency for Research on Cancer). Different heavy metals and their prescribed limits regarding toxicity effects are provided in Fig. 10.2.

Table 10.1 Effects of different heavy metal-induced toxicity in living beings and pollution in environment

Heavy metal	Health effects	Environmental source	Environmental effect
Cadmium	Renal dysfunction, lung diseases, bone defects in humans and animals, lung cancer	Cadmium ores are found in complexation with sulfide ores of zinc, copper, and lead. Vapor and dust emissions during smelting of above sulfide ores release Cd	Contaminate surrounding soil and water
Mercury	Affects brain and central nervous system function Causes abortion, congenital malformation, and development changes when exposed to fetus	Organic mercury in fungicides and herbicides with phenylmercuric salts, and alkyl mercury in methyl mercury, are threat to humans, sediment, anaerobic microorganisms, and fishes	Contaminate surrounding soil and water
Lead	Softening of the bones and kidney failure, lead poisoning	Lead from batteries, additives in gasoline, and paint pigments are major sources Organic lead, for example, tetraethyl lead and tetramethyl lead	Contaminate the soil and water
Chromium	Skin irritation and ulceration by low exposure, whereas long-term exposure effects are damaging of kidney and liver, circulatory and nerve tissues. Blood cancer, blue baby formation are other health effects	Cr-containing rocks, industrial operations, leaching of soils, among others. Ferrochromium slag, chromium plating bath also produce chromium	Water pollution by weathering of direct discharge from rocks Soil pollution: By the dumping of chromium wastes
Arsenic	Damaging effects to kidney and liver along with erythrocyte hemolysis	Insecticide and herbicide or preservative for wood	Soil, water, and air pollution
Aluminum	Bone disorders including fractures, osteopenia, and osteomalacia	Leaching from mine sites, ores	Soil, water, and air pollution
Copper	Anemia, liver and kidney damage, and stomach and intestinal irritation	Leaching from ores, mining activities	Soil and water pollution
Nickel	Lowering of body mass, interrupts in function of heart and liver, irritation in skin are caused by long exposure to Ni	Oxide and sulfide ores of nickel. Nickel mining, industrial processes	Contaminates air, water, and soil, causes greenhouse gas emission, and habitat destruction

Source: Chung et al. (2014), Das et al. (2019), Genchi et al. (2020), Oliveira (2012), Prasad et al. (2021), Rajeswari (2014), Tchounwou et al. (2012), Pachana et al. (2010), Gautam et al. (2016)

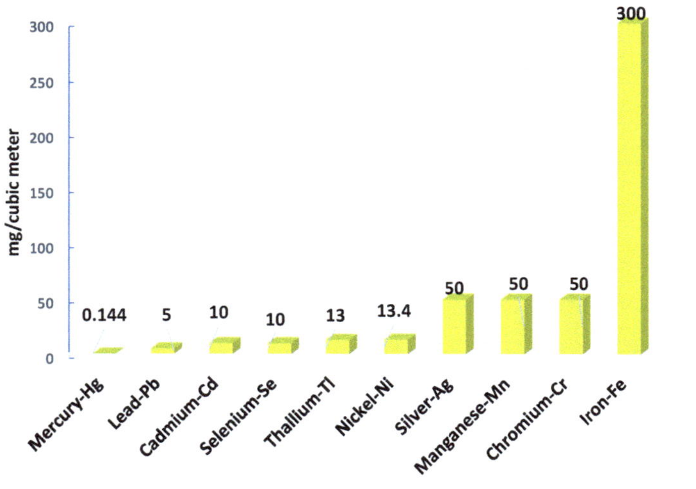

Fig. 10.1 Maximum permissible limits of metal concentration in aquatic ecosystem (USEPA 1987); Federal Register 56 (110): 26460–26564 (1991)

There are severe toxic and poisoning effects (lead poisoning, mercury poisoning, beryllium, arsenic, aluminum phosphide, cadmium, and silver poisoning) of heavy metal ions found in different life forms. So there must be a universal and strict discharge regulation for metal-contaminated wastewater effluents to different aquatic bodies. This will necessitate better, eco-friendly, and sustainable treatment techniques. Environmental biotechnologists and engineers have developed several advanced procedures, such as hemofiltration, phytodetoxification, bioadsorption, bioleaching, and bioremediation using microbes, for the treatment of heavy metal-contaminated wastewater.

10.6 Role of Aquatic Macrophytes as Hyperaccumulators

Accumulation of toxic metals leading to biospheric pollution has been increasing day by day since the dawn of the industrial revolution. Heavy metal pollutants have a great effect on the growth, reproduction, metabolism, and development of both plants and animals. Though the precipitation and absorption of most of these metals occur in soils, only Zn, Cu, and Ni toxicities are found frequently (Biswasi et al. 2014). Phytoremediation is one of the environment-friendly and inexpensive technological ways to remove metal pollutants from contaminated soil or water. This green technique involves the use of plants for decontamination of pollutants from soil or water and makes environmental remediation. In this context, aquaremediation explains the potential use of aquatic plants' macrophytes/hydrophytes to remove toxic environmental contaminants from water bodies to render them usable. A

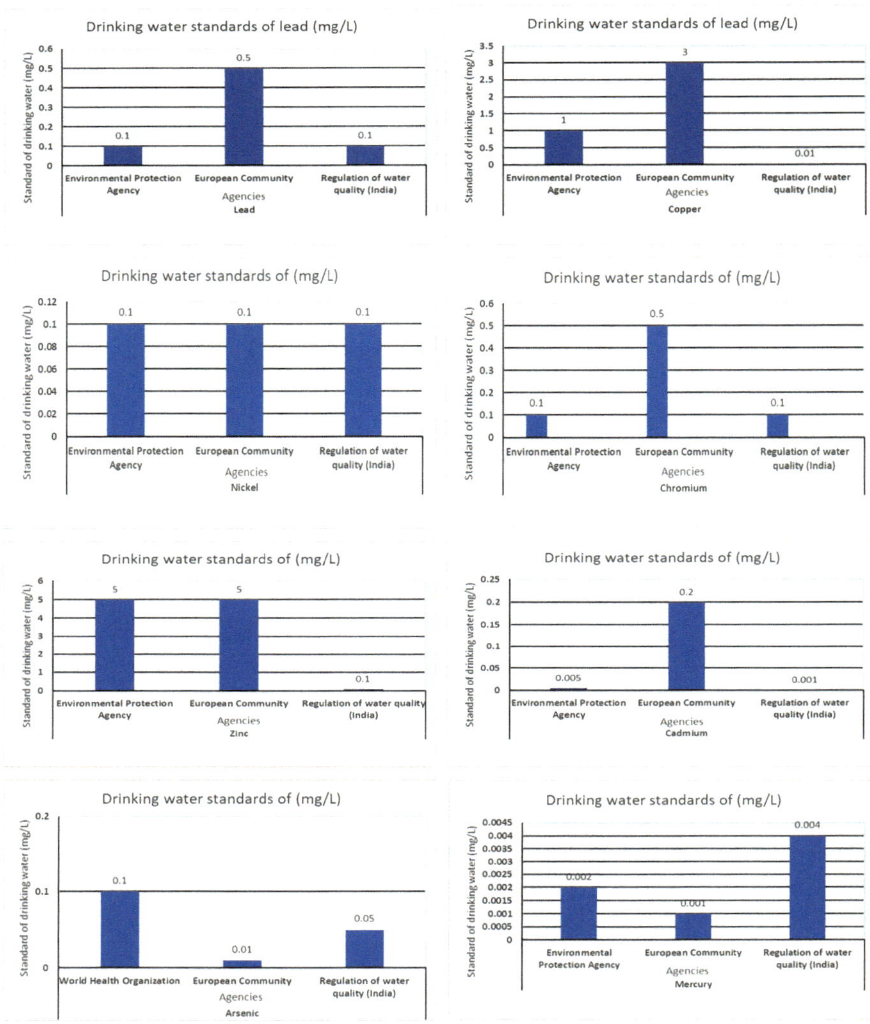

Fig. 10.2 Prescribed permissible limits of heavy metal in drinking water with respect to different toxic impacts and associated health effects (Gautam et al. 2014)

review of a comparative study of aquatic macrophytes in phytoremediation of different heavy metals will give an unbeatable solution to the problem.

Microbe-assisted bioremediation technology is widely used for the treatment of heavy metal-contaminated wastewater. Conventional technologies are expensive and usually produce adverse impacts on aquatic ecosystems. Heavy metal sequestration by aquatic/macrophytes is a unique and cleanup technology for the treatment of contaminated soil and water. Many aquatic plants such as *Phragmites*, *Lemna*, *Eichhornia*, *Azolla,* and *Typha* have been used for the treatment of heavy metal-contaminated water bodies (Gautam et al. 2014). The plants can take up heavy metals through different strategies like phytostabilization, phytoextraction, phytofiltration, or rhizoremediation(Bai et al. 2018; Gautam et al. 2014; Kamel 2013; Newete and Byrne 2016; Pajevic et al. 2003; Wani et al. 2017). Aquatic macrophytes, viz., *Typha*, *Phragmites*, *Eichhornia*, *Azolla*, and *Lemna*, are potential plants for the removal of heavy metals. Phytoextraction follows two strategies: aquaremediation through macrophytes and chemically induced phytoextraction techniques (Ali et al. 2020; Gautam et al. 2014; Mohanty and Patra 2010). Macrophytes include quite a differentiated group of plants like chara (*Charophyta*), mosses (*Bryophyta*), ferns (*Pteridophyta*), and spermatophytes (*Spermatophyta*) (Krems et al. 2013).

Increased biomass-producing plant species (hyperaccumulators) and the application of synthetic and natural chelates will contribute to the efficiency of phytoextraction (Ali et al. 2020; Gautam et al. 2014).

10.6.1 Aquaremediation of Industrial Wastewater

The release of heavy metal-contaminated industrial waste/effluent into different water bodies poses a high risk to human health, living organisms, and the environment. It has been reported that Khan et al. (2009) conducted a study to assess the phytoremediation potential of 12 aquatic plants (cattail, wool grass, water sedge, reeds, jointed rush, coontail, duckweed, water hyacinth, water plantain, knotweed, and waterweed) in a constructed wetland for removal of heavy metals were from industrial wastewater. The efficacies of constructed wetlands for removal of different wetlands were Cd (90%), Cr (89%), Fe (74.1%), Pb (50%), Cu (48.3%), and Ni (40.9%) as reported by researchers (Ali et al. 2020; Khan and Faisal 2018; Khan et al. 2009). As per the study by Miretzky in 2004 in Pampean shallow lakes of Argentina, three commonly found autochthonous floating macrophytes, viz., *Pistia stratiotes*, *Spirodela intermedia*, and *Lemna minor* were used for ex situ remediation of Fe, Cu, Zn, Mn, Cr, and Pb (Miretzky et al. 2004). In a study by Souza and Silva, aquatic macrophytes like *Eichhornia crassipes and Salvinia auriculata* were used for remediation of Cd, Hg, Zn, Ni, and Pb, whereas *Lemna minor* and *Pistia stratiotes* were used as better biomonitoring agents (de Souza and Silva 2019).

10.6.2 Aquaremediation of Municipal Wastewater

Municipal wastewater is potentially risk for the aquatic environment as it possesses noxious heavy metals like Zn, Cu, Ni, Pb, and Hg and may cause chronic and acute health effects. In a hydroponic study by Pedescoll et al. (2015), 14–85% of heavy metals (zinc, lead, arsenic, nickel, iron, copper, aluminum, and magnesium) were removed by *Typha angustifolia* and *Phragmites australis*. Several widely used macrophytes like *Elodea canadensis*, *Ceratophyllum demersum*, *Lemna minor*, and *Potamogeton spp.* and *Myriophyllum sps.* are used for biomonitoring heavy metals in wastewater in Poland. So their possible use for water and sewage phytoremediation was studied by Krems (Krems et al. 2013).

10.7 Phytoremediation Potential of Heavy Metals

Choosing a suitable aquatic macrophyte to remediate particular metals from the wastewater is the need of the hour. A list of macrophytes with different phytoremediation potential for heavy metals are listed in Table 10.2.

The diversity of species and varying distribution of macrophytic vegetation are responsible for the remediation of heavy metals from wastewater. Several research focusing on mechanism of biosorption, effect of factors on the process kinetics and equilibrium, and also on the mutual relations between accumulated pollutants in the living structure and their habitat. Macrophytic resistance to pollution and the possibility of their repeated use in water and sewage phytoremediation, processes of the accumulative properties of water plants, have been carried out for many years to use them in biomonitoring and phytoremediation of wastewater contaminated by heavy metals.

Table 10.2 Aquaremediation potential of macrophytes for wastewater

Heavy metal	Macrophytes/aquatic plants used with their common name
Pb	*Typha latifolia L., Ceratophyllum demersum L*
Pb, Hg, Cu, Cr, Ni, Zn, M	*Eichhornia crassipes*—water hyacinth
K, Ca, Ti, Fe, Cr, Mn, Cu, Zn, Sr, As, Ni, Cr, Cd	*Salvinia* sp. and *Salvinia minima*—water spangles
Cr, Zn, Fe, Mn, Cu	*Pistia stratiotes*—water lettuce
Cd, Cr	*Salvinia herzogii*—water fern
Cd, Pb	*Lemna polyrrhiza L.*
Cr, As, Ni, Cu, Pb	*Lemna minor*—duckweed
Fe, Zn, Mn, Cu, Cr, Pb	*Spirodela intermedia*—duckweed
Cr, Ni, Zn, Cu	*Nasturtium officinale*—watercress
Pb, Cd, Fe, Cu	*Myriophyllum spicatum*—parrot feathers
Cu, Fe, Ni, Zn, M	*Potamogeton crispus*—pondweed
Cd, Pb, Cu, Zn	*Potamogeton pectinatus*—American pondweed
Zn, Mn, Ni, Fe, Pb, Cu	*Typha latifolia*—common cattail
Cu. Cr, Zn, Ni, Mn, Cd, Pb, As	*Spartina alterniflora*—cordgrass
Fe, Cu, Cd, Pb, Zn	*Phragmites australis*—common reed
Cu, Ni, Fe	*Hydrilla verticillata, Elodea canadensis, Salvinia* sp.
Cu, Pb, Zn	*Polygonum hydropiperoides*—smartweed

Source: Ali et al. (2020), Das et al. (2019), Kamel (2013), Miretzky et al. (2004), Mishra et al. (2013)

10.8 Conclusions and Future Prospects

The aquatic plants can be used as low-cost, effective, and potential green tools for the removal of heavy metals from polluted aquatic bodies. This review showed that aquatic plants such as *Pistia, Eichhornia, Salvinia Hydrilla, Lemna*, and others can have phytoremediation potential to attenuate heavy metals from wastewater. Therefore, it is very much essential to utilize the remarkably potential macrophytes for the accumulation of environmental pollutants from wastewater which become a frontier area of research in environmental science and technology. Further research in genetic engineering to enhance the accumulation and tolerance capacity of macrophytes is a perspective approach in phytoremediation technology. Aquaremediation of wastewater through macrophytes are used to treat a huge volume of metal-contaminated wastewater. Treatment of contaminants by macrophytes is a low-cost and feasible advantageous approach for the sustainable development of aquatic ecosystems.

References

Ali S, Abbas Z, Rizwan M, Zaheer IE, Yavaş İ, Ünay A, Abdel-Daim MM, Bin-Jumah M, Hasanuzzaman M, Kalderis D (2020) Application of floating aquatic plants in phytoremediation of heavy metals polluted water: a review. Sustainability 12:1927. https://doi.org/10.3390/su12051927

Bai L, Liu X-L, Hu J, Li J, Wang Z-L, Han G, Li S-L, Liu C-Q (2018) Heavy metal accumulation in common aquatic plants in Rivers and Lakes in the Taihu Basin. Int J Environ Res Public Health 15:2857. https://doi.org/10.3390/ijerph15122857

Biswasi SP, Mohanty M, Pradhan C, Satapathy KB (2014) Phytoremediation of aqueous solutions contaminated with nickel (Ni) by exploitation of Azolla microphylla Kaulf & Salvinia molesta Mitchell: a novel bioseparation process for waste water treatment. In: Pramanik K, Patra JK (eds) Industrial and environmental biotechnology. Chapter-18. Studium Press India Pvt Ltd., pp 289–298. ISBN: 9789380012674

Briffa J, Sinagra E, Blundell R (2020) Heavy metal pollution in the environment and their toxicological effects on humans. Heliyon 6:e04691. https://doi.org/10.1016/j.heliyon.2020.e04691

Chung J-Y, Yu S-D, Hong Y-S (2014) Environmental source of arsenic exposure. J Prev Med Public Health 47:253. https://doi.org/10.3961/jpmph.14.036

Dada EO, Akinola MO, Owa SO, Dedeke GA, Aladesida AA, Owagboriaye FO, Oludipe EO (2021) Efficacy of vermiremediation to remove contaminants from soil. J Health Pollut 11: 210302. https://doi.org/10.5696/2156-9614-11.29.210302

Das KK, Reddy RC, Bagoji IB, Das S, Bagali S, Mullur L, Khodnapur JP, Biradar MS (2019) Primary concept of nickel toxicity – an overview. J Basic Clin Physiol Pharmacol 30:141–152. https://doi.org/10.1515/jbcpp-2017-0171

de Souza CB, Silva GR (2019) Phytoremediation of effluents contaminated with heavy metals by floating aquatic macrophytes species. In: Biotechnology and bioengineering. IntechOpen, pp 1–18. https://doi.org/10.5772/intechopen.83645

Dhir B (2013) Phytoremediation: role of aquatic plants in environmental clean-up. Springer, India. https://doi.org/10.1007/978-81-322-1307-9

Dhir B, Sharmila P, Saradhi PP (2009) Potential of aquatic macrophytes for removing contaminants from the environment. Crit Rev Environ Sci Technol 39:754–781. https://doi.org/10.1080/10643380801977776

Gautam RK, Sharma SK, Mahiya S, Chattopadhyaya MC (2014) Chapter 1: Contamination of heavy metals in aquatic media: transport, toxicity, and technologies for remediation, pp. 1–24. https://doi.org/10.1039/9781782620174-00001

Gautam PK, Gautam RK, Banerjee S, Chattopadhyaya MC, Pandey JD (2016) Heavy metals in the environment: fate, transport, toxicity and remediation technologies. Nova Sci Publ 60:101–130

Genchi G, Carocci A, Lauria G, Sinicropi MS, Catalano A (2020) Nickel: human health and environmental toxicology. Int J Environ Res Public Health 17. https://doi.org/10.3390/ijerph17030679

Home PG, Muthigo KG (2017) Assessment of the efficiency of different mixes of macrophytes in removing heavy metals from wastewater using constructed wetland

Ingole NW, Bhole AG (2003) Removal of heavy metals from aqueous solution by water hyacinth (Eichhornia crassipes). J Water Supply Res Technol AQUA 52:119–128. https://doi.org/10.2166/aqua.2003.0012

Jain P, Andotra A, Aziz A, Kaur P, Mahajan A, Kumar A (2019) Phytoremediation—A miracle technique for waste water treatment. Res J Pharm Technol 12:2009–2016. https://doi.org/10.5958/0974-360X.2019.00341.X

Kamel KA (2013) Phytoremediation potentiality of aquatic Macrophytes in heavy metal contaminated water of El-Temsah Lake, Ismailia, Egypt 14

Khan HN, Faisal M (2018) Phytoremediation of industrial wastewater by hydrophytes. Phytoremediation 6:179–200. https://doi.org/10.1007/978-3-319-99651-6_8

Khan S, Ahmad I, Shah MT, Rehman S, Khaliq A (2009) Use of constructed wetland for the removal of heavy metals from industrial wastewater. J Environ Manag 90:3451–3457

Krems P, Rajfur M, Wacławek M, Kłos A (2013) The use of water plants in biomonitoring and phytoremediation of waters polluted with heavy metals. Ecol Chem Eng S 20:353–370. https://doi.org/10.2478/eces-2013-0026

Kumar S, Deswal S (2020) Phytoremediation capabilities of Salvinia molesta, water hyacinth, water lettuce, and duckweed to reduce phosphorus in rice mill wastewater. Int J Phytoremediat 22:1097–1109

Manorama Thampatti KC, Beena VI, Meera AV, Ajayan AS (2020) Phytoremediation of metals by aquatic Macrophytes. Phytoremediation:153–204. https://doi.org/10.1007/978-3-030-00099-8_6

Masindi V, Muedi KL (2018) Environmental contamination by heavy metals. Heavy metals. IntechOpen. https://doi.org/10.5772/intechopen.76082

Miretzky P, Saralegui A, Cirelli AF (2004) Aquatic macrophytes potential for the simultaneous removal of heavy metals (Buenos Aires, Argentina). Chemosphere 57:997–1005. https://doi.org/10.1016/j.chemosphere.2004.07.024

Mishra S, Mohanty M, Pradhan C, Patra HK, Das R, Sahoo S (2013) Physico-chemical assessment of paper mill effluent and its heavy metal remediation using aquatic macrophytes—a case study at JK Paper mill, Rayagada, India. Environ Monit Assess 185:4347–4359

Mohanty M, Patra HK (2010) Attenuation of chromium toxicity by bioremediation technology. Rev Environ Contam Toxicol 210:1–34

Mohanty M (2015) Phytoremediation – an innovative approach for attenuation of chromium toxicity and rice cultivation in mining areas. Rice Res 03(03). https://doi.org/10.4172/2375-4338.1000e116. (Open Access)

Newete SW, Byrne MJ (2016) The capacity of aquatic macrophytes for phytoremediation and their disposal with specific reference to water hyacinth. Environ Sci Pollut Res 23:10630–10643. https://doi.org/10.1007/s11356-016-6329-6

Oliveira H (2012) Chromium as an environmental pollutant: insights on induced plant toxicity. J Bot 2012:1. https://doi.org/10.1155/2012/375843

Pachana K, Wattanakornsiri A, Nanuam J (2010) Heavy metal transport and fate in the environmental compartments. NU Sci J 7(1):1–11

Pajevic S, Vuckovic M, Kevresan Z, Matavulj M, Radulovic S, Radnovic D (2003) Aquatic macrophytes as indicators of heavy metal pollution of water in DTD canal system. Zb Mat srp prir nauk:51–60. https://doi.org/10.2298/ZMSPN0304051P

Pedescoll A, Sidrach-Cardona R, Hijosa-Valsero M, Bécares E (2015) Design parameters affecting metals removal in horizontal constructed wetlands for domestic wastewater treatment. Ecol Eng 80:92–99

Prasad S, Yadav KK, Kumar S, Gupta N, Cabral-Pinto MMS, Rezania S, Radwan N, Alam J (2021) Chromium contamination and effect on environmental health and its remediation: a sustainable approaches. J Environ Manag 285:112174. https://doi.org/10.1016/j.jenvman.2021.112174

Rai PK (2009) Heavy metal phytoremediation from aquatic ecosystems with special reference to Macrophytes. Crit Rev Environ Sci Technol 39(9):697–753. https://doi.org/10.1080/10643380801910058

Rajeswari TR (2014) Impact of heavy metals on environmental pollution 7

Tchounwou PB, Yedjou CG, Patlolla AK, Sutton DJ (2012) Heavy metals toxicity and the environment. EXS 101:133. https://doi.org/10.1007/978-3-7643-8340-4_6

Ugya AY, Hua X, Ma J (2019) Phytoremediation as a tool for the remediation of wastewater resulting from dyeing activities. Appl Ecol Env Res 17:3723–3735. https://doi.org/10.15666/aeer/1702_37233735

UNEP/GPA (2006) The state of the marine environment: trends and processes, The Hague

US Environmental Protection Agency 1987; Rep.440/5-87-003,207

Wani RA, Ganai BA, Shah MA, Uqab B (2017) Heavy metal uptake potential of aquatic plants through phytoremediation technique - a review. J Bioremed Biodegr 8(404):2

Translocations of Heavy Metals in Aquatic Macrophytes Naturally Grown in the Riverine Ecosystem

11

Sushil Kumar Bharti, Narendra Kumar, and Nandkishor S. More

Abstract

The ever-increasing need to remediate contaminated aquatic ecosystems culminates in identifying economical, readily available, and efficient biological material, and aquatic macrophytes are one of them. To keep in mind, the authors have identified four aquatic macrophytes, viz., *Hydrilla verticillata (L.f.) Royle, Potamogeton pectinatus L., Ranunculus sceleratus L.,* and *Mimulus glabratus Kunth*, which can reduce heavy metals levels in River Gomti. In the current study, surface water and aquatic macrophytes samples were collected from Gaughat to Gomti Barrage in River Gomti. The physicochemical characteristics of surface water were found in the range of pH = 6.35–8.79, dissolved oxygen = 1.58–4.5-mg/L, and fecal coliform = 4.3×10^4 MPN/100 ml, whereas total dissolved solids (TDS) and total suspended solids (TSS) were recorded moderate to high. Whereas metals concentration (mg/L) in surface water ranged from Fe: 1.5–80, Cu: 0.012–0.066, Cd: 0.02–0.24, and As: 0.028–0.40 at different sampling sites. Heavy metals (Cd, Cu, Fe, and As) content was also estimated in order to assess the ability of aquatic macrophytes (Hydrilla verticillata (L.f.) Royle, Potamogeton pectinatus L., Ranunculus sceleratus L., and Mimulus glabratus Kunth) to absorb the metals in their tissues. The highest concentration ($\mu g g^{-1}$) of Fe, Cd, Cu, and As was recorded at Gomti Barrage, Mohan Meakin, and Daliganj Pul. In contrast, other plants have shown selective absorption patterns of heavy

S. K. Bharti
Department of Geology, Babasaheb Bhimrao Ambedkar University, Lucknow, Uttar Pradesh, India

Department of Mining Engineering, Indian Institute of Technology (BHU), Varanasi, India

N. Kumar · N. S. More (✉)
Department of Environmental Science, Babasaheb Bhimrao Ambedkar University, Lucknow, Uttar Pradesh, India

S. Kumar et al. (eds.), *Aquatic Macrophytes: Ecology, Functions and Services*,
https://doi.org/10.1007/978-981-99-3822-3_11

metals in their tissues. Further translocation factors revealed that higher levels of metals were concentrated in roots instead of shoot.

Keywords

Heavy metals · Hyperaccumulator · Macrophytes · Phytoremediation · Translocation Factor

11.1 Introduction

Water bodies have been at the receiving end in the last few decades due to their direct or indirect disposal of industrial, residential, and commercial waste. Direct removal of industrial effluent into water bodies can elevate the level of heavy metals and other contaminants, which directly or indirectly affect the physiology of flora and fauna (Dunbabin and Bowmer 1992; Sinicrope et al. 1992). In addition, heavy metals can be leached through the fracture of rock (fault) and contaminate aquifers' water tables (Chang et al. 2000). Once a water body is contaminated, it can be rejuvenated with the help of aquatic macrophytes, which can accumulate considerable amounts of heavy metals in their tissue (Souza et al. 2013; Abida 2009). In contrast, heavy metal accumulation depends on the contamination level of the growing medium in water bodies from where it can reach their tissue (Deng et al. 2004; Martinez and Nyamboli 2011).

Further, the accumulation of metals in plant tissues varies from species to species, depending on their physiology and adaptation to specific environments (Baldantoni et al. 2004). In plant kingdom, 87 plant species of the Brassicaceae family has the potential to accumulate heavy metals (Zn, Pb, Ni, Cd, Se, and Arsenic) without apparent impairment in plant tissues (Milner and Kochian 2008). Direct consumption of contaminated water may elevate the levels of metals in the kidney, liver, and bones of human beings leading to dysfunction of the cardiovascular, neurological, and renal systems (Johri et al. 2010). Among toxic elements, cadmium and arsenic pose a severe threat to flora and fauna, because they can easily transport to other trophic levels (Ghassemzadeh et al. 2008). During mining and the smelting process in brass manufacturing and electroplating industries, a considerable amount of Cu is added to ambient air by the volatilization process, which ultimately settles down on the earth's surface and may reach water bodies through fractures/leaching.

Similarly, other heavy metals may be added to ambient air, soil, and water bodies when applying fertilizers and agrochemicals in agriculture (Doula et al. 2000). The management practice to reduce the levels of metals in water bodies to a certain extent is phytoremediation. In phytoremediation, phytoextraction is the primary process in which plant species accumulate metals in their tissues. However, plant species that can reduce metal translocation from roots to shoots could be considered phytostabilizers to restore metal-contaminated lands (Deng and Engleherdt 2006). Therefore, identifying the plant tissues that absorb the maximum amounts of trace elements is essential for phytoremediation (Baldantoni et al. 2004). The ability of aquatic plants to remove toxic metals from water depends on (1) sediment geochemistry, (2) water chemistry, (3) plant physiology, and (4) plant genotype. As River Gomti enters the city areas of Lucknow, physicochemical parameters of surface water have deteriorated.

In contrast, biological oxygen demand (BOD), chemical oxygen demand (COD), and levels of trace metals in river water increase, certainly reducing the levels of dissolved oxygen (DO) and promoting fish mortality. This study aimed to assess the physicochemical characteristics of surface water at major sites of Lucknow city. Further metal accumulation potential of four aquatic macrophytes (*P. pectinatus*, *H. verticillata*, *M. glabratus*, and *R. sceleratus*) was estimated by evaluating the translocation factor.

11.2 Material and Method

11.2.1 Site Description

River Gomti is a major tributary of the River Ganga. It originates from GomatTaal, formally known as Fulhaar Jheel in Pilibhit, Uttar Pradesh. It covers about 940 km throughout the 09 districts of Uttar Pradesh and finally merges into Ganga in the Ghazipur district (Singh et al. 2005). The river serves as a foremost source of drinking water, and around 4.58 million (2011 census) are dependent on the River Gomti. Subsequently, it receives the untreated effluent of sugar factories, distilleries, domestic sewage, and wastewater from the household. The study area covers 13 locations, viz., S1 (Gaughat), S2 (Gulalaghat), S3 (Kudiaghat), S4 (Immambara), S5 (Mohan Meakin), S6 (Daliganj Pul), S7 (Hanuman Setu), S8 (Boat Club), S9 (Laxman Mela Ground), S10 (Monkey Bridge), S11 (Parag), S12 (Baikunth Dham), and S13 (Gomti Barrage). Grab samples of water, and aquatic plants were collected from each location.

11.2.2 Water and Aquatic Plant Sampling and Analysis

Water samples were collected 30 cm below the water surface without any air bubbles in polyethylene bottles for chemical analysis from 13 sampling sites of River Gomti. Two sets of water samples were collected from each site. One set of samples was collected without any preservative to analyze physicochemical parameters, and in the other set, 1 ml of concentrated HCl was added to a 250-ml sample bottle to preserve Cd, Cu, Fe, and As. Briefly, for heavy metal extraction, 90 mL of water sample was taken into a conical flask with 10 mL of an acid mixture of HNO_3 and $HClO_4$ (5:1 v/v). Afterward, samples were digested in the acid mixture until the solution became transparent. Afterward, the transparent solution was filtered through the Whatman no.1 filter paper, and metal concentration was estimated using atomic absorption spectrophotometer (AA 240 FS, Varian).

$$\text{Metal concentration}\left(\text{mg L}^{-1}\right)=\frac{\text{X}-\text{Y}}{\text{V2}\times\text{V1}}$$

Where X = reading of sample in mg/L, Y = reading of blank in mg/L, V_1 = final volume of digested sample (mL), and V_2 = volume of a sample taken (mL).

$$\text{Metal concentration}\left(\mu\text{g g}^{-1}\ \text{DW}\right)=\frac{\text{X}-\text{Y}}{\text{W}\times\text{V}}$$

Where X = reading of sample in mg/L, Y = reading of blank in mg/L, V = final volume of digested samples (mL), and W = dry weight of the sample (g).

11.2.2.1 Translocation Factor

The translocation factor was estimated by following Zacchini et al. (2009), Barman et al. (2000), and Gupta et al. (2008).

$$\text{Translocation Factor}=\frac{\text{The concentration of metal in plants in shoots}}{\text{The concentration of metals in corresponding plants root}}$$

11.3 Result and Discussion

11.3.1 Physicochemical Indicators of River Water

The physicochemical properties and heavy metals associated with river water were estimated and presented in Tables 11.1 and 11.2. The results indicate that the pH of river water was slightly acidic to alkaline (6.35 to 8.79), showing the productive nature of the riverine ecosystem (Garg et al. 2010). Further, maximum dissolved oxygen was recorded at Immambara (4.5 mg/L) and minimum at Mohan Meakin (1.58 mg/L). On the other hand, chloride, fluoride, total dissolved solids (TDS), and total suspended solids (TSS) were also recorded as very high. The highest levels of fecal coliform (4.3×10^4 MPN/100 ml) were recorded at Parag, indicating the open defecation in the river. The reason for deteriorating water quality may be the direct disposal of unprocessed sewage from Lucknow city and the discarding of the garbage along the riverbed. However, the finding of this study supports a previous study by (Dutta et al. 2011; Kumar et al. 2013), whereas nitrate and nitrite concentration of all sites were 1–6 folds higher than the approved limit of 10 mg/L and 0.1 mg/L, respectively (WHO 1993). The low concentration of DO can cause fish mortality in a riverine ecosystem (Gower 1980; Chapman 1996). However, a higher concentration of nitrates can cause "blue baby syndrome."

Table 11.1 Concentration of physicochemical parameter of water is expressed as (mg/L) except for pH and total coliform

Sites	pH	Hardness	Do	Nitrate	Nitrite	TSS	TDS	Chloride	Fluoride	T. coliform
S1	7.66 ± 0.17	167.5 ± 3.5	3.4 ± 0.17	45.79 ± 3.28	0.56 ± 0.46	434.5 ± 10.60	350 ± 24.74	3.01 ± 0.22	1.56 ± 0.15	9556.67 ± 801.38
S2	7.45 ± 0.11	178 ± 2.78	2.7 ± 0.267	59.06 ± 5.14	0.14 ± 0.001	523.5 ± 6.36	400 ± 14.41	7.81 ± 1.47	1.56 ± 0.04	25053.3 ± 377.12
S3	7.96 ± 0.24	155.5 ± 1.57	2.45 ± 0.05	49.68 ± 5.58	0.1 ± 0.005	444 ± 15.55	600 ± 27.57	8.9 ± 0.436	1.78 ± 0.08	17,030 ± 424.264
S4	6.35 ± 0.33	139 ± 1.78	4.5 ± 0.187	54.87 ± 2.07	0.47 ± 0.01	489.5 ± 13.43	390 ± 14.14	28.8 ± 2.10	1.8 ± 0.06	12,530 ± 424.264
S5	7.59 ± 0.45	194.5 ± 2.57	1.58 ± 0.45	41.2 ± 2.76	0.28 ± 0.021	570 ± 21.21	558.5 ± 37.5	32.39 ± 1.6	1.3 ± 0.014	17,530 ± 424.264
S6	8.79 ± 0.44	280 ± 2.89	3.15 ± 0.15	57.47 ± 1.05	0.38 ± 0.004	500 ± 13.43	539 ± 8.48	60.6 ± 0.34	1.2 ± 0.028	27,200 ± 2828.43
S7	7.67 ± 0.74	167 ± 2.56	2.9 ± 0.18	74.39 ± 1.48	0.19 ± 0.03	560 ± 10.60	615 ± 28.28	39.5 ± 0.39	1.6 ± 0.042	25266.7 ± 3771.2
S8	7.68 ± 0.84	177.5 ± 1.55	3.65 ± 0.15	66.54 ± 2.96	0.18 ± 0.001	450.5 ± 12.02	500 ± 21.21	22.67 ± 0.6	0.89 ± 0.12	34,610 ± 1414.21
S9	7.45 ± 0.72	160 ± 4.4	3.00 ± 0.05	60.33 ± 1.97	0.04 ± 0.002	539 ± 18.38	404 ± 14.14	16.61 ± 0.6	1.2 ± 0.045	12687.3 ± 6050.9
S10	8.15 ± 0.54	173 ± 11.9	2.5 ± 0.26	60.80 ± 1.16	0.7 ± 0.012	543 ± 13.43	450 ± 7.071	14.65 ± 0.6	1.1 ± 0.04	16059.3 ± 425.94
S11	7.53 ± 0.82	160 ± 14.98	2.7 ± 0.28	55.75 ± 4.55	0.23 ± 0.125	465 ± 12.72	434.5 ± 23.33	17.5 ± 0.85	1.13 ± 0.03	43246.7 ± 4624.7
S12	7.76 ± 0.54	178 ± 3.97	1.9 ± 0.29	49.55 ± 1.69	0.37 ± 0.298	567.5 ± 2.12	520 ± 21.21	4 ± 0.621	1.12 ± 0.106	11807.3 ± 109.8
S13	8.79 ± 0.72	164 ± 1.67	2.5 ± 0.45	67.20 ± 3.58	0.41 ± 0.145	496 ± 12.72	457.5 ± 13.43	18.2 ± 0.34	1.05 ± 0.03	10,390 ± 625.15
BIS	6.5–8.5	300	5.0	45	—	—	500	250	1.0	—

Table 11.2 Heavy metal (Fe, Cu, Cd, and As) concentrations (mg/L) of water collected from River Gomti

Sites	Fe	Cu	Cd	As
S1	6.7 ± 0.15	0.025 ± 0.0007	0.19 ± 0.02	0.028 ± 0.001
S2	7.7 ± 0.07	0.043 ± 0.005	0.19 ± 0.001	0.14 ± 0.003
S3	7.3 ± 0.16	0.028 ± 0.0029	0.20 ± 0.002	0.039 ± 0.002
S4	7.2 ± 0.92	0.066 ± 0.004	0.20 ± 0.009	0.04 ± 0.005
S5	8.5 ± 1.10	0.023 ± 0.001	0.16 ± 0.002	0.06 ± 0.08
S6	10.7 ± 1.16	0.029 ± 0.002	0.23 ± 0.002	0.06 ± 0.004
S7	80 ± 1.54	0.034 ± 0.001	0.21 ± 0.001	0.05 ± 0.004
S8	41 ± 1.91	0.02 ± 0.002	0.24 ± 0.002	0.03 ± 0.004
S9	22 ± 1.07	0.03 ± 0.003	0.18 ± 0.003	0.04 ± 0.005
S10	16.9 ± 1.18	0.013 ± 0.001	0.20 ± 0.006	0.12 ± 0.001
S11	14.9 ± 1.16	0.036 ± 0.002	0.18 ± 0.002	0.4 ± 0.003
S12	15.5 ± 1.51	0.028 ± 0.001	0.14 ± 0.009	0.03 ± 0.002
S13	1.53 ± 0.15	0.012 ± 0.002	0.02 ± 0.003	0.06 ± 0.002

11.3.2 Metal Concentration in River Water

The concentrations of Fe, Cu, Cd, and As in surface water from each sampling site are illustrated in Table 11.2. The highest and lowest Fe concentrations (mg/L) were observed at sites S7 (80 mg/L) and S13 (1.53), respectively. The highest and lowest Cu concentrations (mg/L) were observed in S4 (0.066) and S13 (0.012) sites, respectively. The highest and lowest concentrations (mg/L) of Cd in surface water were observed in S8 (0.24) and S13 (0.02) sites, respectively. The highest (0.40 mg/L) and lowest concentration (0.028 mg/L) of As in surface water were observed in S11 and S1 at sites, respectively.

11.3.3 Metal Accumulation in the Macrophytes

Heavy metal accumulation in the aquatic plant varies from species to species (Alloway et al. 1990). The current study's findings revealed that *H. verticillata* is a hyperaccumulator of metals. However, metal concentration was observed in descending order of Fe > Cu > As > Cd (Figs. 11.1, 11.2, 11.3, and 11.4). *H. verticillata* root accumulated the maximum concentration (μgg^{-1}) (Fe:432, Cd:60, Cu:58.89, and As: 20.8), followed by *R. sceleratus* (Fe:342.67, Cd:49, Cu:46.78, and As:3.32). *M. glabratus* (Fe:68.9, Cd:48.98, and As:4.67), except for Cu, accumulates in shoot up to 21.75 μgg^{-1}. Nevertheless, *P. pectinatus* shoot showed good potential for accumulating Cu 36.45 and As 16.56, Fe 389, and Cd 35.67 in both root and shoot.

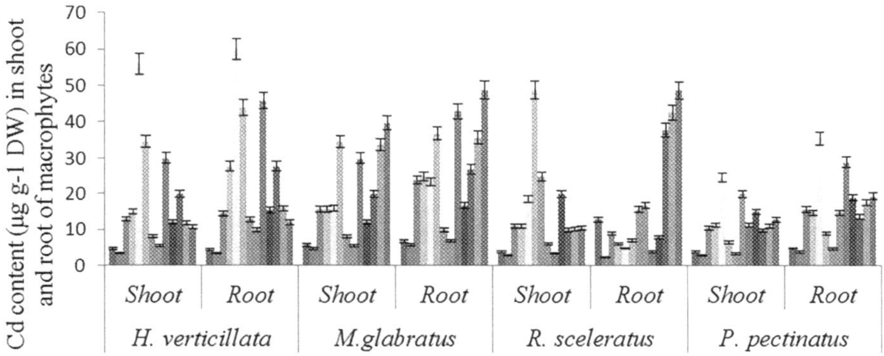

Fig. 11.1 Bioaccumulation of Cd in root and shoot of different aquatic plant species ($n = 6 \pm$ SE)

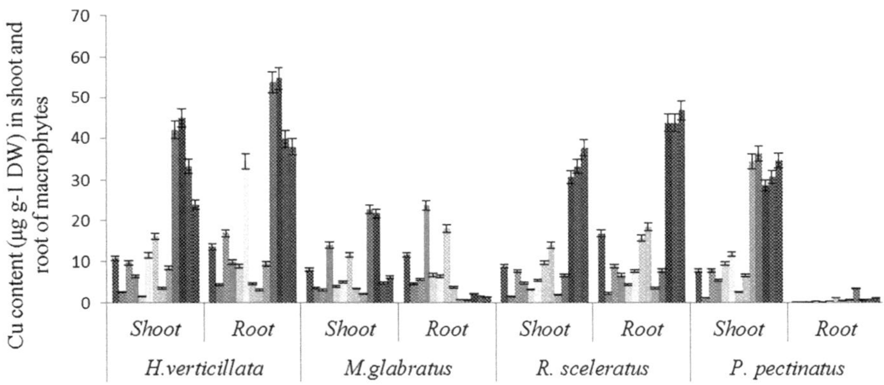

Fig. 11.2 Bioaccumulation of Cu in root and shoot of different aquatic plant species ($n = 6 \pm$ SE)

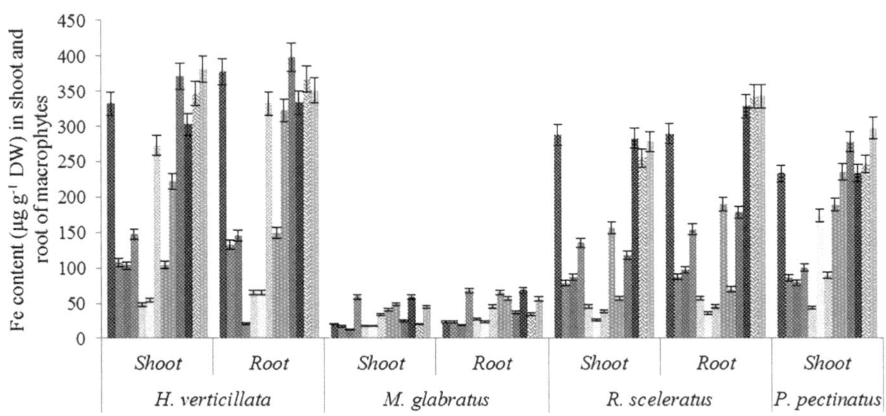

Fig. 11.3 Bioaccumulation of Fe in root and shoot of different aquatic plant species ($n = 6 \pm$ SE)

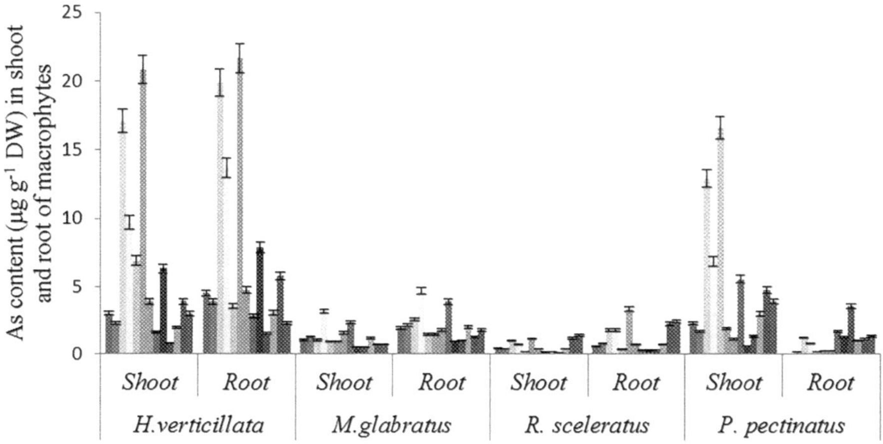

Fig. 11.4 Bioaccumulation of As in root and shoot of different aquatic plant species ($n = 6 \pm$ SE)

11.3.4 Bioaccumulation of Cd in Aquatic Macrophytes

Several investigators have reported the accumulation of Cd by aquatic plants and their consequent toxic effect (Nasu et al. 1983; Sinha and Chandra 1990; Rai et al. 1995). The highest and lowest concentration (μgg^{-1}) of Cd was observed in *H. verticillata* and *R. sceleratus,* respectively (Fig. 11.1). The result indicated that the preferential sequence of Cd accumulation in the studied plant was *H. verticillata* > *P. pectinatus* > *R. sceleratus* > *M. glabratus*. Elevated levels of Cd in all studied plants show that River Gomti is heavily polluted upstream due to untreated waste from the distillery, electroplating, chor-alkali, pigments, plastic, battery, and zinc-refining industries (Holan et al. 1993). Earlier works (Sprenger and McIntosh 1988; Rai et al. 1995) have shown that *H. verticillata* can accumulate more metals than the floating and emergent plants. McLaughlin et al. (1996) also reported that cadmium is a mobile metal in soil and sediment. It is easily transported into the roots of aquatic plants, which support our finding that maximum concentration was found in the root.

11.3.5 Bioaccumulation of Cu in Aquatic Macrophytes

The levels of Cu varied significantly with the variation in sites, plant species, and tissue (Table 11.3). Generally, aquatic macrophytes' roots accumulate higher levels of Cu than shoots. The highest & lowest concentration was observed in the root of *H. verticillata* and *P. pectinatus* (Figure 11.2). The concentration (μgg^{-1}) of Cu in plant tissue ranged from 0.6 to 60 dry weight. A higher pH value (Fig. 11.1) may be attributed to elevated levels of copper in plant tissue (Weis and Weis 2004; Sundby et al. 1998). However, excessive concentrations of this metal are considered highly

Table 11.3 Pearson correlation between physicochemical parameter of river water

Parameter	pH	Hardness	DO.	Nitrate	Nitrite	TSS	TDS	Chloride	Fluoride	Coliform	Fe	Cu	Cd	As
pH	1													
Hardness	−0.1115	1												
DO	0.08068	−0.2497	1											
Nitrate	−0.4277	0.26227	−0.3183	1										
Nitrite	−0.2579	−0.0907	0.16448	**0.5534**	1									
TSS	0.1274	−0.0266	−0.4443	0.24091	−0.0161	1								
TDS	−0.0604	0.44047	−0.3889	0.29988	−0.1831	0.46243	1							
Chloride	−0.1295	**0.71941**	0.05666	0.24232	0.24318	−0.1771	0.30083	1						
Fluoride	0.21155	0.34319	−0.4757	0.03178	−0.1844	0.70042	0.4572	0.06689	1					
Coliform	0.23368	−0.2932	0.341	0.12915	0.48776	0.07746	−0.4076	−0.0551	−0.2204	1				
Fe	0.15921	−0.0765	−0.0685	0.35863	−0.1333	−0.1131	−0.0122	−0.1675	−0.4845	0.19726	1			
Cu	0.09319	**0.67216**	0.03001	−0.0508	−0.1646	−0.1994	0.3434	0.33894	0.11097	−0.0594	−0.0143	1		
Cd	−0.2371	0.15077	0.06607	−0.0525	−0.2884	−0.2216	−0.1616	0.07339	−0.3278	0.33202	0.37214	0.32772	1	
As	0.02871	0.03759	−0.526	−0.1163	−0.3352	**0.70579**	0.4001	−0.3594	**0.50404**	−0.2568	−0.1024	0.1152	−0.0006	1

toxic. The probable source of Cu is mining, smelting, and manufacturing fertilizers and algaecide (Peng et al. 2005a, b).

11.3.6 Bioaccumulation of Fe in Aquatic Macrophytes

Iron toxicity is associated with increased iron uptake and translocation by plant tops (Clements et al. 1974). Fe availability in plants depends on alkaline pH, a characteristic of River Gomti. In the current study, all submerged macrophytes accumulate considerable iron in their corresponding tissue (Table 11.4). The highest concentration (432.56 μgg^{-1} dry weight) of Fe was attained in *H. verticillata* (Table 11.4), followed by (378.41 $\mu g\ g^{-1}$ DW) in *R. sceleratus* (Fig. 11.3), and the minimum concentration (8.89 $\mu g\ g^{-1}$ DW) of Fe was found in *M. glabratus* (Table 11.4). The primary source of iron may be attributed to the disposal of colored clay statues during Dussehra and Ganesh Chaturdashi in the River Gomti.

11.3.7 Bioaccumulation of As in Aquatic Plants

Arsenic has low mobility for translocation from roots to shoots in aquatic macrophytes (Sun et al. 2012). *Pteris vittata*, a fern species, was first discovered as a hyperaccumulator of arsenic that accumulates more than 20 μgg^{-1} in their tissues (Ma et al. 2001). In the present study, *H. verticillata* accumulated the highest concentration (20.8 μgg^{-1} DW), followed by *P. pectinatus* (16.56 μgg^{-1} DW). A minimum arsenic concentration (0.11 μgg^{-1} dry weight) was found in *R. sceleratus* (Fig. 11.4). The probable sources of arsenic are untreated waste from insecticides, pesticides, fungicides, herbicides, glass, and preservatives. *H. verticillata* is a bioindicator of metal pollution and has been tested and validated to remove arsenic, copper, lead, zinc, and chromium from surface water (Lee et al. 1991; Elankumaran et al. 2003; Dixit and Dhote 2010; Srivastava et al. 2010).

11.3.7.1 Statistical Analysis

To see the correlation between heavy metals in submerged macrophytes and surface water, paired sample correlation coefficients (R) were estimated. It was observed that arsenic in the surface water is highly correlated ($R = 0.679$; $p < 0.01$) with total coliform. Moreover, similar results were also reported by Green et al. (2011). Cd in *H. verticillata* shoot and root is highly correlated ($R = 0.968$; $p < 0.01$). However, arsenic in *R. sceleratus* root and *H. verticillata* root was highly correlated ($R = 0.754$; $p < 0.01$). However, Cu and Fe were highly correlated ($R = 0.891$; $p < 0.05$) with fluoride and nitrate content of surface water (Chase et al. 2002). Pearson's correlation revealed that the levels of heavy metals in the same plant's roots and shoots were highly correlated. The data analysis of the current study was carried out using the statistical package IBM SPSS (20) and MS-Excel (2007).

Table 11.4 Translocation factor of the aquatic plants collected at 13 sites

Sites	*H. verticillata*				*M. glabratus*				*R. sceleratus*				*P. pectinatus*			
	Cd	Fe	Cu	As	Cd	Fe	Cu	As	Cd	Fe	Cu	As	Cd	Fe	Cu	As
1	1.051	0.883	0.834	0.674	0.844	0.882	0.696	0.552	0.386	0.993	0.533	0.757	0.794	0.784	23.214	46.183
2	0.963	0.816	0.676	0.593	0.798	0.743	0.785	0.593	1.367	0.965	0.712	0.515	0.773	0.593	4.534	10.675
3	0.894	0.717	0.583	0.862	0.666	0.767	0.568	0.445	1.224	0.956	0.873	0.554	0.672	0.845	24.863	10.733
4	0.548	3.868	0.668	0.717	0.643	0.864	0.594	0.683	1.846	0.882	0.724	0.395	0.771	0.573	12.875	8.712
5	1.073	0.733	0.183	1.952	0.688	0.638	0.623	0.627	3.824	0.876	0.735	0.524	0.897	0.755	15.813	27.563
6	0.791	0.831	0.336	0.968	0.944	0.755	0.793	0.645	7.027	0.724	0.722	0.343	0.694	0.663	28.124	57.621
7	0.637	0.828	3.374	0.819	0.831	0.744	0.655	0.887	1.582	0.843	0.636	0.588	0.722	0.742	26.223	7.323
8	0.558	0.776	1.127	0.573	0.793	0.638	0.923	0.634	0.363	0.836	0.762	0.534	0.723	0.677	1.994	5.074
9	0.654	0.694	0.923	0.814	0.767	0.865	2.643	0.546	0.923	0.861	0.566	0.543	1.355	0.773	11.333	3.334
10	0.799	0.939	0.748	0.534	0.724	0.677	29.251	0.578	2.534	0.667	0.868	0.342	0.393	0.743	43.266	0.496
11	0.727	0.916	0.824	0.642	0.742	0.854	9.937	0.595	0.265	0.863	0.734	0.564	0.792	0.877	10.243	0.374
12	0.751	0.955	0.843	0.674	0.573	0.598	2.374	0.577	0.385	0.757	0.758	0.523	0.724	0.734	36.624	2.993
13	0.919	0.875	0.654	0.767	0.815	0.794	2.823	0.414	0.231	0.815	0.763	0.567	0.673	0.767	28.983	4.414

11.4 Translocation Factor of Aquatic Macrophytes

The ratio of metal concentrations in the shoot to root is called the translocation factor (TF). According to Zacchini et al. (2009), the plant can translocate metals from the root to the shoot. Generally, plants absorb and store pollutants into root biomass and transport them to the shoot. They absorb impurities until they are harvested and disposed of safely (Barman et al. 2000; Gupta et al. 2008). Both processes must be repeated frequently to downgrade the contamination to acceptable levels for phytoextraction purposes. Consequently, we have to identify the plants that can absorb large amounts of contaminants and also have the capacity to transport contaminants, especially in the aerial part, for proper removal. The plants that can accumulate more than 0.1% of contaminants and transport to shoot for proper removal are called hyperaccumulator. According to Baker and Brooks (1989), hyperaccumulator can accumulate more than 1000 $\mu g g^{-1}$ of Cu, Cd, Cr, Pb, Ni, Co, Zn, and Mn in dry matter. Compared to ordinary plants, hyperaccumulator plants can concentrate the 10–500 times pollutants in their corresponding tissue. In recent decades, scientific communities identified hyperaccumulator by estimating their translocation factors (TF). More than one TF corresponds to hyperaccumulator and less for the ordinary plant (Raskin and Ensley 2000; Yanqun et al. 2005).

11.5 Discussion

The hypothesis of the current study is to identify the most appropriate hyperaccumulator of metals among four aquatic macrophytes taken for the current study. It was found that *H. verticillata* is the hyperaccumulator of four metals taken for the current study. In contrast, the highest translocation factor was accredited to *P. pectinatus*. The heavy metals concentration in aquatic macrophytes and their corresponding tissues were found in the order of Fe > Cu > As > Cd. Very low fractions of essential metals were required for plant growth and their physicochemical activities prescribed by regulatory authorities. However, the concentration of heavy metals beyond the permissible limit is toxic to all life forms. Due to industrial effluent, the highest concentration of Cd was observed in the root of *H. verticillata*, that is, 60 mg/L at Mohan Meakin. Apart from Mohan Meakin, several small-scale electroplating, battery, ceramic, and glaze industries produce effluent that may produce Cd in the surface water of the river. The lowest concentration (1.45 mg/L) of Cd was found in the shoot of *R. sceleratus* at Daliganj Pul. The highest concentration (58.89 mg/L) of Cu was observed in the root of *H. verticillata* at Gomti Barrage. The lowest concentration of Cu was found in the roots of *P. pectinatus* at Hanuman Setu. The Cu source may be electroplating, paints, steel industry, fertilizer industry, and results from municipal and industrial effluent discharge. The highest concentration (432.56 mg/L) of Fe in all aquatic plants was observed in the root of *H. verticillata* at Gomti Barrage.

The lowest concentration (8.89 mg/L) of Fe was found in the stem of *M. glabratus* at Gulalaghat. The probable sources of Fe in surface water may be due to the disposal of colored clay statues in the river during the festival, especially Dussehra and Ganesh Chaturdashi. The highest concentration of As was observed in the root of *H. verticillata,* that is, 19.87 mg/L at Kudiaghat site. In contrast, the lowest concentration of As was found in the stem of *R. sceleratus*, that is, 0.11 mg/L at Monkey Bridge. The probable sources of As may be attributed to pesticides, fungicides, insecticides, herbicides, and preservatives. Translocation is the leading factor for metal translocation from roots to shoots, and *P. pectinatus* is more capable of translocating the metal than other aquatic macrophytes species.

11.6 Conclusion

Heavy metal concentration in four aquatic macrophytes was in descending order of Fe > Cu > As > Cd. The *H. verticillata* root can accumulate almost all metal. Still, in the case of metal translocation from root to shoot, it was found that *P. pectinatus* is more capable of translocating the metal than other plant species. In the case of cadmium, *H. verticillata* is the most effective plant because it can accumulate the highest cadmium in the stem. Therefore, these aquatic plants could be employed for phytoremediation strategies.

References

Abida B (2009) Concurrent removal and accumulation of Fe^{2+}, Cd^{2+} and Cu^{2+} from waste water using aquatic macrophytes. Der Pharma Chemical 1(1):219–224

Alloway BJ, Jackson AP, Morgan H (1990) The accumulation of cadmium by vegetables grown on soils contaminated from a variety of sources. Sci Total Environ 91:223–236

Baker AJM, Brooks RR (1989) Terrestrial higher plants which hyper accumulate metal elements: a review of their distribution, ecology, and photochemistry. Biorecovery 1:81–126

Baldantoni D, Alfani A, Tommasi PD, Bartoli G, Santo AVD (2004) Assessment of macro and microelement accumulation capability of two aquatic plants. Environ Pollut 130:149–156

Barman SC, Sahu RK, Bhargava SK, Chaterjee C (2000) Distribution of heavy metals in wheat, mustard, and weed grown in field irrigated with industrial effluents. Bull Environ Contam Toxicol 64:489–496

Chang CP, Zhang Y, Li T (2000) Interannual and interdecadal variations of the East Asian summer monsoon and tropical Pacific S.S.T.s, Part I: Roles of the subtropical ridge. J Clim 13:4310–4325

Chapman D (1996) Water quality assessment: a guide to the use of biota, sediments and water in environmental monitoring, 2nd edn. Chapman and Hall, London

Chase Z, Green AV, Kosro PM, Marra J, Wheeler PA (2002) Iron, nutrient and phytoplankton distribution in Oregon coastal waters. J Geophys Res 107(C10):38–31. https://doi.org/10.1029/200/JC000987

Clements HF, Putnam EW, Suehisa RG, Yee GLN, Wehling ML (1974) Soil toxicities as causes of sugarcane leaf freckle, macadamia leaf chlorosis (Keaau) and Maui sugarcane growth failure. Hawaii Agric Exp Stn Tech Bull 88:52

Deng Y, Engleherdt JD (2006) Treatment of landfill leachate by the Fenton process. Water Res 40: 3683–3694

Deng H, Ye ZH, Wong MH (2004) Accumulation of lead, zinc, copper and cadmium by 12 wetland plant species thriving in metal-contaminated sites in China. Environ Pollut 132:29–40

Dixit S, Dhote S (2010) Evaluation of uptake rate of heavy metals by *Eichhornia crassipes* and *H. verticillata*. Environ Monit Assess 169:367–374

Doula M, Ioannou A, Dimirkou A (2000) Thermodynamics of copper adsorption-desorption by ca-kaolinite. Adsorption 6:325–335

Dunbabin JS, Bowmer KH (1992) Potential use of constructed wetlands for treatment of industrial wastewaters containing metals. Sci Total Environ 111:151–168

Dutta V, Srivastava RK, Yunus M, Ahmad S, Pathak VV, Rai A, Prasad N (2011) Restoration plan of Gomti River with designated best use classification of surface water quality based on river expedition, monitoring and quality assessment. J Earth Sci India 4:80–104

Elankumaran R, Raj MB, Madhyastha MN (2003) Biosorption of copper from contaminated water by *H. verticillata* Casp. And Salvinia sp. Green Pages, Environmental News Sources, Karnataka

Garg RK, Rao RJ, Uchchariya D, Shukla G, Saksena DN (2010) Seasonal variations in water quality and major threats to Ramsagar reservoir, India. Afr Environ Sci Technol 4(2):61–76

Ghassemzadeh F, Yousefzadeh H, Arab-Zavar MH (2008) Arsenic phytoremediation by *Phragmites australis*; Green technology. Int J Environ Stud 65:587–594

Gower AM (1980) Ecological effects of changes in water quality. In: Gower AM (ed) Water quality in catchment ecosystems. John Wiley and Sons Ltd., New York, pp 145–172

Green, AV., Ahmed, KM., Akita, Y., Alam, MJ., Culligan, PJ., Emch, M., Escamilla, V., Feighery, J., Ferguson, AS., Knappett, P., Layton, AC., Mailloux, J.OB., McKay LD., Mey JL., Serre ML., Streatfield PK., Wu J., Yunus M., 2011. Fecal contamination of shallow tubewells in Bangladesh inversely related to arsenic. Environ Sci Technol 45, 1199–1205 https://doi.org/10.1021/es103192b

Gupta S, Nayek S, Saha RN, Satpati S (2008) Assessment of heavy metal accumulation in macrophyte, agriculture soil and crop plant adjacent to discharge zone of sponge iron factory. Environ Geol 55:731–739

Holan ZR, Volesky B, Prasetyo I (1993) Biosorption of cadmium by biomass of marine algae. Biotechnol Bioeng 41:819–825

Johri N, Jacquillet G, Unwin R (2010) Heavy metal poisoning: the effects of cadmium on the kidney. Biometals 23:783–792

Kumar D, Verma A, Dhusia N, More NK (2013) Water quality assessment of river Gomti in Lucknow. Univers J Environ Res Technol 3:337–344

Lee CK, Low KS, Hew NS (1991) Accumulation of arsenic by aquatic plants. Sci Total Environ 103:215–227

Ma LQ, Komar KM, Tu C, Zhang W, Cai Y, Kennelley ED (2001) A fern that hyper accumulating arsenic. Nature 409:579

Martinez EA, Nyamboli CS (2011) Determination of selected heavy metal concentrations and distribution in a southwestern stream using macrophytes. Ecotoxicol Environ Saf 74:1504–1511

McLaughlin MJ, Tiller KG, Naidu R, Stevens DP (1996) Review: the behaviour and environmental impact of contaminants in fertilizers. Aust J Soil Res 34:1–54

Milner MJ, Kochian LV (2008) Investigating heavy-metal hyperaccumulation using Thlaspi caerulescens as a model system. Ann Bot 102(1):3–13

Nasu Y, Kugimoto M, Tanaka O, Takimoto A (1983) Comparative studies on the absorption of cadmium and copper in *Lemna paucicostata*. Environ Pollut Ser A Ecol Biol 32:201–209

Peng HY, Yang XE, Jiang LY, He ZL (2005a) Copper Phytoavailability and uptake by Elsh {o} ltiza Splendus from contaminated soil as affected by soil amendments. J Environ Sci Health 40: 839–856

Peng HY, Yang X, Tian S (2005b) Accumulation and ultra structural distribution of copper in Elsholtzia splendens. J Zhejiang Univ Sci B 6:311–318

Rai UN, Sinha S, Tripathi RD, Chandra P (1995) Wastewater treatability potential of some aquatic macrophytes: removal of heavy metals. Ecol Eng 5:5–12

Raskin I, Ensley BD (2000) Phytoremediation of toxic metals. John Wiley and Sons, Inc, New York

Singh KP, Malik A, Sinha S (2005) Water quality assessment and apportionment of pollution sources of Gomti river (India) using multivariate statistical techniques—cases study. Anal Chim Acta 538:355–374
Sinha S, Chandra P (1990) Removal of cu and cd from water by Bacopa monnieri L. Water Air Soil Pollut 51:271–276
Sinicrope TL, Langis R, Gersberg RM, Busnardo MJ, Zedler JB (1992) Metal removal by wetland mesocosms subjected to different hydroperiods. Ecol Eng 1:309–322
Souza MS, Hallgren P, Balseiro E, Hansson L-A (2013) Low concentrations, potential ecological consequences: synthetic estrogens alter life-history and demographic structures of aquatic invertebrates. Environ Pollut 178:237–243. https://doi.org/10.1016/j.envpol.2013.03.038
Sprenger M, McIntosh AW (1988) Relationship between concentrations of aluminium, cadmium, leads and zinc in water sediments and aquatic macrophytes in six acidic lakes. Achiev Environ Contaminat Toxicol 18:225–232
Srivastava S, Mishra S, Dwivedi S, Tripathi RD (2010) Role of thiol metabolism in rsenic detoxification in *H. verticillata (L.f.) Royle*. Water Air Soil Pollut 212:155–165
Sun GX, Wiele TVD, Alava P, Tack F, Laing GD (2012) Arsenic in cooked rice: effect of chemical enzymatic and microbial process on bioaccessibility and speciation in the human gastrointestinal tract. Environ Pollut 162:241–246
Sundby B, Vale C, Cacador I, Catarino F, Madureira MJ, Caetano M (1998) Metal-rich concretions on the roots of salt marsh plants: mechanisms and rate of formation. Limnol Oceanogr 43:245–252
Weis JS, Weis P (2004) Metal uptake, transport and release by wetland plants: implications for phytoremediation and restoration. Environ Int 30:685–700
WHO (1993) Guidelines for drinking-water quality. Volume 1 Recommendations, 2nd edn. World Health Organization, Geneva
Yanqun Z, Yuan L, Jianjun C, Haiyan C, Li Q, Schvartz C (2005) Hyperaccumulation of Pb, Zn and Cd in herbaceous grown on lead–zinc mining area in Yunnan, China. Environ Int 31:755–762
Zacchini M, Pietrini F, Scarascia Mugnozza G, Iori V, Pietrosanti L, Massacci A (2009) Metal tolerance, accumulation and translocation in poplar and willow clones treated with cadmium in hydroponics. Water Air Soil Pollut 197:23–34

Enhanced Effluent Treatment and Bioelectricity Generation Using Coupled Constructed Wetland-Microbial Fuel Cell (CW-MFC) Technology: Challenges and Opportunities

12

Himani Chandel, Kashika Keshari, Sibiraj Murugesan, Rahul Yadav, Kirti Shyam, Navneet Kumar, Geetansh Sharma, and Gaurav Saxena

Abstract

The depletion of natural resources for freshwater generation and energy production is occurring at an exceptional pace. Constructed wetlands (CWs) combined with microbial fuel cells (MFCs) are possibly the creator of clean energy while also remediating effluents for pollution prevention and control. The exudates from the roots acting as native substrate in combination with the metabolic processes in microbes, this environmental biotechnology technique is a very promising bioelectrochemical technology for conversion of sunlight into clean energy. The presence of organic matter in CWs as a result of wastewater characteristics with the redox gradient occurring naturally in midst of CWs treatment bed layer under aerobic conditions and the layer under anaerobic conditions giving synergistic effect to MFC and CW. This review focuses over providing an incisive information on the current position of CW-MFC systems,

H. Chandel · K. Shyam · N. Kumar · G. Sharma · G. Saxena (✉)
EMBL-Environmental Microbiology and Biotechnology Laboratory, EERG-Ecotoxicology and Environmental Remediation Group, School of Biotechnology, Shoolini University, Solan, Himachal Pradesh, India

MATER-Microalgae Technology for Environmental Resources, EERG-Ecotoxicology and Environmental Remediation Group, School of Biotechnology, Shoolini University, Solan, Himachal Pradesh, India
e-mail: gauravsaxena@shooliniuniversity.com

K. Keshari · S. Murugesan
EMBL-Environmental Microbiology and Biotechnology Laboratory, EERG-Ecotoxicology and Environmental Remediation Group, School of Biotechnology, Shoolini University, Solan, Himachal Pradesh, India

R. Yadav
School of Biological and Environmental Sciences, Shoolini University, Solan, Himachal Pradesh, India

S. Kumar et al. (eds.), *Aquatic Macrophytes: Ecology, Functions and Services*,
https://doi.org/10.1007/978-981-99-3822-3_12

including a brief about the characteristics of the technology's evolution, such as clean energy generation, designs, plant species, rhizodeposits, electrode materials, wastewater treatment, factors influencing, merits, demerits, challenges, and future perspectives.

Keywords

Wastewater · Environmental pollution · Constructed wetlands · Microbial fuel cell · Bioelectricity · Environmental safety

12.1 Introduction

At an unprecedented rate, natural resources for freshwater generation and energy production are being depleted. Water scarcity and contamination from agriculture and industry are major threats to domestic freshwater supplies in the twenty-first century, making a reliable supply a top priority (Albert et al. 2020). A variety of wastewater treatment methods have been used to deal with the environmental threat posed by wastewater. Reverse osmosis, activated sludge, membrane filters, and trickling filters are among the wastewater treatment methods now in use cleaning the industrial and municipality wastewater discharge. But when it comes to their operational costs and energy requirements, they are not very productive, and they also have a huge carbon footprint (Singh et al. 2019).

Wetland engineering, also known as constructed wetland or reed bed treatment, is an ecologically friendly and thrifty wastewater treatment technology that can be put to good to establish a water-energy nexus like that of a natural wetland (Bharagava et al. 2017; Vymazal 2010; Valipour and Ahn 2016). While the old technologies that need a large investment in both time and resources, constructed wetland (CW) technology utilizes natural processes to remove contaminants efficiently and is environmentally safe (Rabbani et al. 2021). Traditional technologies are also more expensive to create, operate, and maintain than CW technology (Li et al. 2014). Treatment in CWs is the result of ecological interactions occurring simultaneously (Chandra and Kumar 2015). As a result of this system's operation and the involvement of numerous species, a variety of adsorption and retention mechanisms as well as filtering processes such as redox, precipitation, microbial decomposition, and transformation have all played a role. In shallow basins, a filter media (typically gravel) is put, and the basins are filled to the top with water (macrophytes).

Wastewater from a wide range of sources has been successfully treated by CW systems: agricultural (Liu et al. 2019a, b), municipal (Ma et al. 2019), septic tank wastewater (de Rozari et al. 2018), urban (Benvenuti et al. 2018), industrial (Saeed and Khan 2019), mining effluents (Sheridan et al. 2018), runoff stormwater (Schmitt et al. 2015), as well as the percolation of water from solid wastes (Mulamoottil et al. 2018). Subsurface flow wetlands, horizontal and vertical flow CWs, and other forms of CWs exist. The first type of CW is one of the most often used (García et al. 2010). It has a porous gravel base that allows wastewater to flow beneath the surface.

Macrophyte plants cover the gravel bed's surface, while a diverse microbial population lives in biofilms connected to the gravel and roots (Corbella et al. 2015). Although CWs have been demonstrated to be effective for conventional and nonconventional pollutants, they have some inherent limitations such as substrate clogging and limited pollutant removal effectiveness, which limit their further deployment. The limited efficacy of the CW technique in removing nitrogen is perhaps its most critical flaw. Relatively low rates of total nitrogen (TN) removal within the range of 40–50% has been observed in conventional wastewater (CW) systems where nitrogen inputs ranged from 0.6 to 2 g N m^{-2} d^{-1} density (Liu et al. 2019a, b; Li et al. 2014).

Potentially, increased nitrogen elimination can be achieved with constructed wetlands in conjunction with various biological processes (Srivastava et al. 2018; Liu et al. 2015). The Bioengineered device called the microbial fuel cell, which substantially produces energy while also remediating wastewater, proves to be a promising equipment for implementation. Traditional treatment methods like activated sludge or septic tanks have significant drawbacks in terms of aeration and waste disposal expenses, as well as important outputs like hydrogen or power (Ramirez-Vargas et al. 2019). A cathode, an anode, a separator, and a circuit external integrate a bioelectrochemical system that generates energy by oxidizing organic matter. MFCs are based on the redox gradient shift that occurs during substrate digestions in microorganisms (Xu et al. 2016; Wu et al. 2020). Since MFCs can process a wide variety of materials, they can remove contaminants including heavy metals, nonmetallic inorganics, and poorly biodegrading organic compounds (Saxena et al. 2020; Zhang et al. 2020; Wu et al. 2019). By using MFCs in WW treatment systems, electron acceptors in anaerobic conditions can be increased, facilitating the decomposition of organic pollutants (De et al. 2008). As per the study by Tender et al. 2002, anode sulphide oxidation may enhance organic material oxidation rates due to the presence of MFCs in an environment lushed with inorganic anions of sulfur. Acetate is also a substrate for exoelectrogenic microbes, which reduces the carbon supply for methane-producing bacteria. Methane emissions (another greenhouse gas) may be reduced during anaerobic wastewater treatment due to competition between exoelectrogenic bacteria and methane-producing bacteria.

Even though CWs and MFCs have both been extensively explored, the CW-MFC fusion has appeared recently as an emerging technology. Microorganisms in the plant's rhizosphere produce bioelectricity in the CW-MFC, which is an electrochemical device (Nitisoravut and Regmi 2017). Redox gradients are innate in constructed wetland (CW) systems and are quite similar to those found in MFC systems, such as anaerobic zones at air-water interfaces, and aerobic zones deeper in. CW-MFCs differ from MFCs in the design of the system with the minimal difference includes the supply of organics and the primary aim of the reactors. According to Yadav et al. 2012, CW-MFCs utilize both rhizodeposits and wastewater to feed the electrogenic bacteria.

There are only a few publications on CW-MFCs, and this chapter provides the existing state of CW-MFC succinctly and clearly. The functions of microbes, wetland plants, electrodes, and substrates in the mechanism were extensively reviewed as well as debated. Challenges as well as possible future avenues for improvement in operational efficiency and practical application were also examined and presented. CW-MFC operation mechanisms were also examined in this work, to shed light on their significance and provide direction for CW-MFC development.

12.2 Fundamentals of CW-MFC System

Lakes, ponds, reservoirs, and marshes are naturally self-cleaning. Several microbial activities that are dependent on redox potential collaborate to facilitate pollution removal through the simultaneous oxidation or reduction of several components. Electrochemical reactor design and inclusion into a wetland, for example, is influenced by research into plants, microorganisms, and electrochemical processes. It is therefore necessary to combine these aspects into a biological organization consisting biological as well as inorganic components for the generation of clean energy. There are two parts to the biosystem of a CW-MFC: the biocontrol and the bioprocess. The plant is a part of the biocontrol structure since it gets sunlight as an input for its photosynthetic processes, which produce electricity (exudates). Because it uses sunlight to power its photosynthetic functions, the plant is an integral member of the biocontrol system (exudates). It is the microbial population, which consumes the secretion of fluids from the roots then building energy accompanied by the microbial metabolism, that constitutes the bioprocess structure (Nitisoravut and Regmi 2017). Several electron donors and acceptors found in wastewater are utilized by microbes for their growth. There are only a finite number of electron acceptors, therefore they eventually run out. Anode and cathode addition to deficient zones within CW-MFC, in contrast, adjusts redox chemistry, which in turn influences microbial activity, by providing an infinite supply of electron acceptors (anode) (Wang et al. 2016b). According to Fang et al. (2017), this stratified redox gradient is analogous to the MFC two-chambered cell's redox reactions in one of its two-chambered cells, which have an aerobic layer on the surface and an anaerobic layer underneath.

In a traditional integrated system, proton exchange membranes, fibrous materials, or lands separate the anode and the cathode (Santoro et al. 2017). During biochemical events in the anodic chamber, rhizodeposition-released root exudates from aquatic plants are taken up by anodic bacteria that create electrons and protons (Corbella et al. 2017; Singh et al. 2018; Wetser 2016). Cathodic chamber receives protons via the separators. An external resistance completes the electrical circuit, allowing the negative sub-atomic particles to flow into the anode fragment. The negative and positive subatomic particles are collected in the fragment of cathode whereby reduction of O takes place, producing vaporized water, which is then let outside (Helder et al. 2012; Regmi et al. 2018). Exudates or carbon compounds are released by the submerged plant's root system in a process known as "rhizodeposition," which serves as a carbon source for denitrifiers to remove nitrate from the water (Aguirre Sierra 2017; Doherty et al. 2015a, b). The attachment surface that wetland plants provide for bacterial breakdown is also critical in sustaining a diverse range of microorganisms.

CW-MFC has been used with a variety of aquatic plants, including *Pennisetum setaceum* (Chiranjeevi et al. 2012), *Cyperus involucratus* (Nattawut and Holasut 2015), *Phragmites australis* (Villaseñor et al. 2013), *Ipomoea aquatica* (Fang et al. 2013a, b), *Acorus calamus* (Yan et al. 2015), *Echinola glabrescens* (Bombelli et al. 2013), *Iris pseudacorus* (Wu et al. 2015a), *Typha angustifolia* (Saz et al. 2018), *Echinorriea crassipes* (Mohan et al. 2011), *Lolium perennee* (Habibul et al. 2016), *Asparagus fern* (Manohar et al. 2017), *Canna indica* (Regmi et al. 2018), *Sporobolas arabicus* (Gilani et al. 2016), *Typha latifolia, Elodea nuttallii* (Oon et al. 2015a, b), *Alocasia macrorrhiza* (Zaman and Wardhana 2018), *Glyceria maxima* (Strik et al. 2008a, b), *Typha orientalis, Scripus Validus* (Wang et al. 2017a, b, c, d), *Chlorophytum inornatum, Oriza sativa* (Kouzuma et al. 2014), *Juncus effuscus* (Ramírez et al. 2018), and *Sedum kamschaticum* (Tapia et al. 2018).

12.3 Configurations in CW-MFC System

The hybrid CW-MFC system utilizes ecological interactions among the substrates, microorganisms, as well as the plants within a constructed wetland to integrate MFCs into the ecosystem for remediating water and production of clean energy (Sierra et al. 2017; Yang et al. 2018a; Yan et al. 2018). Hybrid systems that combine the advantages of the two methods can achieve significant levels of water reuse and bioenergy production (Yang et al. 2018a). MFCs, like standard integrated system, have two main regions: anaerobic and aerobic, where the cathode and anode materials functioning as electrodes are located (Yang et al. 2018a, b). Titanium wires with external resistance are used to link the electrodes and generate energy. As electrode materials, carbon and graphite are frequently used because of their nonoxidative properties, high electrical conductivity, and the fact that they act as an excellent medium for microbial attachment and growth (Doherty et al. 2015a, b; Yadav et al. 2018). Separators commonly act as separators between the lower anaerobic zone and the upper cathodic region. Redox gradients are formed in the system so that electron transport can take place (Figs. 12.1 and 12.2).

A glass wool separator was utilized in the initial CW-MFC built by Yadav et al. (2012) in a regime of upflow, and other investigations by Villaseor et al. (2013, b), Doherty et al. (2015a), and Yang et al. (2018a) used a bentonite layer as separators as used previously. The use of separators has been avoided in recent research because of the system's jamming up vulnerability as well as increased system's internal resistance, which results in poorer energy generation. In practice, CW-MFC systems consist of two groups based on the flow pattern: vertical flow CW-MFC (Fang et al. 2015; Doherty et al. 2015a; Oon et al. 2015a, b, 2017a, b; Zhao et al. 2013) and horizontal flow CW-MFC (Corbella et al. 2015; Villaseňo et al., 2013, b; Wei et al. 2015). There have, however, been experiments with alternative types of flow patterns such as vertical and the horizontal subsurface flow (VSSF, HSSF), hybrid subsurface flow CW and flow CW (Corbella et al. 2016; Türker and

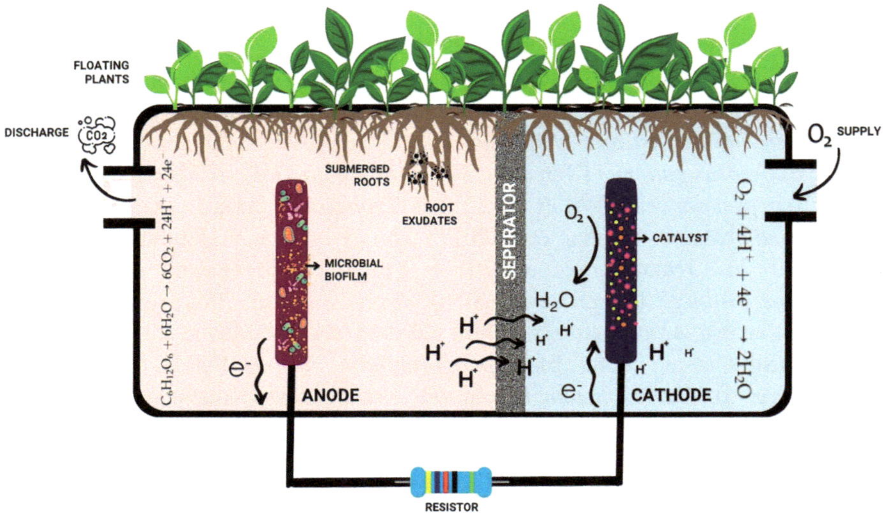

Fig. 12.1 An illustration of a constructed wetland-microbial fuel cell (CW-MFC). (Adopted from Guadarrama et al. 2019)

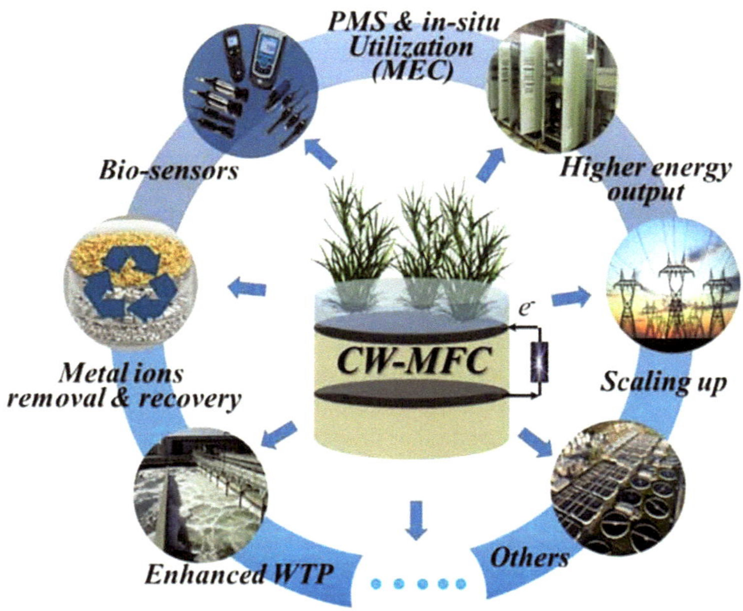

Fig. 12.2 Future research needs in CW-MFC system (PMS, power management system; MEC, microbial electrolysis cell). (Adopted from Wang et al. 2017a, b, c, d)

Yakar 2017; Wu et al. 2016). The one used most often is a CW-MFC hybrid flow, in which wastewater is initially routed through a horizontal CW (HFCW) before being routed vertically (VFCW).

12.4 Operation in CW-MFC System

Generally, an integrated system of CW-MFC has two compartments: an anodic chamber and a cathodic chamber, which are partitioned using lands, fibrous materials, or proton exchange membranes. As eq.1 shows, the anodic chamber, microorganisms break down exudates released from roots or contaminants to release protons and electrons through anaerobic digestion (Kabutey et al. 2019). External circuits and current collector send electrons to the cathode, while the separator separates protons from electrons. In eq. 2, according to Santoro et al. (2017), water and bioelectricity are produced in the fragment of cathode by reducing oxygen involving electrons and protons. O acts most often as the acceptor for negative subatomic particles at cathodic fragment because of the increased redox potential and long-term stability. Biochemically, microorganisms around the cathode of an MFC can convert nitrate to nitrite, making it a viable electron acceptor (Zhang and He 2012). In CWL-MFCs, further research should be done into the use of nitrate as an oxidizer. The generation of the current occurs by the movement of negative and positive subatomic particles through a circuit, which generates electricity and allows for the creation and restoration of power.

Acetate reduction at anodic fragment:

$$CH_3COO^- + 4H_2O \rightarrow 2HCO_3 - + 9H^+ + 8e^- \tag{12.1}$$

O_2 reduction at cathodic fragment:

$$O_2 + 4H^+ + 4e^- \rightarrow 2H_2O \tag{12.2}$$

It is not uncommon for the anode electrode of a CW-MFC reactor similar to a CW to be supported in its lower anaerobic zone by a soil layer or gravel or modern substrates, for instance, activated carbons, zeolites, or alum sludge. Additionally, this support matrix generates the ideal conditions for microorganisms to carry out oxidation reactions, which remove contaminants and transfer electrons (Yang et al. 2018a; Yadav et al. 2012, 2018; Doherty et al. 2015a). The integrated system must have all of these components to function. Planting macrophytes in the upper cathode compartment allows the cathode to get oxygen via plant root respiration, hence boosting the rate at which pollutants are removed.

12.5 Performance in CW-MFC System

Wastewater is frequently contaminated with a large selection of stubborn and dangerous contaminants, for example, textile dyes, personal care products (PCPs), and antibiotics. Wetland plants and organic loads have a direct impact on CW-MFC energy production and sewage purification efficacy (Liu et al. 2014; Sharma and Li 2010; Yoongling et al. 2015; Srivastava et al. 2015; Fang et al. 2013a, b). The performance of CW-MFCs is significantly impacted by COD loading. There must be an adequate supply of organics at the anode, but restriction on COD at the cathode to maintain a good equilibrium. The rise in the concentration of COD in the influent from 50 mg/L to 250 mg/L power densities improved within the vertical upflow type CW-MFC built by Liu et al. (2014). Further increases from 250 to 500 mg/L and 1000 mg/L achieved PDs of 33.7 and 21.3 mW/m^2 rather than 44.63 mW/m^2 at 250 mg/L COD concentrations, as previously reported in the literature. With an increase in cathodic COD, heterotrophic biofilms on the electrode may impede the diffusion of reactants and the products (Freguia et al. 2008).

Yadav et al. (2012) successfully enhanced the elimination of synthetic wastewater dye while achieving 1.57 mW/m^2 power density, using a combination of MFC and CW technologies. Similarly, in an investigation of Fang et al. (2013a, b), bioelectrodes improved dye decolorization and generated a 302 mW/m^3 power density within CW-MFC. Aromatic amines (a refractory chemical) can affect the environment when the azo dye is decolored in an anaerobic atmosphere. Amines have previously been considered to be destroyed aerobically, hence CW-MFCs can only be employed for amine treatment because of the aerobic layer. HY99, a new bacterium discovered by Kahng et al. (2000), can be used in either an aerobic or anaerobic setting to cure anilines. The anode chamber of MFCs generated energy that eliminated p-nitroaniline (Liu et al. 2014). In CW-MFCs, HRT has a considerable impact on treatment efficacy (Wu et al. 2015b; Zhang et al. 2018). Numerous parameters, including DO, HRT, filtering media, external resistances, and the CW-MFC's recirculation ratio were discovered to vary the CW performance. HRT was determined to be the most important design component for treatment efficiency, influencing TN removal by more than 45% and COD, TP, and NH4 + -N removal by more than 50%. Increasing the HRT in CW-MFC allows for more time for bacteria and substrate to come into touch with each other, allowing for improved adsorption and degradation of pollutants.

Design variables such as flow regime, the quantity of organic matter, plant impacts, filter medium, and reaction speed in oxidation and reduction inside the column all have an impact on the dissolved oxygen concentration. This concentration in turn influences the efficiency of CW-MFCs. In an upflow CW-MFC column, the amount of DO changes greatly, concentrating heavily in the upper parts and sparsely in the lower ones. Liu et al. (2013) with his studies on CW-MFC discovered that the depth varied by a V shape with a height of 50 cm, which is consistent with these variations. The DO gradually decreased up the column from 3.10 mg/L at reactor's base to 0.15 mg/L at 30 cm, leading to a sharp spike and continued to rise to the column's surface (4.31 mg/L). The CW-MFC's DO discrepancies at the bottom and top can be attributed to one of two things. One is photosynthesis and reoxygenation, and the other is organic matter decomposition by microorganisms. Organic matter concentration, type of plant, and chemical response of half-cell all played a significant role in the DO variance, according to the research.

The effectiveness of the integrated system can be varied due to the size and type of electrodes employed. While developing a CW-MFC system, it is also critical to choose appropriate electrode materials. The optimal electrode material must have fine conductance capacity for the current as well as the strength, higher hardness, biocompatibility, and electrochemical stability, while providing an attachment surface for the microorganisms. Electrodes made of C are commonly utilized because they both have excellent electrical conductivity, cannot be oxidized spontaneously, also give microbial adsorption spots. They outperform stainless steel, brass, Ni, and Al in every way. In bioelectrochemical systems like CW-MFCs, a higher cathodic oxidation-reduction reaction (ORR) and lower internal resistance are achieved with a huge surface area of the electrode which has been extensively documented in the literature (Ghangrekar and Shinde 2007; Ebrahimi et al. 2017; Tang et al. 2019). According to the study's findings, the surface area of electrode in MFCs is negatively correlated with the internal resistance of the cells (Fan et al. 2008). Increasing the size of electrodes reduces the system's internal opposition by eventually exceeding the energy generation.

Macrophytes play in wastewater treatment is another crucial one for the plant component of CW-MFC. Species of wetland plants significantly impacts the pollutant removal capacity, according to the research by Liu et al. (2016). Liu et al. (2013) employed the fluid secreted from *Ipomoea aquatica* plant's roots, powers a photosynthetic MFC in their research. Both the efficiency with which pollutants eliminated from the nature and the composition of the microorganisms in wetlands are strongly influenced by the plant species present there, according to an investigation by Liu et al. (2013). Since *Phragmites australis* has greater radial oxygen loss (ROL) than *Iris pseudacorus*, it can improve the nitrification process in the rhizosphere. Researchers at the University of Waterloo in Canada used an *Elodea nuttallii* CW-MFC to remove 98% of the COD in their sample.

Electroactive bacteria (EAB) and electrons are microorganisms that are critical to the operation of CW-MFCs, and substrate or filter media play an important role in supporting and storing various pollutants in these devices. Substrates and filtration media affect the performance of CW-MFCs. The anodic compartment is oxidized by EABs, while the cathodic compartment is reduced by chemical or microbiological reactions. The CW- MFC's wastewater treatment and energy-generating performance is mostly influenced by the microbial community makeup (Wang et al. 2016a). Electron transfer and pollution elimination depend heavily on the metabolic processes of microbes, which are encoded by their functional genes (Xu et al. 2018a). It is widely accepted that EABs are a critical component of superior CW-MFC's performance.

Bacteroidetes, Proteobacteria, Spirochaetes, Firmicutes, Actinobacteria, Lentisphaerae, Acidobacteria, and Chloroflexi are the bacterial phyla most commonly reported in CW-MFCs (Li et al. 2019, 2016; Aguirre-Sierra et al. 2016; Xu et al. 2018a; Corbella et al. 2015; Lu et al. 2015; Wang et al. 2017a; Ramírez-Vargas et al. 2018). The members of levels of different class and family, *Gammaproteobacteria, Alphaproteobacteria, Anaerolineae, Deltaproteobacteria,* and *Betaproteobacteria Anaerolineaceae, Desulfovibrionaceae, Bacteroidales, Xanthomonadaceae, Rhodocyclaceae, Spirochaetaceae, Weeksellaceae, Oxalobacteraceae, Flavobacteriaceae,, Comamonadaceae, Clostridiaceae, Desulfobulbaceae, Planctomycetaceae, Bacteroidaceae, Rhizobiales,* and *Nocardioidaceae* have been shown to exist in high numbers (Xu et al. 2018a; Wang et al. 2016a, b, 2019). As far as electricity-generating bacteria go, *Geobacter* and *Dechloromonas* are household one (Corbella et al. 2015; Wang et al. 2016b). In the anode region, dye-decolorizing microorganisms *Desulfovibrio* and *Clostridium* have been found (Rathour et al. 2019).

According to Liu et al. (2020a) investigation, the differing amounts of EAB found in the cathode and anode fragments of CW-MFCs are indicative of their respective roles in the two operating modes. As an illustration, the anaerobic anode-respiring bacterium *Comamonas* has been shown to generate electricity when placed at the anode section instead of the cathode section. *Thauera and Bacillus* are two genera of denitrifying bacteria have been found in the cathodic section according to reports (Liu et al. 2020a). Xu et al. (2018a) found that *Geobacter, Desulfovibrio, and Bacillus* were the most common bacteria in the cathodic section. CW-MFCs rely on these microbes to enhance the entire process by transferring electrons directly or indirectly. More research into CW-MFC microorganisms and their roles in various processes are required for these systems to work to their full potential.

12.6 Applications in Effluent Treatment and Bioelectricity Generation in CW-MFC System

These systems were initially developed to generate bioelectricity from rhizodeposition but have now been applied to wastewater treatment because of their promising performance in this area despite their modest power output (Nattawut and Holasut 2015). Organic matter oxidation and energy production synergistically clearly depicts the significant role of CW-MFCs in wastewater treatment compared to traditional CWs (Wang et al. 2017a, b, c, d). For COD and total nitrogen removal, the efficiency of the three macrophytes used in a CW-MFC was 5.8% higher than that of an unplanted one (*T. orientalis*, *S. validus*, and *Juncus effusus*) (Rincon et al. 2013). Under various operating conditions, Wu et al. (2015a) reached the most compact density of power with the value of 9.6 mW/m^2 by employing a run of a small-scale CW on a membrane-less MFC. Batch flow mode removed 91.2% of COD and removed 95 to 99% of ammonium nitrogen, TN, and TP. Bioelectricity output was affected by the starting COD concentration and the temperature. Wetland microbial fuel cells planted with a variety of plant species like *Typha latifolia, Ipomoea aquatica, cattail, Phragmites australis, and Canna indica* produced more power than sediment-based systems alone in one experiment. Table 12.1 summarizes research of utilization of integrated system for cleaning effluents.

Table 12.1 CW-MFC system for wastewater treatment and bioelectricity generation

Plant	Cathode	Anode	Wastewater/ contaminants	Wastewater characteristics	Process Type	Removal efficiency (%)	Maximum power density	Maximum current density (%)	Reference
Typha latifolia	Carbon felt	A and C	Dye WW	Sodium benzoate: 107, sodium acetate: 205, NH_4NO_3: 176, AR18: 100–500	VCW-MFC	COD: 77, dye removal: 96	1.58 mW/ m^2	CE: 1.38	Oon et al. (2018, 2020)
Fimbristylis dichotoma	Stainless steel plates	A and C	Dye WW	COD: 1580, BOD: 705, TDS: 2170, TS: 2248, nitrate: 193, Sulphate: 867, ADMI: 1712	VCW-MFC	COD: 70, NO_3: 61, TDS: 91, BOD: 94, ADMI: 82	199 mW/ m^2	235 mA/m^2	Rathour et al. (2019)
Ipomoea aquatica	SSM	GAC	Dye WW	Glucose: 50–210, NH_4Cl: 310, ABRX3: 150–425	VCW-MFC	COD: 86, Decolorization rate: 91	910 mW/ m^3	850 mA/ m^2, CE: 5.1	Fang et al. (2013a, b, 2017)
Ipomoea aquatica	SSM	GAC	Dye WW	Glucose: 140, X-3B: 166.7, COD: 200	VCW-MFC	Decolorization rate: 93	200 mW/ m^3	NA	Li et al. (2016)
Canna indica	NA	NA	Dye WW	Sucrose: 8000, NH_4Cl: 5000, MB: 2000	VCW-MFC	COD: 75, MB: 80	15.7 mW/ m^2	69.8 mA/m^2	Yadav et al. (2012)
Canna indica	SSM	GAC	Pharmaceutical WW	Glucose: 200, NH_4Cl: 25, SMX: 5–100 μg/L, TC: 5–50 μg/L	VCW-MFC	Glucose: 200, NH_4Cl: 25, SMX: 5–100 μg/ L, TC: 5–50 μg/L	125 mW/ m^3	NA	Wen et al. (2020)
I. aquatica, A. calamus, O. javanica	Carbon fiber felt and SSM	GAC	Pharmaceutical WW	Glucose: 200–400, NH_4Cl: 25, IBP:10, BPA:10, SDZ: 2, CIP: 2, SMX: 0.95	VCW-MFC	COD: 98, NH_4: 64, IBP: 96, BPA: 91, SDZ: 87, SMX: 75	18.4 mW/ m^2	49.3 mA/m^2	Ebrahim et al. (2021)
Typha latifolia L.	Graphite node	Graphite node	Boron-containing WW	Boron;12.3	HCW-MFC	NO_3: 47.9, NO_2: 19.1, boron: 63.4	78 mW/ m^{-2}	105 mA m^{-2}	Türker and Yakar (2017)
Ipomoea aquatica	SSM	GAC	Pharmaceutical WW	Glucose: 400, NH_4Cl: 135, SDZ: 4	VCW-MFC	COD: 94, SDZ: 99.8	15.4 mW/ m^2	NA	Song et al. (2018)
Oenanthe javanica	SSM	GAC and TM	Pharmaceutical WW	Glucose: 125–250, NH_4Cl: 25–130, SMX;1.74–30, TC;02–1.6	VCW-MFC	SMX: 97	57.8 mW/ m^2	1200–5800 mA/ m^3	Zhang et al. (2016, 2018)
				–				–	

(continued)

Table 12.1 (continued)

Plant	Cathode	Anode	Wastewater/ contaminants	Wastewater characteristics	Process Type	Removal efficiency (%)	Maximum power density	Maximum current density (%)	Reference
Ipomoea aquatica	Granular activated carbon (with stainless steel)	Granular activated carbon (with stainless steel)	Bisphenol A and ibuprofen		UCW-MFC	95% for ibuprofen, 91% for bisphenol	28.63 mW/m^2		Li et al. (2019)
Canna indica	Plate graphite	Plate graphite	Saline WW	COD: 3040, TDS: 2650	VCW-MFC	COD: 88, TDS: 82	1120 mWh/m	NA	Das et al. (2019)
Phragmites australis	Titanium mesh	Activated sludge, biochar and titanium mesh	Saline WW	NaCl: 1000–5000	VCW-MFC	COD: 65, TN: 71, TP: 86, NH_4: 80	16.4 mW/m^2	65.6 mA/m^2	Xu et al. (2016)
Canna indica	Carbon felt	Activated carbon and SSM	Saline WW	COD: 505–1045, NH_4: 215–306	VCW-MFC	COD: 88, NH_4: 77.5, NO_3: 49	456 mW/m^3	22.5 mA/m^2	Liu et al. (2020a, b)
Phragmites australis	Graphite plate	Graphite plate	Saline WW	COD: 1060, TN: 147, NH_4: 89.9	VCW-MFC	COD: 77, TN: 50, TP: 66, NH_4: 77, SS: 93, PO_4: 46	9.4 mW/m^2	CE: 0.1–0.6	Zhao et al. (2013)
Iris tectorum	Activated carbon with titanium mesh	ActivateD carbon with titanium mesh	Pb(II)-containing WW	–	VCW-MFC	COD: 85.4, NH_4: 78.3, TN: 76.7, TP: 89.2, Pb(II): 84.9	7.43 mW/m^2	37.2 mA/m^2	Zhao et al. (2020)
Phragmites australis	GAC and graphite rod	GAC and graphite rod	Saline WW	COD: 411–854, TN: 63, NH_4: 40, TP: 8.9, RP: 6.2	HSSSSW-MFC	COD: 64, TN: 58, TP:85, NH_4: 75, PO_4: 90	276 mW/m^2	CE: 0.1–0.36	Doherty et al. (2015a)
Acorus calamus	MnO_2	Blank carbon felt	Oil-contaminated WW	COD: 520, BOD: 109, oil content: 235, TOC: 33	VCW-MFC	COD: 73, TOC: 57, oil removal: 95.7	3868 mW/m^3, 102 mW/m^2	NA	Yang et al. (2016)

Nutrients in plant root exudates promote microbial growth, which in turn enhances power generation (Liu et al. 2013; Lu et al. 2015; Oon et al. 2015a, b). Macrophytes, according to Brix (1997), increase nitrification and the decomposition of organic compounds in the rhizosphere through their root systems aerobically. Nitrogen elimination in CWs is a major benefit of the CW and MFC combination. In MFC, bioelectrochemical reactions are primarily responsible for nitrogen removal (Ucar et al. 2017). NO_3 and NO_2 might be employed in accepting electrons at cathodic section in MFC via the biocathode (Kelly and He 2014; Clauwaert et al. 2007; Puig et al. 2011; Virdis et al. 2010, 2008) so that combined system's nitrogen removal can be improved. An electrochemical denitrification process utilizing EAB electrons transforms nitrate into nitrogen gas, which helps remove nitrogen from these systems (Wang et al. 2016b). Cathodic reduction techniques, such as those used to remove nitrogen from CW-MFCs, can use NO_3 as an electron acceptor, according to these studies (Wang et al. 2017a, b, c, d; Logan 2008; Gonzalez et al. 2021; Ge et al. 2020). Nitrate's electron-accepting potential for usage in a chemical reaction is shown in eq. 3 (Oon et al. 2018). In a case where nitrate is the lone acceptor of electron, the maximal cell potential may drop because of the decrease in the reduction potential as $ENO_3 = 0.7$ V in case of NO_3 (Ucar et al. 2017).

$$2NO^{-3} + 12H^{+} + 10e^{-} \rightarrow N_2 + 6H_2O \qquad (12.3)$$

12.7 Suggestions to Enhance Bioelectricity Generation in CW-MFC System

Integrated systems can be essentially used for treatment of wastewater while simultaneously generating energy. Concerns about low power densities must be addressed. The bioelectrochemical CW-MFC system's capacity has been increased from tens of mW/m^2 to hundreds of mW/m^2 during the various experiments. The CW-MFCs have lower power densities and volumetric power densities than most traditional MFCs. Since the exposed electrode (conductive material) surface area is substantially larger than the volume of wastewater, this is the primary cause.

Some ideas have been floated for increasing CW-capacity MFCs to produce bioelectricity (Table 12.2). For CW-MFC power density increases, the most notable advance has been the investigation of various electrode materials and rhizodeposition. As a result, the anode surface area expanded, allowing more microbes and graphite to come into direct contact with each other, resulting in lower internal resistance and better power densities.

The CW-MFC system's PD (power density) can be increased by including metallic materials. With graphite and nickel used as electrodes in the combined system, Gilani et al. (2016) looked at PD production. Graphite and nickel electrodes generated 23/10.7 mW/m^2 PDs respectively, in this study. For CW-MFCs, the material selection of electrode is a major factor in their ability to produce bioelectricity efficiently.

Table 12.2 Factors affecting the efficiency of the CW-MFC system. (Adopted from Ebrahimi et al. 2021)

Factor	Effect on system's efficiency	Suggestions for improvement
Operation time	The long-term operation harms CW-MFC performance	Investigating various inoculation sources, quality, quantity, and pretreatment requirements can help reduce inoculation and acclimatization time, which can help combat the negative effects of operating time
CW-MFC scale	Laboratory investigations offer information for designing larger pilot-scale systems	A growing need for additional pilot-scale research to comprehend and address the problems that are only seen at a bigger scale
Electrode spacing	A large gap increases internal resistance, whereas little spacing results in oxygen diffusion from cathode to anode	Creating a model that takes other aspects into account while addressing the link between electrode distance and power generation
Filtration media	The large-size inorganic filter media produces more voltage and has a bigger population of EAB. The risk for clogging is increased by the small-size inorganic filter media has greater COD and $NO^{3-}N$ removal rates	The impact of the filters' type, particle size, shape, homogeneity, and absorption coefficient
Plant	Plants gain the EAB community is improved by CW-MFC, which also releases DO, lowers internal resistance, and produces biomass	Plant harvesting needs, plant durability, evapotranspiration rates, and the likelihood of system clogging with various plant varieties, comparison of the effectiveness of floating, emergent, and submerged plants in CW-MFC
Separator	Separator in CW-MFC: Prevents oxygen from entering the anode zone from the cathode and prevents short circuits when electrodes are in proximity	Comparing the different types, prices, and thicknesses of separators used in CW-MFC with designs without separators
Nutrient concentration	The high nitrate concentration competes with oxygen in the cathode chamber to take part in the reduction reaction, and the high concentration of ammonia is damaging to plant life	More research should be done to determine the ideal C/N ratio for getting the best electrical performance, as well as how planted and unplanted systems react to various C/N ratios
Microorganism	Exoelectrogenic bacteria have a significant role in both biodegradation and power generation in CW-MFC	A thorough review of the main microbial community of the CW-MFC and its purpose
Recirculation	Positive aspects (enhancing nitrification-denitrification efficiency, decreasing organic load and risk of wetland blockage, and	The effect of recirculation injection points (RIP)

(continued)

Table 12.2 (continued)

Factor	Effect on system's efficiency	Suggestions for improvement
	raising the DO content in the cathode zone) and negative aspects (energy consumption, disturbing anaerobic conditions in the anode zone)	

Comparing CW-MFC to MFC, there is a significant difference in their internal resistances. The internal resistance of these bioelectrochemical systems must be reduced to get a higher power density. Carbon paper anodes with plasma modifications have been proven to considerably enhance CW-MFC performance (He et al. 2012). The hydrophobicity and roughness of the carbon paper's surface were altered by this preparation, which increased the anode's ability to transport electrons. For future applications of CW-MFC, the voltage reversal unpredictability that occurs when some of the unit cells are used to generate power for the battery and others are utilized to consume the battery's output can be explored in greater detail because of this issue (Xu et al. 2017b; Tamta et al. 2020; Gajda et al. 2020). As compared to concentrations greater than 200 mg/l, Xu et al. (2017a) found that the reduced response time (about 2 h), which was found to be reliable at 5 h for COD concentrations less than 100 mg/L was reliable for the first time in their investigation into CW-MFC biosensing for COD concentrations below or above 1000 mg/L. Because the system is saturated with substrate, the signals become even more similar at COD values above 700 mg/L. With a dosage of 200–700 mg/L voltage signals were distinct and reproducible.

MFC-based devices suffer from the fact that the capacitor's charging and discharging limits are controlled by a relatively modest input voltage (Xu et al. 2018b). When the charge storage is low in both MFC as well as supercapacitor, a charge pump was added between them to increase the storage capacity and regulate the charging pace. PMS is increasingly being used to power electronic devices. There are three components to the PMS, which are the DC/DC converter, a supercapacitor for storing and dispensing charge, and the CW-MFC, which provides the power supply. For example, the integration of PMS with biocathode CW-MFC was examined by Xu et al. (2018b) utilizing the electrons stored in the anode capacitor to gain more power. There were three operational ways used to compare the electron harvesting efficiency of this device: CW-MFC with CDC, DC, and CL which means the capacitor duty cycling (i.e., three operational approaches), intermittent external resistance, and continuous external resistance. The electron-harvesting efficiency based on the CDC, DC, and CL operational ways were 91.16%, 84.4%, and 76.1%, respectively. An increase in electron-harvesting efficiency of 91.16% and 95.1% was demonstrated by the D values, 31.6% and 20%, respectively. It is the sole work in CW-MFC with PMS that has been published, and more research into its electrical applications may provide new paths.

In addition, electrocoagulation-driven enhanced treatment methods can be used with CW-MFC. The EC technique may successfully remove resistant contaminants, such as dye and metals, from industrial effluent. Several pollutants like, nitrate, sulfamethoxazole, phosphate, COD, ammonium, and sulphate have been removed using this technology, which has also been integrated into CW (Liu et al. 2020b; Gao et al. 2017; Ju et al. 2014).

12.8 Limitations, Challenges, and Opportunities

According to all previous CW-MFC research, the system's efficacy and effectiveness for effluent treatment and clean energy production are strongly influenced because of wetlands, substrate, electrode material, and HRT. In contrast, most research specifically designed to evaluate the treatment efficiency effects of macrophytes concentrated on nitrate, phosphorus, and COD removal with little attention paid to utilize appropriate macrophytes to eliminate heavy metals. It is also well-known that wetlands play an essential role in the effectiveness and functioning of electrogenic bacteria. Radial oxygen loss (ROL), ecological requirements, chemical composition, and physiological features vary by CW-MFC species and effects both the elimination efficiency and bioenergy creation. Not many studies have examined the effect of substrate material and HRT on CW-MFCs, along with electrode materials.

Despite the advances in material science, there is still an opportunity for CW-MFC electrode development. The use of materials that are both very porous and electrically conductive could be investigated. Using this method in municipal wastewater treatment has significant potential. For some organic and inorganic contaminants, CW-removal MFC's efficiency is commendable, but a study is needed to improve the technology's ability to generate energy. The indirect use of CWL-MFCs is a severe problem because of their limited practical power output. For typical CW-MFCs, PDs (power densities) vary from 9 to 72 mWm^{-2} and 0.05% to 10.48%, respectively (Liu et al. 2017).

Capacitors, as studied and proposed by Xu et al. (2018a, b), have the potential to store dissipated energy and boost the collection of charges in general. Future investigations may involve the usage of combined CW-MFC's with supercapacitive electrodes connected to harvesting equipment present outside for increasing the system's energy generation. Future studies must include a report on the CW-MFC's total energy recovery based on the flow rate or decreased COD of the wastewater to further understand the energy generation potential of CW-MFC. Given that the parameter of energy is less sensitive to the cell dimension, comparisons between CW-MFC and MFC can be made with more accuracy.

Understanding the beneficial interaction of CW combined with MFC regarding the management of various waste streams is still a work in progress. CW-MFC coupling, LCA-based CW (life cycle assessment-based constructed wetlands), and cold temperature operation of CW-based study all appear to have research gaps that should be filled to push CW performance even higher in the future. According to the conclusions of this study, significant research effort is required to maintain the temperature levels necessary for CW efficiency in cold areas. It may be beneficial to incorporate some design methods for extended retention times to maintain consistent performance throughout the year (Jenssen et al. 1993).

More research into the ideal ratio of nitrogen to organic materials for treatment effectiveness is required. As a result of the decreased thermal dissipation from the layer of air between the frozen surface and water level, horizontal subsurface flow CW has been reported to work well in a cold climate. With the temperature-sensitive phase of nitrification being particularly difficult to maintain in freezing temperatures, a bioaugmentation strategy using microorganisms with specific activities in the right proportions can provide some benefit. Ammonia-oxidizing bacteria and nitrite-reducing bacteria appear to be excellent bioaugmentation options for low-temperature CW (Uggetti et al. 2010). Since their introduction, CW and MFC synergy has been primarily driven by a desire to maximize energy production. Other benefits of MFC implementation in CW technology, for instance, process monitoring, increased effectiveness of treatment, and the lowering of methane emissions and jamming up can be fully addressed by investigations on the optimum use of CW and MFC systems to know the further benefits of MFC utilization.

12.9 Conclusion

With the CW-MFC technology, wastewater treatment efficiency can be improved while also generating power. Combined CW and MFC treatment employ electrogenic bacteria to maximize phytoremediation efficiency and to boost power generation through plant impacts. Electrogenic degradation of COD is made more efficient by the layer of anode and electrodes in combined systems (CW-MFC), so it is reasonable to assume that CW-MFC is more effective at removing COD in comparison to CW. Additionally, it enhances the CWs' ability to remove nitrogen completely. It is needed to be emphasized that the CW-MFCs' maximal pollutant removal can vary based on their design settings. In terms of greenhouse gas emissions, CW-MFCs outperform CWs. When it comes to extracting creating energy and removing organic materials from highly complicated effluent, CW-MFCs are competent. However, their weak power output per unit area is their biggest drawback. Additionally, more macrophyte species need to be studied for comparing distinguished species efficiency within the system. It is also imperative to probe into the harms of pollutants on the CW-MFC plant component, namely the phytotoxicity produced by contaminants.

According to recent research, the fundamental objective of future CW-MFC research should not be energy harvesting. Consequently, more synergistic effects between MFCs and CWs should be investigated over time. Future CW-MFC research and development should focus on the creation of in situ biosensors, which hold great promise. To sum it all up, CW-MFCs have a great deal of potential for removing metal pollutants, emerging micro-pollutants, and recovering metals, all of which have not yet been studied. Shortly, we expect that this brand-new integration will evolve into a usable kind of innovation.

References

Aguirre Sierra MA (2017) Integrating Microbial Electrochemical Systems in Constructed Wetlands, a New Paradigm for Treating Wastewater in Small Communities. PhD thesis, Departamento de Química Analítica, Química Física e Ingeniería Química, Universidad de Alcalá, Alcalá de Henares, Spain

Aguirre-Sierra A, Bacchetti-De Gregoris T, Berná A, Salas J, Aragón C, Esteve-Núñez A (2016) Microbial electrochemical systems outperform fixed-bed biofilters in cleaning up urban wastewater. Environ Sci Water Res Technol 2(6):984–993

Albert JS, Destouni G, Duke-Sylvester SM, Magurran AE, Oberdorff T, Reis RE, Winemiller KO, Ripple WJ (2020) Scientists' warning to humanity on the freshwater biodiversity crisis. Ambio 50:1–10

Benvenuti T, Hamerski F, Giacobbo A, Bernardes AM, Zoppas-Ferreira J, Rodrigues MA (2018) Constructed floating wetland for the treatment of domestic sewage: a real-scale study. J Environ Chem Eng 6:5706–5711

Bharagava RN, Saxena G, Chowdhary P (2017) Constructed wetlands: an emerging phytotechnology for degradation and detoxification of industrial wastewaters. In: Environmental pollutants and their bioremediation approaches. CRC Press, pp 397–426

Bombelli P, Iyer DMR, Covshoff S et al (2013) Comparison of power output by rice (Oryza sativa) and an associated weed(Echinochloa glabrescens) in vascular plant bio-photovoltaic (VP-BPV) systems. Appl Microbiol Biotechnol 97(1):429–438

Brix H (1997) Do macrophytes play a role in constructed treatment wetlands? Water Sci Technol 35(5):11–17

Chandra R, Kumar V (2015) Mechanism of wetland plant rhizosphere bacteria for bioremediation of pollutants in an aquatic ecosystem. Advances in biodegradation and bioremediation of industrial waste. CRC Press, Boca Raton, FL, pp 329–380

Chiranjeevi P, Mohanakrishna G, Mohan SV (2012) Rhizosphere mediated electrogenesis with the function of anode placement for harnessing bioenergy through CO2 sequestration. Bioresour Technol 124:364–370

Clauwaert P, Rabaey K, Aelterman P, de Schamphelaire L, Pham TH, Boeckx P, Boon N, Verstraete W (2007) Biological denitrification in microbial fuel cells. Environ Sci Technol 41:3354–3360

Corbella C, Guivernau M, Viñas M, Puigagut J (2015) Operational, design and microbial aspects related to power production with microbial fuel cells implemented in constructed wetlands. Water Res 84:232–242

Corbella C, Garfí M, Puigagut J (2016) Long-term assessment of best cathode position to maximise microbial fuel cell performance in horizontal subsurface flow constructed wetlands. Sci Total Environ 5634:48–455

Corbella C, Puigagut J, Garfí M (2017) Life cycle assessment of constructed wetland systems for wastewater treatment coupled with microbial fuel cells. Sci Total Environ 58:355–362

Das B, Thakur S, Chaithanya MS, Biswas P (2019) Batch investigation of constructed wetland microbial fuel cell with reverse osmosis (RO) concentrate and wastewater mix as substrate. Biomass Bioenergy 122:231–237

de Rozari P, Greenway M, El Hanandeh A (2018) Nitrogen removal from sewage and septage in constructed wetland mesocosms using sand media amended with biochar. Ecol Eng 111:1–10

De SL, Rabaey K, Boeckx P, Boon N, Verstraete W (2008) Outlook for benefits of sediment microbial fuel cells with two bio-electrodes. Microb Biotechnol 1:446–462

Doherty L, Zhao YQ, Zhao XH, Wang WK (2015a) Nutrient and organics removal from swine slurry with simultaneous electricity generation in an alum sludge based constructed wetland incorporating microbial fuel cell technology. Chem Eng J 266:74–81

Doherty L, Zhao X, Zhao Y, Wang W (2015b) The effects of electrode spacing and flow direction on the performance of microbial fuel cell-constructed wetland. Ecol Eng 79:8–14

Ebrahimi A, Yousefi Kebria D, Najafpour Darzi G (2017) Enhancing biodegradation and energy generation via roughened surface graphite electrode in microbial desalination cell. Water Sci Technol 76(5):1206–1214

Ebrahimi A, Sivakumar M, McLauchlan C (2021) A taxonomy of design factors in constructed wetland-microbial fuel cell performance: a review. J Environ Manag 291:112723

Fan Y, Sharbrough E, Liu H (2008) Quantification of the internal resistance distribution of microbial fuel cells. Environ Sci Technol 42(21):8101–8107

Fang Z, Song HL, Cang N, Li XN (2013a) Performance of microbial fuel cell coupled constructed wetland system for decolorization of azo dye and bioelectricity generation. Bioresour Technol 144:165–171

Fang Z et al (2013b) Performance of microbial fuel cell coupled constructed wetland system for decolorization of azo dye and bioelectricity generation. Bioresour Technol 144(6):165–171

Fang Z, Song H-L, Cang N, Li X-N (2015) Electricity production from azo dye wastewater using a microbial fuel cell coupled constructed wetland operating under different operating conditions. Biosens Bioelectron 68:135–141

Fang Z, Cheng S, Wang H, Cao X, Li X (2017) Feasibility study of simultaneous azo dye decolorization and bioelectricity generation by microbial fuel cell-coupled constructed wetland: substrate effects. RSC Adv 7(27):16542–16552

Freguia S, Rabaey K, Yuan ZG, Keller J (2008) Sequential anode-cathode configuration improves cathodic oxygen reduction and effluent quality of microbial fuel cells. Water Resour 42:1387

Gajda I, Obata O, Jose Salar-Garcia M, Greenman J, Ieropoulos IA (2020) Long-term biopower of ceramic microbial fuel cells in individual and stacked configurations. Bioelectrochemistry 133: 107459

Gao Y, Xie Y, Zhang Q, Wang A, Yu Y, Yang L (2017) Intensified nitrate and phosphorus removal in an electrolysis-integrated horizontal subsurface-flow constructed wetland. Water Res 108:39–45

García JL, Rousseau PD, Morató J, Lesage E, Matamoros V, Bayona JM (2010) Contaminant removal processes in subsurface-flow constructed wetlands, a review. Crit Rev Env Sci Technol 40:561–661

Ge X, Cao X, Song X, Wang Y, Si Z, Zhao Y, Wang W, Tesfahunegn AA (2020) Bioenergy generation and simultaneous nitrate and phosphorus removal in a pyrite-based constructed wetland-microbial fuel cell. Bioresour Technol 296:122350

Ghangrekar MM, Shinde VB (2007) Performance of membrane-less microbial fuel cell treating wastewater and effect of electrode distance and area on electricity production. Bioresour Technol 98(15):2879–2885

Gilani SR, Yaseen A, Zaidi SRA, Zahra M, Mahmood Z (2016) Photocurrent generation through plant microbial fuel cell by varying electrode materials. J Chem Soc Pakistan 38(1):17–27

Gonzalez T, Puigagut J, Vidal G (2021) Organic matter removal and nitrogen transformation by a constructed wetland-microbial fuel cell system with simultaneous bioelectricity generation. Sci Total Environ 753:142075

Guadarrama PO, Gutiérrez MT, García SL, Guadarrama PVH, Estrada AEB (2019) Recent advances in constructed wetland-microbial fuel cells for simultaneous bioelectricity production and wastewater treatment: a review. Int J Energy Res 43(10):5106–5127

Habibul N, Hu Y, Wang YK, Chen W, Yu HQ, Sheng GP (2016) Bioelectrochemical chromium (VI) removal in plant-microbial fuel cells. Environ Sci Technol 50(7):3882–3889

He YR, Xiao X, Li WW et al (2012) Enhanced electricity production from microbial fuel cells with plasma-modified carbon paper anode. Phys Chem Chem Phys 14(28):9966–9971

Helder M, Strik DP, Hamelers HV, Buisman CJ (2012) The flat-plate plant-microbial fuel cell: the effect of a new design on internal resistances. Biotechnol Biofuels 5(70):1–10

Jenssen PD, Mehlum T, Krogstad T (1993) Potential use of constructed wetlands for wastewater treatment in northern environments. Water Sci Technol 28:149–157

Ju X, Wu S, Zhang Y, Dong R (2014) Intensified nitrogen and phosphorus removal in a novel electrolysis-integrated tidal flow constructed wetland system. Water Res 59:37–45

Kabutey FT, Zhao Q, Wei L, Ding J, Antwi P, Quashie FK, Wang W (2019) An overview of plant microbial fuel cells (PMFCs): configurations and applications. Renew Sust Energ Rev 110:402–414

Kahng HY, Kukor JJ, Oh KH (2000) Characterization of strain HY99, a novel microorganism capable of aerobic and anaerobic degradation of aniline. FEMS Microbiol Lett 190:215–221

Kelly PT, He Z (2014) Nutrients removal and recovery in bioelectrochemical systems: a review. Bioresour Technol 153:351–360

Kouzuma A, Kaku N, Watanabe K (2014) Microbial electricity generation in rice paddy fields: recent advances and perspectives in rhizosphere microbial fuel cells. Appl Microbiol Biotechnol 98(23):9521–9526

Li F, Lu L, Zheng X, Ngo HH, Liang S, Guo W, Zhang X (2014) Enhanced nitrogen removal in constructed wetlands: effects of dissolved oxygen and step-feeding. Bioresour Technol 169: 395–402

Li T, Fang Z, Yu R, Cao X, Song H, Li X (2016) The performance of the microbial fuel cell-coupled constructed wetland system and the influence of the anode bacterial community. Environ Technol 37(13):1683–1692

Li H, Zhang S, Yang X-L, Yang Y-L, Xu H, Li X-N, Song H-L (2019) Enhanced degradation of bisphenol A and ibuprofen by an up-flow microbial fuel cell-coupled constructed wetland and analysis of bacterial community structure. Chemosphere 217:599–608

Liu S, Song H, Li X, Yang F (2013) Power generation enhancement by utilizing plant photosynthate in microbial fuel cell coupled constructed wetland system. Int J Photoenergy 2013

Liu T et al (2014) Bio-cathode materials evaluation and configuration optimization for power output of vertical subsurface flow constructed wetland—microbial fuel cell systems. Bioresour Technol 166(166C):575–583

Liu R, Zhao Y, Doherty L, Hu Y, Hao X (2015) A review of incorporation of constructed wetland with other treatment processes. Chem Eng J 279:220–230

Liu J, Yi NK, Wang S, Lu LJ, Huang XF (2016) Impact of plant species on spatial distribution of metabolic potential and functional diversity of microbial communities in a constructed wetland treating aquaculture wastewater. Ecol Eng 94:564–573

Liu S, Feng X, Li X (2017) Bioelectrochemical approach for control of methane emission from wetlands. Bioresour Technol 241:812–820

Liu T, Xu S, Lu S, Qin P, Bi B, Ding H, Liu Y, Guo X, Liu X (2019a) A review on removal of organophosphorus pesticides in constructed wetland: performance, mechanism and influencing factors. Sci Total Environ 651:2247–2268

Liu Q, Zhou B, Zhang S, Xu D, Pan R, Xia S (2019b) Embedding microbial fuel cells into the vertical flow constructed wetland enhanced denitrogenation and water purification. Pol J Environ Stud 28:1799–1804

Liu F, Sun L, Wan J, Shen L, Yu Y, Hu L, Zhou Y (2020a) Performance of different macrophytes in the decontamination of and electricity generation from swine wastewater via an integrated constructed wetland-microbial fuel cell process. Int J Environ Sci 89:252–263

Liu X, Wang Y, Lu S, Liu Y, Zhao B, Xi B, Guo X, Guo W, Zhang J (2020b) Intensified sulfamethoxazole removal in an electrolysis-integrated tidal flow constructed wetland system. Chem Eng J 390:124545

Logan BE (2008) Microbial fuel cells. John Wiley & Sons, New York

Lu L, Xing D, Ren ZJ (2015) Microbial community structure accompanied with electricity production in a constructed wetland plant microbial fuel cell. Bioresour Technol 195:115–121

Ma Y, Zhai Y, Zheng X, He S, Zhao M (2019) Rural domestic wastewater treatment in constructed ditch wetlands: effects of influent flow ratio distribution. J Clean Prod 225:350–358

Manohar K, Shide A, Supriya S (2017) Green electricity production from living plant and microbial fuel cell. Int J Adv Res Sci Eng 6(09):459–466

Mohan SV, Mohanakrishna G, Chiranjeevi P (2011) Sustainable power generation from floating macrophytes based ecological microenvironment through embedded fuel cells along with simultaneous wastewater treatment. Bioresour Technol 102(14):7036–7042

Mulamoottil G, de Rozari P, Greenway M, El Hanandeh A, Constructed Wetlands for the Treatment of Landfill Leachates, Routledge (2018) Nitrogen removal from sewage and septage in constructed wetland mesocosms using sand media amended with biochar. Ecol Eng 111:1–10
Nattawut K, Holasut K (2015) Electricity generation of plant microbial fuel cell (PMFC) using Cyperus involucratus R. Eng Appl Sci Res 42(1):117–124
Nitisoravut R, Regmi R (2017) Plant microbial fuel cells: a promising biosystems engineering. Renew Sust Energ Rev 76:81–89
Oon YL, Ong SA, Ho LN et al (2015a) Hybrid system up-flow constructed wetland integrated with microbial fuel cell for simultaneous wastewater treatment and electricity generation. Bioresour Technol 186:270–275
Oon Y-L, Ong S-A, Ho L-N, Wong Y-S, Oon Y-S, Lehl HK, Thung W-E (2015b) Hybrid system up-flow constructed wetland integrated with microbial fuel cell for simultaneous wastewater treatment and electricity generation. Bioresour Technol 186:270–275
Oon YL, Ong SA, Ho LN et al (2017a) Role of macrophyte and effect of supplementary aeration in up-flow constructed wetland- microbial fuel cell for simultaneous wastewater treatment and energy recovery. Bioresour Technol 224:265–275
Oon YL, Ong SA, Ho LN, Wong YS, Dahalan FA, Oon YS, Nordin N (2017b) Role of macrophyte and effect of supplementary aeration in up-flow constructed wetland-microbial fuel cell for simultaneous wastewater treatment and energy recovery. Bioresour Technol 224:265–275
Oon YL, Ong SA, Ho LN, Wong YS, Dahalan FA, Oon YS, Lehl HK, Thung WE, Nordin N (2018) Up-flow constructed wetland-microbial fuel cell for azo dye, saline, nitrate remediation and bioelectricity generation: from waste to energy approach. Bioresour Technol 266:97–108
Oon YL, Ong SA, Ho LN, Wong YS, Dahalan FA, Oon YS, Teoh TP, Lehl HK, Thung WE (2020) Constructed wetland–microbial fuel cell for azo dyes degradation and energy recovery: influence of molecular structure, kinetics, mechanisms and degradation pathways. Sci Total Environ 720:137370
Puig S, Serra M, Vilar-Sanz A, Cabre M, Baneras L, Colprim J, Balaguer MD (2011) Autotrophic nitrite removal in the cathode of microbial fuel cells. Bioresour Technol 102:4462–4467
Rabbani A, Zainith S, Deb VK, Das P, Bharti P, Rawat DS, Kumar N, Saxena G (2021) Microbial technologies for environmental remediation: potential issues, challenges, and future prospects. In: Microbe mediated remediation of environmental contaminants. Woodhead Publishing, pp 271–286
Ramírez-Vargas CA, Arias CA, Zhang L, Brix H (2018) Microbial community function in electroactive biofilm-based constructed wetlands. Biogeosci Discuss 428:1–28
Ramirez-Vargas CA, Arias CA, Carvalho P, Zhang L, Esteve-Nunez A, Brix H (2019) Electroactive biofilm-based constructed wetland (EABB-CW): a mesocosm-scale test of an innovative setup for wastewater treatment. Sci Total Environ 659:796–806
Rathour R, Patel D, Shaikh S, Desai C (2019) Eco-electrogenic treatment of dyestuff wastewater using constructed wetland-microbial fuel cell system with an evaluation of electrodeenriched microbial community structures. Bioresour Technol 285:121349
Regmi R, Nitisoravut R, Ketchaimongkol J (2018) A decade of plant- assisted microbial fuel cells: looking back and moving forward. Biofuels 9(5):605–612
Rincon VA, Carvalhais LC, Schenk PM (2013) Culture-independent molecular tools for soil and rhizosphere microbiology. Diversity 5(3):581–612
Saeed T, Khan T (2019) Constructed wetlands for industrial wastewater treatment: alternative media, input biodegradation ratio and unstable loading. J Environ Chem Eng 7:103042
Santoro C, Arbizzani C, Erable B, Ieropoulos I (2017) Microbial fuel cells: from fundamentals to applications. A review. J Power Sources 356:225–244
Saxena G, Thakur IS, Kumar V, Shah MP (2020) Electrobioremediation of contaminants: concepts, mechanisms, applications and challenges. In: Combined application of physico-chemical & microbiological processes for industrial effluent treatment plant, pp 291–313
Saz C, Türe C, Türker OC, Yakar A (2018) Effect of vegetation type on treatment performance and bioelectric production of constructed wetland modules combined with microbial fuel cell (CW-MFC) treating synthetic wastewater. Environ Sci Pollut Res 25(9):8777–8792

Schmitt N, Wanko A, Laurent J, Bois P, Molle P, Mosé R (2015) Constructed wetlands treating stormwater from separate sewer networks in a residential Strasbourg urban catchment area: micropollutant removal and fate. J Environ Chem Eng 3:2816–2824
Sharma Y, Li B (2010) Optimizing energy harvest in wastewater treatment by combining anaerobic hydrogen producing biofermentor (HPB) and microbial fuel cell (MFC). Int J Hydrog Energy 35(8):3789–3797
Sheridan C, Akcil A, Kappelmeyer U, Moodley I (2018) A review on the use of constructed wetlands for the treatment of acid mine drainage. Constructed Wetlands for Industrial Wastewater Treatment, pp. 249–262
Sierra MA, Esteve Núñez A, Salas Rodriguez JJ (2017) Integrating microbial electrochemical systems in constructed wetlands, a new paradigm for treating wastewater in small communities. Ph.D. Thesis, Universidad de Alcalá, Madrid, Spain, pp. 100–165
Singh HM, Pathak AK, Chopra K, Tyagi VV, Anand S, Kothari R (2018) Microbial fuel cells: a sustainable solution for bioelectricity generation and wastewater treatment. Biofuels:1–21
Singh HM, Pathak AK, Chopra K, Tyagi VV, Anand S, Kothari R (2019) Microbial fuel cells: A sustainable solution for bioelectricity generation and wastewater treatment. Biofuels 10:11–31
Song HL, Li H, Zhang S, Yang YL, Zhang LM, Xu H, Yang XL (2018) Fate of sulfadiazine and its corresponding resistance genes in up-flow microbial fuel cell coupled constructed wetlands: effects of circuit operation mode and hydraulic retention time. J Chem Eng 350:920–929
Srivastava P, Yadav AK, Mishra BK (2015) The effects of microbial fuel cell integration into constructed wetland on the performance of constructed wetland. Bioresour Technol 195:223–230
Srivastava P, Yadav AK, Abbassi R, Garaniya V, Lewis T (2018) Denitrification in a low carbon environment of a constructed wetland incorporating a microbial electrolysis cell. J Environ Chem Eng 6:5602–5607
Strik DPBTB, Hamelers HVM, Snel JFH, Buisman CJN (2008a) Green electricity production with living plants and bacteria in a fuel cell. Int J Energy Res 32:870–876
Strik DP, Snel JF, Buisman CJ (2008b) Green electricity production with living plants and bacteria in a fuel cell. Int J Energ Res 32(9):870–876
Tamta P, Rani N, Yadav AK (2020) Enhanced wastewater treatment and electricity generation using stacked constructed wetland–microbial fuel cells. Environ Chem Lett 18:871–879
Tang C, Zhao Y, Kang C, Yang Y, Morgan D, Xu L (2019) Towards concurrent pollutants removal and high energy harvesting in a pilot-scale CW-MFC: insight into the cathode conditions and electrodes connection. Chem Eng J 373:150–160
Tapia N, Rojas C, Bonilla C, Vargas I (2018) A new method for sensing soil wáter content in green roofs using plant microbial fuel cells. Sensors 18(1):71
Tender LM, Reimers EC, Stecher AH, Holmes ED, Bond RD, Lowy AD, Pilobello K, Fertig JS, Lovley RD (2002) Harnessing microbially generated power on the seafloor. Nat Biotechnol 20: 821–825
Türker OC, Yakar A (2017) A hybrid constructed wetland combined with microbial fuel cell for boron (B) removal and bioelectric production. Ecol Eng 1024:11–421
Ucar D, Zhang Y, Angelidaki I (2017) An overview of electron acceptors in microbial fuel cells. Front Microbiol 8:643
Uggetti E, Ferrer I, Molist J, Garcı J (2010) Technical , economic and environmental assessment of sludge treatment wetlands. Water Res 45:573–582
Valipour A, Ahn YH (2016) Constructed wetlands as sustainable ecotechnologies in decentralization practices: a review. Environ Sci Pollut Res 23:180–197
Villaseñor J, Capilla P, Rodrigo MA, Canizares P, Fernandez FJ (2013) Operation of a horizontal subsurface flow constructed wetland–microbial fuel cell treating wastewater under different organic loading rates. Water Res 47(17):6731–6738
Villaseňor J, Capilla P, Rodrigo MA, Caňizares P, Femάndez FJ (2013) Operation of a horizontal subsurface flow constructed wetland—microbial fuel cell treating wastewater under different organic loading rates. Water Res 47:6731–6738

Virdis B, Rabaey K, Yuan Z, Keller J (2008) Microbial fuel cells for simultaneous carbon and nitrogen removal. Water Res 42:3013–3024
Virdis B, Rabaey K, Rozendal RA, Yuan Z, Keller J (2010) Simultaneous nitrification, denitrification and carbon removal in microbial fuel cells. Water Res 44:2970–2980
Vymazal J (2010) Constructed wetlands for wastewater treatment. Water 2:530–549
Wang J, Song X, Wang Y, Abayneh B, Ding Y, Yan D, Bai J (2016a) Microbial community structure of different electrode materials in constructed wetland incorporating microbial fuel cell. Bioresour Technol 221:697–702
Wang J, Song X, Wang Y, Abayneh B, Li Y, Yan D, Bai J (2016b) Nitrate removal and bioenergy production in constructed wetland coupled with microbial fuel cell: establishment of electrochemically active bacteria community on anode. Bioresour Technol 221:358–365
Wang J, Song X, Wang Y et al (2017a) Bioenergy generation and rhizodegradation as affected by microbial community distribution in a coupled constructed wetland-microbial fuel cell system associated with three macrophytes. Sci Total Environ 607-608:53–62
Wang Y, Zhao Y, Xu L et al (2017b) Constructed wetland integrated microbial fuel cell system: looking back, moving forward. Water Sci Technol 76(2):471–477
Wang J, Song X, Wang Y, Zhao Z, Wang B, Yan D (2017c) Effects of electrode material and substrate concentration on the bioenergy output and wastewater treatment in air-cathode microbial fuel cell integrating with constructed wetland. Ecol Eng 99:191–198
Wang J, Song X, Wang Y, Bai J, Li M, Dong G, Lin F, Lv Y, Yan D (2017d) Bioenergy generation and rhizodegradation as affected by microbial community distribution in a coupled constructed wetland-microbial fuel cell system associated with three macrophytes. Sci Total Environ 607: 53–62
Wang G, Guo Y, Cai J, Wen H, Mao Z, Zhang H, Wang X, Ma L, Zhu M (2019) Electricity production and the analysis of the anode microbial community in a constructed wetlandmicrobial fuel cell. RSC Adv 9(37):21460–21472
Wei M, Rakoczy J, Vogt C, Harnisch F, Schumann R, Richnow HH (2015) Enhancement and monitoring of pollutant removal in a constructed wetland by microbial electrochemical technology. Bioresour Technol 196:490–499
Wen H, Zhu H, Yan B, Xu Y, Shutes B (2020) Treatment of typical antibiotics in constructed wetlands integrated with microbial fuel cells: roles of plant and circuit operation mode. Chemosphere 250:126252
Wetser K (2016) Electricity from Wetlands: Technology Assessment of the Tubular Plant Microbial Fuel Cell with an Integrated Biocathode (Doctoral Thesis). Netherlands: Wageningen University
Wu D, Yang L, Gan L et al (2015a) Potential of novel wastewater treatment system featuring microbial fuel cell to generate electricity and remove pollutants. Ecol Eng 84:624–631
Wu H, Zhang J, Ngo HH, Guo W, Hu Z, Liang S, Fan J, Liu H (2015b) A review on the sustainability of constructed wetlands for wastewater treatment: design and operation. Bioresour Technol 175:594–601
Wu S, Lv T, Lu Q, Ajmal Z, Dong R (2016) Treatment of anaerobic digestate supernatant in microbial fuel cell coupled constructed wetlands: evaluation of nitrogen removal, electricity generation, and bacterial community response. Sci Environ 580:339–346
Wu D, Lu D, Sun F, Zhou Y (2019) Process optimization for simultaneous antibiotic removal and precious metal recovery in an energy neutral process. Sci Total Environ 695:133914
Wu Q, Jiao S, Ma M, Peng SP (2020) Microbial fuel cell system: a promising technology for pollutant removal and environmental remediation. Environ Sci Pollut Res 27(7):6749–6764
Xu L, Zhao Y, Doherty L, Hu Y, Hao X (2016) The integrated processes for wastewater treatment based on the principle of microbial fuel cells: a review. Crit Rev Environ Sci Technol 46(1): 60–91
Xu L, Zhao Y, Wang T, Liu R, Gao F (2017a) Energy capture and nutrients removal enhancement through a stacked constructed wetland incorporated with microbial fuel cell. Water Sci Technol 76(1):28–34

Xu L, Zhao Y, Fan C, Fan Z, Zhao F (2017b) First study to explore the feasibility of applying microbial fuel cells into constructed wetlands for COD monitoring. Bioresour Technol 243: 846–854

Xu F, Cao F-Q, Kong Q, Zhou L-L, Yuan Q, Zhu Y-J, Wang Q (2018a) Electricity production and evolution of microbial community in the constructed wetland-microbial fuel cell. Chem Eng 339:479–486

Xu L, Wang B, Liu X, Yu W, Zhao Y (2018b) Maximizing the energy harvest from a microbial fuel cell embedded in a constructed wetland. Appl Energy 214:83–91

Yadav AK, Dash P, Mohanty A, Abbassi R, Mishra BK (2012) Performance assessment of innovative constructed wetland-microbial fuel cell for electricity production and dye removal. Ecol Eng 47:126–131

Yadav AK, Srivastava P, Kumar N, Abbassi R, Mishra BK (2018) Constructed wetland microbial fuel cell: an emerging integrated technology for potential industrial wastewater treatment and bio-electricity generation. Constr Wetl Ind Wastewater Treat:493–510

Yan Z, Jiang H, Cai H, Zhou Y, Krumholz LR (2015) Complex interactions between the macrophyte Acorus calamus and microbial fuel cells during pyrene and benzo[a]pyrene degradation in sediments. Sci Rep 5(10709):1–12

Yan D, Song X, Weng B, Yu Z, Bi W, Wang J (2018) Bioelectricity generation from air cathode microbial fuel cell connected to constructed wetland. Water Sci Technol 78:1990–1996

Yang Q, Wu Z, Liu L, Zhang F, Liang S (2016) Treatment of oil wastewater and electricity generation by integrating constructed wetland with microbial fuel cell. Materials 9(11):885

Yang Y, Zhao Y, Liu R, Morgan D (2018a) Global development of various emerged substrates utilized in constructed wetlands. Bioresour Technol 261:441–452

Yang Q, Gao C, Wu Z-X, Liang S-N, Liu M-H (2018b) Activated carbon clogging analysis in an integration of constructed wetland with microbial fuel cell. E3S Web Conf 53:01025

Yoongling O et al (2015) Hybrid system up-flow constructed wetland integrated with microbial fuel cell for simultaneous wastewater treatment and electricity generation. Bioresour Technol 186: 270–275

Zaman B, Wardhana IW (2018) Potential of electric power production from microbial fuel cell (MFC) in evapotranspiration reactor for leachate treatment using Alocasia macrorrhiza plant and Eleusine indica grass. E3S Web Conf 31:02010

Zhang F, He Z (2012) Integrated organic and nitrogen removal with electricity generation in a tubular dual-cathode microbial fuel cell. Process Biochem 47(12):2146–2151

Zhang S, Song HL, Yang XL, Yang YL, Yang KY, Wang XY (2016) Fate of tetracycline and sulfamethoxazole and their corresponding resistance genes in microbial fuel cell coupled constructed wetlands. RSC Adv 6(98):95999–96005

Zhang S, Song HL, Yang XL, Li H, Wang YW (2018) A system composed of a biofilm electrode reactor and a microbial fuel cellconstructed wetland exhibited efficient sulfamethoxazole removal but induced sul genes. Bioresour Technol 256:224–231

Zhang J, Wang W, You S, Qi D, Liu Z, Liu D, Ma M, Cui F, Ren N, Chen X (2020) Photothermal Janus anode with photosynthesis-shielding effect for activating low-temperature biological wastewater treatment. Adv Funct Mater 30(7):1909432

Zhao Y, Collum S, Phelan M, Goodbody T, Doherty L, Hu Y (2013) Preliminary investigation of constructed wetland incorporating microbial fuel cell: batch and continuous flow trials. Chem Eng J 229:364–370

Zhao C, Shang D, Zou Y, Du Y, Wang Q, Xu F, Ren L, Kong Q (2020) Changes in electricity production and microbial community evolution in constructed wetland microbial fuel cell exposed to wastewater containing Pb(II). Sci Total Environ 732:139127

Role of Microbial Communities and Aquatic Macrophytes in Constructed Wetlands for Tannery Wastewater Treatment: Challenges and Opportunities

13

Navneet Kumar, Prachi Bhatnagar, Rahul Yadav, Himani Chandel, Sibiraj Murugesan, Geetansh Sharma, and Gaurav Saxena

Abstract

Tannery wastewaters are notoriously difficult to manage because of its complicated chemical makeup. There is a significant amount of chromium-laden wastewater produced by industries like the tannery because industrial water use is less than the abstraction rate (Cr). Toxic tannery effluent is used to irrigate food crops, posing a global threat to human health. Instead of using traditional treatment methods, constructed wetlands (CWs) be effective in cleaning up chromium-tainted wastewater. A possible alternative solution for treating wastewater from tanneries is CWs, they have been successfully utilized to treat a variety of wastewater. CWs are also economical and kind to the environment. This chapter provides evidence that CWs are an excellent method for cleaning up wastewater from tanning operations. Through investigating the function of microorganisms and aquatic macrophytes in CWs, improving wetland design and management, and analyzing the impact of microbes. The wetland structure, operation parameters, water, pH, temperature, and mechanism for chromium remediation are all mentioned as factors affecting the efficacy of CWs for wastewater treatment.

N. Kumar · P. Bhatnagar · H. Chandel · S. Murugesan · G. Sharma · G. Saxena (✉)
MATER-Microalgae Technology for Environmental Resources, EERG-Ecotoxicology, and Environmental Remediation Group, School of Biotechnology, Shoolini University, Solan, Himachal Pradesh, India

EMBL-Environmental Microbiology and Biotechnology Laboratory, EERG-Ecotoxicology and Environmental Remediation Group, School of Biotechnology, Shoolini University, Solan, Himachal Pradesh, India
e-mail: gauravsaxena@shooliniuniversity.com

R. Yadav
School of Biological and Environmental Sciences, Shoolini University, Solan, Himachal Pradesh, India

S. Kumar et al. (eds.), *Aquatic Macrophytes: Ecology, Functions and Services*,
https://doi.org/10.1007/978-981-99-3822-3_13

Keywords

Tannery wastewater treatment · Constructed wetlands · Bioremediation · Aquatic macrophytes · Microbial communities · Environmental remediation

13.1 Introduction

Since the tannery business has expanded, tannery effluent has become a significant contributor to industrial water pollution. The annual volume of tannery effluent, around 300 million metric tonnes, is similar to the volume of cowhides processed in the world's leather markets. Leather is made from animal hides and skin through a chemical process known as tanning (Dargo and Ayalew 2014; Lofrano et al. 2013). The tannery sector in China uses roughly 1.4 billion cubic meters of water per year, with about 1.2 billion cubic meters being discharged. As a result, pollution is one of the main challenges that slow the growth of the tannery sector (Ma 2004). One of the biggest causes of pollution that is raising environmental concerns is the discharge of effluents from tanneries into bodies of water (Bosinc et al. 2000). The effluents from the tanneries harm the air, land, and water, causing a wide range of health issues (WHO 2002). The tanneries in Brazil, a country with a sizable leather industry, use as much water per year as 5.5 million people. Daily, each resident consumes about 150 L of water. The challenge of figuring out how to manage tannery effluent effectively, in other words, is growing (Streit et al. 2014). Historically, communities would coordinate their tanning operations to satisfy the need for leather goods like shoes and instruments (Durai and Rajasimman 2011). Pollutants from tannery wastewater discharge are currently subject to extremely stringent regulations. Standards for the proper disposal of wastewater from tanning operations have been established in Europe and the United States based on the most recent scientific findings in this area. Most nations have specified limits for things like pH, BOD, COD, oil, sulfide, and total chromium. Ammonia is a major harmful by-product of wastewater treatment, yet few nations have declared limitations for the nitrogen it produces (Chen et al. 2011a, b). Most tanneries switched their methods of production in the twentieth century from vegetable tanning to chromium tanning. Since each receiving body of water can only receive limited amounts of contaminants without degradation, regulations from numerous regulatory authorities, such as the World Health Organization (WHO), emphasize minimizing or eliminating organic matter, sediments, minerals, chromium, and other pollutants. Wastewater from tanneries undergoes a multistep purification procedure called effluent treatment before it is released into the environment (Buljan and Kral 2011).

Wastewater treatment plants often operate either over or under capacity due to insufficient planning or shifts in production. Remediation of Cr-contaminated wastewater has been accomplished through a wide range of treatment technologies, from physicochemical methods (such as adsorption, membrane filtration reverse osmosis, activated carbon, ion exchange, precipitation) to biological methods (using fungi, bacteria, algae, yeast, etc.) (Bibi et al. 2018; Ashraf et al. 2016; Pradhan et al. 2017;

Shahid et al. 2017; Papaevangelou et al. 2017). Although effective, methods like the activated sludge process, membrane bioreactors, and membrane separation for treating wastewater are expensive and may not be practical for widespread use in rural regions (Chen et al. 2014). In addition, they are inadequate and constrained in light of the ever-increasingly severe water and wastewater treatment regulations (Wu et al. 2013b). Consequently, it is crucial, especially in developing regions, to pick low-cost and efficient alternative technologies for wastewater treatment. Constructed wetlands (CWs) are a cost-effective and low-maintenance alternative to traditional wetlands for wastewater treatment, and as such, they are receiving considerable attention (Bharagava et al. 2017; Rai et al. 2013). There are physical, chemical, and biological mechanisms at work in a man-made wetland to filter out pollutants. To be more specific, the physical process that results in the removal of contaminants is the sedimentation of suspended particles in the wastewater. Sedimentation rises as wastewater is held for longer periods (DBT 2019). Some believe that constructed wetlands (CWs) can be used to treat wastewater at the secondary or tertiary level using biological processes. The goal is to create the same biogeochemical conditions seen in natural wetlands within a specially designed man-made facility. Additionally, they are being researched as a phytoremediation method to improve wastewater treatment (Schröder et al. 2007). By utilizing the energy from the sun, plants, microbes, and algae produce carbon dioxide and water through a process called photosynthesis. Microorganisms including bacteria, fungi, algae, and protozoa that carry out photosynthesis populate man-made wetland ecosystems. When compared to other methods, built wetlands (CWs) are superior at treating wastewater. Interest in these has increased recently, in developing countries, because of their inexpensive pricing, ease of use, eco-friendliness, and aesthetic worth (Li et al. 2007; Katsenovich et al. 2009; Zhi and Ji 2012; Zhang et al. 2009). Since the late 1990s, several different treatment centers have opened, although surprisingly little information about the effectiveness of Portuguese CWs has been published (Duarte et al. 2010). The first extensive study of the application of CWs in the tannery industry for the treatment of effluent was published by Calheiros et al. (2007, 2008a, b, 2009a, b, c). Since 2001, researchers in Portugal have examined these systems from a wide variety of angles, including plant composition, substrate composition, treatment intensity, cell alignment, microbial diversity, and toxicological concerns (Calheiros et al. 2007, 2008a, b, 2012b). This chapter is an attempt to compile the findings of these investigations and provide guidelines for improving the CW technology for treating wastewater from tanning operations. Furthermore, the chapter mainly discussed the significance of microorganisms and aquatic macrophytes in CWs for the purification of tannery effluent.

13.2 Nature and Characteristics of Tannery Wastewater

There are a wide variety of industrial operations that rely on water, which is essential to life. It takes a lot of chemicals and water to treat raw hides and skins in the tanning process, and for every tonne of raw hides and skins handled, there is a corresponding amount of effluent, on average 30–35 m^3 (Islam et al. 2014; Lofrano et al. 2008). However, the type of raw materials, the final product, and the manufacturing procedures are used in wastewater formation (Lofrano et al. 2013; Tunay et al. 1995). These results in two significant challenges for the leather industries (LIs): ensuring access to high-quality water and properly treating a substantial amount of highly contaminated wastewater. Wastewater and sludge from the tanning business may have significant concentrations of Cr. Additional possible contaminants found in this water include 0.025–3.3 gL^{-1} of sulfide, 0.0–4.1 gL^{-1} of Cr^{3+}, and 0.08–3.8 gL^{-1} of nitrate. The TSS (total suspended solids) range from 2.5 to 70 gL^{-1}, whereas the TDS ranges from 22 to 67 gL^{-1} (Borba et al. 2018; Wang et al. 2014).

The amount of water required to treat a given type of leather varies substantially due to the wide range of methods employed. The amount of water used in the manufacturing process varies widely, even for the same type of leather, based on the specific technologies employed at each stage of production. Tannery effluent is a dark brown, acidic waste with high concentrations of chromium (III), phenolics, COD, and total dissolved solids (Saxena 2020; Goutam et al. 2018; Dixit et al. 2015; Suganthi et al. 2013; Durai and Rajasimmam 2011). However, TWW features may change depending on the LIs' chosen sector of the economy, the raw materials and chemicals they utilize, the nature of the ultimate product they create, and the methods of production they employ (Lofrano et al. 2013; Apaydin et al. 2009; Rameshraja and Suresh 2011). High amounts of salts (including chloride, ammonium, chromium, and sulfate) are released into TWW during the tanning process, which is the second most polluting stage in the leather manufacturing process after the beam house (Rameshraja and Suresh 2011; Cooman et al. 2003). That is why you will find an alkaline pH in beam house wastewater and a very acidic pH and higher COD value in tanning wastewater (Lofrano et al. 2013). Tannery effluent typically has high levels of nitrogen, especially organic nitrogen, but low levels of phosphorus (Durai and Rajasimmam 2011). Although most of the salt in TWW comes from the hides/skins soaked in liquor, the retanning streams have high COD and low biological oxygen demand and total suspended solids, and they contain tannins, wasted colors, sulfonated oils, and trivalent chromium (III) (Lofrano et al. 2013; USEPA 1986). The effluents from the beamhouse (deliming/bating, liming, splitting machines, and water from fleshing) are the most different of the three primary types of tannery effluent because they include sulphides, have a high pH, and lack chromium. The acidic, high-Cr effluents from the tanning and retanning processes are used in the sammying. Low Cr content is typical of soaking water and other general effluents, especially those produced after tanning processes (fat-liquoring and dying) (Buljan et al. 2011). In addition, the biodegradation study of TWW makes use of the BOD5/COD (as a result of inhibitors) or BOD5/TOC (as a result of high sulphide and chloride concentration) ratio (Lofrano et al. 2013).

13.3 Environmental Concerns Related with Tannery Wastewater

13.3.1 Pollution Profile

When compared to other types of industrial wastewaters, TWW is considered a major pollutant (Gupta et al. 2012; Verma et al. 2008). A scheme of leather processing operation and release of tannery TWW is depicted in Fig. 13.1. The widespread use of toxic and dangerous chemicals in TWW, including chromium, detergents, biocides, chlorophenols, STs, formaldehyde, oils, resins, and phthalates, etc., gives LIs a bad name (Saxena et al. 2020a, b; Lofrano et al. 2013; Dixit et al. 2015). CETP (common effluent treatment plant) wastewater is hazardous to human and animal health due to its elevated levels of TDS, COD, BOD, and several toxic heavy metals, most notably chromium (Saxena et al. 2020c; 2017; Dixit et al. 2015; Lofrano et al. 2013; Mondal et al. 2012). In addition, tannery effluent contains a variety of chemical compounds utilized in the leather-processing industry but which do not decompose adequately even after normal treatment, harming both living creatures and the environment (Saxena and Bharagava 2015; Oral et al. 2007; Tigini et al. 2011; Shakir et al. 2012; Alvarez-Bernal et al. 2006; Kumar et al. 2008; Lofrano et al. 2013; Siqueira et al. 2011). Tannery effluent contributes greatly to the contamination of both water and soil. This dark brown tint prevents light from reaching the water's surface, which is bad for aquatic life because it lowers photosynthesis and oxygenation in receiving water bodies (Carpenter et al. 2013; Song et al. 2000; Bakare et al. 2009; Kongjao et al. 2008; Mwinyihija 2010). In addition, the anaerobic condition is encouraged by the decrease in dissolved oxygen, which results in the putrid smell of receiving water bodies (Sahu et al. 2007; Rai et al. 2005; Verma et al. 2008). Polluted water bodies become even more eutrophic due to TWW, which has a negative impact on aquatic ecosystems (Dixit et al. 2015; Durai and Rajasimmam 2011; Rai et al. 2005; Schilling et al. 2012). Heavy metals have been observed to be abundant in the Ganga river and its tributaries' sediments (Tare et al. 2003; Singh et al. 2003; Bhatnagar et al. 2013). Soil fertility has declined and drinking water quality has deteriorated in Tamil Nadu, India as a result of the salinization of rivers and groundwater (Money 2008). More than 55,000 acres of land have been contaminated due to TWW, and it is estimated that five million people are adversely affected by the low quality of their drinking water and social environment (Sahasranaman and Jackson 2005; CSIRO 2001). In addition to causing a major foaming problem on surface waters, TWW has been shown to hinder the nitrification process (Lofrano et al. 2013; Trujillo-Tapia et al. 2008; Szpyrkowicz et al. 2001; Schilling et al. 2012).

13.3.2 Ecotoxicity

As heavy metals accumulate in aquatic plants, they trigger a cascade of biochemical and physiological responses that stunt the development of various plant parts

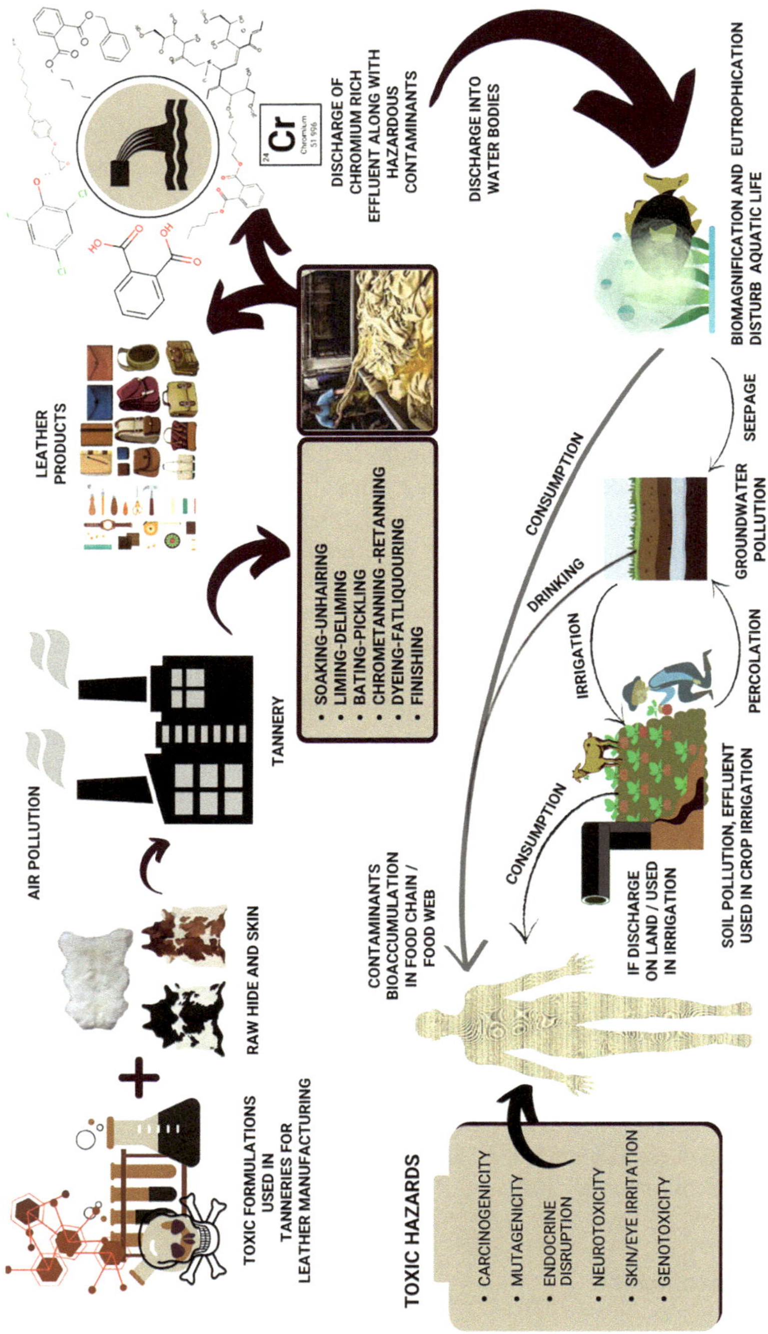

Fig. 13.1 The leather manufacturing process and release of wastewater cause toxicity in living beings

(Shanker et al. 2005; Chandra and Kulshreshtha, 2004; Satyakala and Jamil 1992; Singh and Sinha 2005). To add insult to injury, fish and other aquatic species are severely harmed by the treated/partially treated TWW. *Oreochromis niloticus* fish were used in a micronucleus test and a comet assay to determine the mutagenicity and genotoxicity of TWW-polluted water (Matsumoto et al. 2006). According to research conducted by De Nicola et al. (2007), 1 mg L^{-1} of vegetable tannins and syntan water extracts negatively impacted sea urchin embryogenesis (*Sphaerechinus granularis* and *Paracentrotus lividus*) and marine algae cell proliferation (*Dunaliella tertiolecta*). As a rule, phytotoxicity limits are denoted as a percentage of growth reduction. Its phytotoxicity is a result of Cr′s high soluble form and powerful oxidizing ability. Damage to cell membranes caused by Cr (VI) is substantial since it is a potent oxidizing agent (Shahandeh and Hossner, 2000; Vazquez et al. 1987; Mei et al. 2002). Due to its ability to form complexes with proteins, nucleic acids, and chemical molecules, Cr (III) is also a potent phytotoxin (Goutam and Saxena 2021; Su et al. 2005). It was shown by Afaq and Rana (2009) that teleost treated with leather dyes had a significant drop in total protein content (acid leather brown and Bismarck brown). Researchers in this study looked at how leather dyes affected the protein-making processes of the freshwater teleost *Cirrhinus mrigala* (Ham.). In addition, the histopathological characteristics and survival rates of *Channa punctatus* and *Oreochromis mossambicus* fish were examined concerning the harmful effects of TWW (Navaraj and Yasmin 2012; Mohanta et al. 2010). Choo et al. discovered similar symptoms in *Nymphaea spontanea* plants after 7 days in Cr concentrations of 5 and 10 mg/L (2006). They also found that when exposure time and Cr concentration increased, protein and chlorophyll concentrations decreased, suggesting a linear link between the two. Zayed and Terry (2003), Shanker et al. (2005), and Chandra and Kulshreshtha (2004) all found a similar decline in photosynthetic pigments in a variety of aquatic plants. *Labeo rohita* (Hamilton) and *Tilapia mossambica*, two common freshwater fish, have had their hematological parameters studied concerning tannery effluent in recent studies (Praveena et al. 2013; Lesley Sounderraj et al. 2012). Typically, LIs release their sewage into surrounding canals or rivers that are used by farmers for irrigation purposes (Gupta et al. 2012; Trujillo-Tapia et al. 2008). The employment of this method results in the transfer of chromium and other potentially hazardous metals from water to crop plants, which can then be ingested by humans and other animals (Chandra et al. 2009; Sinha et al. 2008). Furthermore, TWWs are extremely rich in organic and inorganic contents, which means they may offer an opportunity for a range of dangerous bacteria to develop and contaminate the receiving water bodies (Bharagava et al. 2014; Verma et al. 2008).

As an early indicator of metal toxicity, Unnikannan et al. (2011) found that plants produced less chlorophyll than normal after being exposed to 50 mg/L of Cr (VI) for 21 days. They also found that biomass decreased by 50%. It was determined that the decrease in biomass was the result of stunted root development and inadequate nutrition transfer to the shoots. Chandra et al. (2011), who studied a CETP in India that degrades and detoxifies TWW, revealed recently that they had discovered several OPs (organic pollutants) and bacterial communities in two aeration lagoons.

TWW toxicity was also evaluated by observing how it affected the germination and development of mung bean (*Phaseolus mungo*) seeds. As a result of being tapped for energy, plants' sugar reserves are depleted during times of stress, resulting in less sugar being present in the biomass of the plants themselves (Choo et al. 2006; Outridge and Noller 1991). Plants' protein levels drop because Cr exposure causes the formation of complex chemicals that inhibit enzyme action (Choo et al. 2006; Satyakala and Jamil 1992). Further, many researchers have evaluated TWW's bacteriological quality and found that harmful microorganisms persist in it even after secondary treatment (Bharagava et al. 2014; Ramteke et al. 2010; Verma et al. 2008). Chromium's toxicity and the health impacts connected with it are largely dependent on the chemical speciation to which the metal is exposed (Rameshraja and Suresh 2011). Chromium (VI) is a known carcinogen for humans, plants, animals, and microbes alike because it enters cells through the surface transport system, is converted to chromium (III), and then causes a variety of genotoxic consequences (Raj et al. 2014; Ackerley et al. 2004; Matsumoto et al. 2006; Aravindhan et al. 2004; Tripathi et al. 2011). To investigate Cr accumulation and toxicity in the aquatic plants *Vallisneria spiralis* and *Hydrilla verticillata*, Gupta et al. (2011) employed treated tannery effluent containing 2.16 mg/L of Cr. They found that both of these aquatic plants had dangerously high levels of Cr deposited in their tissues. As a result, irrigating crops with Cr-loaded TWW reduces seed germination, shoot, root growth, seedling growth, and biomass (Hussain et al. 2010; Lopez-Luna et al. 2009), induces chlorosis, photosynthetic impairment, and ultimately kills the plant (Chidambaram et al. 2009; Shanker et al. 2005; Gupta et al. 2012). (Asfaw et al. 2012; Akinici and Akinci 2010). Shoot and root lengths start to shorten at 50 mg/L of Cr, according to research by Calheiros et al. (2008a). This result indicates that plants are vulnerable to Cr toxicity during growth. Mishra and Tripati (2009) employed *Eichhornia crassipes* with varying Cr concentrations to study Cr removal from solutions, and they found that at Cr concentrations of 10 and 20 mg/L, the plants exhibited some morphological indications of toxicity. Metal toxicity symptoms, such as leaf chlorosis and root loss, were observed in the plants after 7 days of exposure. As the scientists increased the time and concentration of Cr exposure, they also saw a decrease in protein, sugar, and chlorophyll content in the plants. Haddad and Mizyed (2011) found similar things (2009). TWW has a concentration-dependent and crop-specific effect on germination and early growth of plants. Recently, Raj et al. (2014) did a study on mung bean (Vigna radiate (L.) wilczek) and found that 25% untreated and treated TWW significantly reduced germination rates (by 90% and 75%, respectively). Furthermore, it is possible to be exposed to chlorinated phenols, particularly pentachlorophenol (PCP), which is highly carcinogenic, teratogenic, and mutagenic. PCP does this via reducing the efficiency of oxidative phosphorylation, blocking the function of respiratory enzymes, and damaging the structure of mitochondria (Jain et al. 2005; Verma and Maurya 2013; Tripathi et al. 2011). Pentachlorophenol (PCP) and other chlorinated phenols are harmful to living organisms by multiple pathways, including suppression of oxidative phosphorylation, inactivation of respiratory enzymes, and damage to mitochondrial structure (Dixit et al. 2015; Tewari et al. 2011; USDHHS 2001).

Concern has also grown over the widespread environmental release of endocrine-disrupting chemicals (EDCs) that occur simultaneously with TWW. Endometrial and ovarian cancer (ERC) can develop when EDCs disrupt the body's delicate hormonal balance and reduce an organism's ability to reproduce (Dixit et al. 2015). For example, Kumar et al. (2008) found a variety of EDCs, including nonylphenol (NP), hexachlorobenzene, benzidine, and 4-aminobiphenyl, in TWW from northern India and investigated the effects of these chemicals on male rat reproductive systems. In contrast, it has also been reported that EDCs such bis (2-methoxyethyl) phthalate (BDEHP), dibutyl phthalate (DBP), and bis (2-Ethylhexyl) phthalate (DEHP) can be found in TWW (Alam et al. 2009, 2010). Numerous aquatic vascular plants exposed to a Cr solution over an extended period showed decreased photosynthetic pigments, sugar, and protein levels, according to research by Hasan et al. (2007) and Cheng et al. (2007). Due to damaged chloroplasts, Cr reduces the production of chlorophyll, which lowers photosynthetic activity (Mishra and Tripati 2009). To properly dispose of TWW in the environment, it must first undergo suitable treatment.

13.4 Treatment Technologies Available for Tannery Wastewater

Tannery wastewater is a significant source of soil and water contamination, so it must be properly treated before being safely disposed of into the environment. Utilizing physical, chemical, and biological processes can accomplish this (Metcalf and Eddy 1979). Many organic chemicals found in wastewater have been reported to be resistant to standard chemical and/or biological treatment (Schrank et al. 2004; Rameshraja and Suresh 2011). However, efficiency, affordability, and environmental competence all play a role in the decision of which effluent treatment method to use (Costa and Olivi 2009). Additionally, the properties of the wastewater should be taken into account when selecting the appropriate procedure (Costa and Olivi 2009). There have been numerous reviews of the various methods for treating tannery effluent by authors in the past (Cassano et al. 2001; Aravindhan et al. 2004). As a result, there are numerous techniques to remediate the tannery effluent. There are instances where a tannery performs each of the aforementioned procedures for treating wastewater on-site. Sending the effluent to a centralist effluent treatment facility, a particular tannery may only apply pretreatment, only pretreatment in part, or no pretreatment at all (Dargo and Ayalew 2014). The processes used to treat tannery effluents are described below.

13.5 Constructed Wetlands for Tannery Wastewater Treatment

CWs (constructed wetlands) are ecologically sound systems that have been man-made to filter contaminants out of heavily contaminated industrial and wastewater. The use of CWs to treat industrial wastewater has progressed rapidly in recent years, and they are now widely implemented for the effective removal of numerous toxins from these waters. A precise balance must be struck between the plants, microorganisms, soil, wastewater quality, and operating parameters for a wetland system to function properly (Aguilar et al. 2008). Constructed wetlands (CWs) are eco-friendly systems that are man-engineered to eliminate pollutants from severely contaminated industrial and municipal wastewater. Selecting plant species that can withstand and filter out these toxins has involved a lot of work. Industrial wastewater treatment has recently experienced fast growth, which has made it possible to successfully use CWs to remove a variety of contaminants from wastewater. The intricate interplay between plants, bacteria, soil, wastewater properties, and operating settings determines how effective a wetland system is (Rabbani et al. 2021; Aguilar et al. 2008). Several studies on the use of constructed wetlands for tannery wastewater treatment are listed in Table 13.1.

13.5.1 Role of Plants

Constructed wetlands (CWs) are eco-friendly, man-made systems used to treat highly contaminated wastewater from industries and municipalities. Rapid expansion in recent years has seen an increase in the usage of CWs for the removal of various pollutants from industrial effluent. To work properly, a wetland system depends on many factors, including plant life, microbial communities, soil composition, wastewater characteristics, and operational settings (Aguilar et al. 2008). There have been numerous attempts to identify plant species with the tolerance and removal of excessive salt ions into vacuoles or older tissues, the creation of pressure-equalizing suitable solutes through biosynthesis, the activation of antioxidant enzymes, and the synthesis of antioxidant compounds. Catalase, superoxide dismutase, glutathione peroxidase, ascorbate peroxidase, and glutathione reductase are all examples of antioxidant enzymes that are made in response to high levels of salt stress. Reactive oxygen species are neutralized, and cellular oxidation is halted (Kumari et al. 2015). In times of adversity for plants, they produce more of the phytohormone abscisic acid, which limits cell development and reduces the inhibitory effect of salts (Gupta and Huang 2014; Deinlein et al. 2014). Based on their salt accumulation and transportation mechanisms, halophytes can be broken down into three broad categories: pseudohalophytes, recretohalophytes, and euhalophytes. Recretohalophytes are separated into two groups: exo-recretohalophytes, such as Limonium Mill, which emits salts through the salt gland, and endo-recretohalophytes, which store salts in vesicles (e.g., Atriplex Linn.). Euhalophytes (i.e., stem succulents and leaf succulents) absorb salts through the vacuole of the leaf or stem as opposed to pseudohalophytes (such as Phragmites Adans) doing so through their roots (Robinson et al. 1997; Deinlein et al. 2014; Zhao et al. 1999, 2002). Furthermore, depending on their position concerning the water, halophytes can be classified as submerged, floating, or emergent plants (Zhou and Xiang 2013; Sun et al. 2013). In most cases, emerging halophytes are used in CWs, including *Scirpus validus*, *Phragmites australis*, and *Typha orientalis*. In certain research, the

Table 13.1 Application of constructed wetlands for treatment of tannery wastewater

CW type	CW surface area (m^2)	Substrate	Plant species	Pre-treatment	Operation parameters	Contaminants load ($g/m^2/d$) and removal efficiency (%)	Reference
HF	1.92	Gravel and pumice substrate	*Chrysopogon zizaniodes*	–	Effective size (ES) (d_{10}) of the Media: 1.5–4.5; uniformity coefficient (UC) (d_{60}/d_{10}): 3.5–4; HRT: (3,5,7 and 9 days)	Cr (18.33): 98.91%	Aregu et al. (2021)
VF	0.14	Gravel, sand, and soil	*Leptochloa fusca, T. Domingensis, B. Mutica*	Collected from CETP receiving wastewater from tanneries in the industrial area of Kasur, Pakistan.	The concentration of tannery effluent used with 7 days time intervals: 25%, 50%, 75% & 100%; HLR: 300 mL/min; HRT: 7 days	Cr (134): 55%	Ashraf et al. (2020a, b)
HF	0.15	Gravel and sand	*HF-I. cernua, FSW-I. aquaticum*	Physico-chemical treatment method	The concentration of tannery effluent: 50%, 75% & 100%; constant Flow Rate:0.016 m^3 day^{-1}; for the first stage of horizontal subsurface flow: HLR: 14 mmd^{-1}; HRT: 4 days; For the second stage of free flow water: HLR: 18 mm d^{-1}; HRT: 2 days	Cr (8.11): 99.22%	Zapana et al. (2020)

(continued)

Table 13.1 (continued)

CW type	CW surface area (m^2)	Substrate	Plant species	Pre-treatment	Operation parameters	Contaminants load ($g/m^2/d$) and removal efficiency (%)	Reference
HF	67.5	Gravel and sand	*Phragmites australis*	Physico-chemical and biological treatment	Two HRT_S were followed: 3 days and 7 days	Cr (0.36): 38.9%	García-Valero et al. (2020)
VF	–	Gravel, sand, and soil	*Leptochloa fusca*	–	All the CWs were operated batch-wise with a single exposure to tannery effluent and bacterial inoculation (200 mL). During the experimental period (27 days), wastewater samples from each treatment reactor were collected every 72 h regularly till no further degradation was observed.	Cr (247): 98.66%	Ashraf et al. (2018a, b, c)
HF	0.31	Pea gravel	*Typha spp.*	–	Two wetland system was incorporated; Hourly Flow Rate: 8 h each day; Daily Flow Rate: 11–25 L/day; HLR: 2.67 cm/day; Fertilizer used as nutrient on monthly basis: 14 g per event	Cr (5): 90–99%	Dotro et al. (2011a, b)

HF	4.5	Granitic rock	*Typha latifolia*	Primary treatment	Batch System with batch reactors; Cr content: 1800 mg/L/ pH 4; HLR: 0.1 m/d and Wetland Loading received: 0.048 m/d; HRT in Wetlands H and L: 2.4 and 5 days	Cr (10–20): 50–95%	Dotro et al. (2012)
HSF	–	Expanded clay and sand	*A. donax*	Biological treatment	TWW: 4 $m^3 d^{-1}$; HLR_S: 60 mmd^{-1} and 210-mm d^{-1}; HRTS: 2 days and 0.6 days	COD (38–885): 23–615% and BOD_5 (9–458): 6–363%	Calheiros et al. (2012a)
VF-HF-VF	0.65–1.3	organic coco-peat, cupola slag and pea grave	*P. australis*	–	Wastewater Flow Rate: 6 cm d^{-1}; HRT_S: 4.8, 12.5 & 2.4 d	COD (11,500 ± 6000): 98% and BOD_5 (4200 ± 2800): 98%	Saeed et al. (2012)
VF	0.096	Gravel	*Penisetum purpureum, Brachiaria decumbens, Phragmites australis*	Primary treatment (precipitation, sieving, and decantation)	Water residence time: 7 to 12 days; Hydraulic conductivity: 10^7 ms^{-1}	Cr (6): 78.10%	Maine et al. (2009)
HSF	1.2	Filtralite MR 3–8 (FMR), with particle size ranging from 3 to 8 mm.	*Phragmites australis, Typha latifolia*	Primary treatment	HLR: 8 cm d^1; HRT: 7, 2, & 5 days	TN (3.7–12.3): 53–60%, TSS (2.4–15.1): 88–90%, and Cr: 30–100%	Calheiros et al. (2009)
VF	0.0254	Round washed gravel	*Phragmites australis*	–	HRT (based on the average void volume of the bed): 3 d Average flow rate: 1.04 mL/min	TN (3.3): 67% and TP (0.40): 26%	Ong et al. (2009)
HSF	1.2	Expanded clay	*P. australis*	Primary treatment	HLR: 8 cm d^1; HRT: 7, 2, & 5 days	COD (242–1925): 141–1283% and BOD_5 (126–900): 72–529%	Calheiros et al. (2009)

(continued)

Table 13.1 (continued)

CW type	CW surface area (m^2)	Substrate	Plant species	Pre-treatment	Operation parameters	Contaminants load ($g/m^2/d$) and removal efficiency (%)	Reference
FWS	450	Sediment	*Typhs spp., Scirpus americanus*	Treatment with alkali and sedimentation	Wastewater effluent from the sedimentation pond arrived at a rate of 20 L min^{-1}. Influent totalled 80 m^3; HRT: 2d	Cr (22–31): 99%	Aguilar et al. (2008)
HSF	1.2	Clay and fine gravel-FG	*Typha latifolia, Phragmites australis*	–	The systems operated for 31 months under different hydraulic conditions and interruptions in feed; HLR: 18, 8, and 6 cm d^1	TN (7.32–14): 31–66%, TP (0.025–0.6): 16–33%, TSS (4.8–30.2): 69–85%, and Cr (0.022–0.66): 55–57%	Calheiros et al. (2008)
VF, HSF	80	Gravel and sand	*Phragmites australis*	–	Wastewater flow rate: 1.44, 7.2, and 14.4 m^3/day; HRT: 2, 0.7, and 0.4 day	TN (5–7.4): 28–52% and TSS (129): 80–93%	Bulc et al. (2008)
HF	1.2	Expanded clay	*T. latifolia*	Primary treatment	Hydraulic loading rates: 3 and 6 cm d^1	Cr (10–0.027): 001–0.080 COD (1755–2669): 54–73% and BOD_5 (740–1300): 41–58%	Calheiros et al. (2007)
HSSF	378	Gravel	*Phragmites australis*	Biological treatment	Hydraulic retention times between 5–11 days	Cr (0.2): 43–55%	Kucuk et al. (2004)
HSSF	378	Gravel	*Phragmites australis*	Biological treatment	Hydraulic retention times between 5–11 days	TP (0.16): 20% and Cr (0.007): 43–55%	Kucuk et al. (2003)

efficacy of halophytes in CWs has been assessed. For instance, in a pilot-scale subsurface wetland, Lymbery et al. (2006) tested how well *Juncus kraussii* performed in filtering out nutrient-, phosphate-, and sodium-rich wastewater from aquaculture (6900–24,500 mg L^{-1}). When evaluating eight distinct halophyte species in 2005, they discovered that CWs treating simulated saline water with an EC of 14–16 mS cm-1 removed a significant amount of TN (69%) and TK (85.5%). The research team discovered that *Digitaria bicornis* had the highest 5-day biochemical oxygen demand (BOD5) and elimination capacity, while *T. angustifolia* performed the best in terms of growth and nutrient assimilation (removal percentage of 78.9%). Additionally, both species thrived in CWs with salty water, which aided in the treatment procedure. According to Gao et al. (2015), Canna indica, P. australis, and *Scirpus validus* showed promising removal rates for COD (61.5–70.5%), NH4 + -N (59.3–68.4%), TN (61.9–70.4%), and TP (61.5–70.5%) when the salt concentration of the influent was less than 2.0%. (40.4–47.3%). Heavy metal removal was also an impressive ability displayed by several halophyte species in CWs. At a high rate, native halophytes like *T. latifolia* and *P. australis* removed heavy metals from saline industrial effluent in the Gadoon Amazai Industrial Estates, including 50% Pb, 91.9% Cd, 74.1% Fe, 40.9% Ni, 89.1% Cr, and 48.3% Cu (Khan et al. 2009). To treat salty wastewater from the tool industry with high EC, pH, and high Zn, Ni, Cr, and nutrient content, Hadad et al. (2006) transplanted numerous natural macrophytes (including *T. domingensis* and *Cyperus alternifolius*) to a pilot-scale CW. One of the macrophytes, *T. domingensis*, was proposed as a means of purifying wastewater. Numerous tests have demonstrated that CWs planted with halophytes can improve the treatment of saline wastewater. However, in contrast to their removal effectiveness for nonsaline or low-saline wastewater, their removal performance for excessively salty wastewater is constrained and erratic. When CWs were fed a mixture consisting of increasing ratios of saltwater and residential sewage (where the ratio is to be 30–40% (v/v)), Qiu et al. (2009) observed a severe decrease in COD removal percentages. In addition, the presence of both salts and heavy metals or nonmetallic ions often leads to a complicated interaction between those components, which in turn impacts plant absorption in certain wastewater types (e.g., salts with a dominant NaCl or Na_2SO_4 ion content). For instance, Liu and Liu (2008) provided evidence that salts in marine water can prevent halophytes from absorbing nutrients and metals. According to Fritioff et al. (2005), *Elodea canadensis* and *Potamogeton natans* both saw less metal buildup as salt levels rose. Salinity results in nutrient imbalances or deficits in plants because Na + and Cl– compete with nutrients like Ca^{2+}, K^{+}, and NO^{3-}, according to Hu and Schmidhalter (2005). These antagonistic effects may be the result of competition for ion channel uptake or a change in the ion carrier type (Kumari et al. 2015). Soybean germination can be negatively impacted by sodium chloride, but Patil et al. (2012) demonstrated that mixes containing boron (B) and NaCl were able to mitigate these effects. These salt, nutrient, and metalloid/metal interactions appear to be element and species-specific, even though the interactive processes are still unclear. Future studies should focus on the selection of halophytes for particular wastewater based on the interaction of certain pollutants and salts (Calheiroset al. 2008a).

13.5.2 Role of Microbial Communities

The bacteria in CWs have been proven in multiple studies to have an essential role in the degradation of various contaminants. Few attempts have been made to introduce halophilic microorganisms into CWs for the treatment of saline wastewater, but Margesin and Schinner (2001) hypothesized that these organisms have a wide variety of actual or potential applications in numerous areas of CW remediation. To supply knowledge for upcoming CW investigations, we studied the use of halophilic bacteria in biological treatment facilities. Organics are typically contaminated in saline water that comes from sources like industrial and aquacultural operations. For the elimination of COD, microorganisms are very important. Due to salt stress on microorganisms, biological technologies have poor removal effectiveness, which has restricted their widespread use for treating saline wastewater (Abou-Elela et al. 2010). Yang et al. (2013) demonstrated that salts inhibit the microbial activities that degrade organic matter. Due to their interaction with inorganic ions, the solubility of organic molecules, which are normally nonpolar or weakly polar and provide a food source for microorganisms, can decrease; Dissolved ions can lower gas solubility, making it harder for oxygen-dependent organisms to function; High salinity can cause eucryons to lose water, which in turn causes microbial cells to shrink or die off. Due to the aforementioned methods, soluble organic compounds cannot be effectively broken down and utilized by microbes. A decrease in pollution abatement rates was seen when salt concentrations in an activated sludge process exceeded 4% (w/v) NaCl, as reported by Aloui et al. (2009). Salinity could greatly reduce the ability of CWs to remove pollutants from municipal wastewater, particularly organics and N, according to Wu et al. (2008). As salinity increased from 0 to 30 ngL^{-1}, the percentages of dissolved organic carbon (DOC), ammonia-N, and inorganic-N removed decreased from 91% to 71%, 98% to 83%, and 78% to 56%, respectively. Although many microorganisms are inhibited by salts, halophilic (or salt-tolerant) bacteria allow for the biological and/or ecological treatment of saline wastewater. In environments high in salinity or hypersaline conditions, halophilic microorganisms flourish. Halophilic bacteria were divided into communities that were only moderately (0.5–2.5 M), highly tolerant (2.5–5.5 M), and very tolerant (0.2–0.5 M) of NaCl (Larsen 1986). There are many halophilic bacteria in places like pickled food industrial locations. They can also be divided into eukaryotes, bacteria, and archaea, with bacteria and archaea making up the majority of these divisions. The ability of halophilic bacteria to maintain an osmotic balance under stress is largely responsible for their ability to live in saline and hypersaline environments. To explain the mechanisms of salt tolerance in halophilic bacteria, the following three major ideas were developed (DasSarma and DasSarma 2001; Lanyi 1974; Zhuang et al. 2010; Russell 1989): The cell membrane alters to adapt to salt stress; more anions are produced by moderately halophilic groups than by non-halophilic groups; and typical archaebacterial ether lipids are found in severely halophilic groups rather than the proteins seen in non-halophilic groups. In saline environments, the halophilic groups are so made more soluble and malleable. More anions are produced by moderately halophilic groups than by non-halophilic groups, and typical archaebacterial ether lipids are found in severely halophilic groups rather than the proteins seen in non-halophilic groups. To keep the concentration of ions inside the cell stable, they use mechanisms called Na+/H+ antiporters to actively excrete Na+. Some halophilic bacteria, when subjected to high levels of salt, can biosynthesize essential solutes (such as amino acids and carbohydrates) from materials they have ingested or manufactured themselves. The use of halophilic bacteria in various biotechnological applications is widespread. Using halophilic bacteria obtained from Romanian salt lakes to decontaminate dichlorvos may be a useful approach to dealing with nonbiodegradable chemicals in the wild (Oncescu et al. 2007). For NaCl concentrations above 2%, a halophilic bacterium (*Staphylococcus xylosus*) isolated from a vegetable pickling business has been proven to be more effective than either activated sludge alone or a combination of *S. xylosus* plus activated sludge. That was the conclusion reached by researchers Abou-Elela et al. (2010). The addition of the halophilic bacterium *Halobacter halobium* to the activated sludge reactor for the treatment of saline wastewater (1–5% salt) has been shown to significantly increase the removal percentage of COD, particularly at salinities above 2% (w/v) (Kargi 2002). A Bacillus sp. SCUN strain was employed in an anaerobic method by Wu et al. (2013a, b) to remove 58.3% of the COD from extremely saline wastewater from pretreated ethyl chloride manufacturing, which had a COD of 20,000 mg $L^{-1}$and 4% salt(w/v). These results are much above those obtained by conventional anaerobic techniques. Other CWs use inoculation halophilic bacteria, which could provide useful data for future research. Nutrient removal was maximized at a NaCl concentration of 1.5% (w/v) in a pilot-scale subsurface CW by inoculating the substrate with halophilic microorganisms (isolated from a saltern), as shown by Karaji'cet al. (2010). These results show that CWs infused with targeted halophilic bacteria offer a promising strategy for handling salty water waste. Experimental results in biological processes and CWs lead us to the conclusion that inoculating CWs with halophilic microorganisms presents promising new avenues for the treatment of saltwater runoff. To effectively remove contaminants, future research should concentrate on isolating and identifying certain halophilic bacteria.

13.5.3 Role of Substrates

In CWs, the substrates play a potential role in the elimination of certain types of pollutants (e.g., heavy metals and P). Using Liebig's law of the minimum, we may conclude that phosphorus is more important than nitrogen in creating eutrophication (Wang et al. 2013). Substrate adsorption (Xu et al. 2006; Hill et al. 2000), microbial action (mostly by poly-P bacteria), plant uptake, chemical precipitation, and integration into organic matter all contribute to P removal from CWs. Similar mechanisms, such as adsorption, precipitation, plant absorption, complexation, and microbially driven processes like oxidation and reduction, remove and/or immobilize heavy metals in CWs. Several other processes, such as complexation and precipitation, are linked to the adsorption process. Adsorption of metal cations occurs either on the substrate's external face through physical mechanisms or on the substrate's interior face through ligand exchange with lattice ions or chemical mechanisms (Lee and Scholz 2006; Dunbabin and Bowmer 1992; Haynes 2015; Xu et al. 2006). Certain metals (such as Zn and Cu) may undergo sulphide precipitation as the primary process in CWs over extended periods. Machemer and Wildeman (1992) found that metals in acid mine drainage are primarily and rapidly adsorbed onto organic sites in substrate materials. These metals include copper, manganese, iron, and zinc. Allende et al. (2012). propose employing planted column reactors with a variety of substrate media to remove Fe, As, and B. According to their findings, as was eliminated via coprecipitation with Fe, which typically involved an alkaline substrate, while B removal was linked to the organic components of the latter. Different metals required different substrates due to their unique reaction mechanisms; however, the removal of heavy metals was still closely tied to the kind of substrate, making substrate selection critical to guarantee effective removals of phosphorus (P) and metals in CWs. In particular, many studies have demonstrated that substrate optimization can greatly increase CWs' removal effectiveness, particularly for P and heavy metals. Particularly effective at removing soluble P from CWs (With a 40-h hydraulic residence time (HRT), 96% of the contaminants were removed) is the calcium metasilicate mineral wollastonite, which is mined in upstate New York (Brooks et al. 2000). The removal efficiency of hydrated oil-shale ash sediment in CWs for P removal is estimated to be between 67% and 85% (loading 5–300 mgL^{-1} of PO_4^{3-} per 1.5 g of sediment) (Kaasik et al. 2008). Additionally, various target pollutants may call for varying substrates. A study by Allendeet al. looked into the efficacy of four substrates (limestone, gravel, zeolite, and cocopeat) in removing Fe, As, and B from acidic mine wastewater in vertical flow CWs (2012). Average removal percentages for As (99%) and Fe (98%) were highest for limestone, followed by the zeolite unit (86% and 92% for Fe and As, respectively). With a clearance percentage of roughly 6.3%, coco peat stood as the sole possible substrate for B removal. When compared to the other substrates used in the experiment, Gravek proved inefficient at removing any of the components. These results show that screening and using the most effective substrates may improve CWs' ability to remove certain pollutants from wastewater. Very few researchers have analyzed the role and/or effects of the substrate in CWs in the treatment of salty wastewater. Using models of surface and subsurface flow wetland systems, Mitchell and Karathanasis (1995) investigated the removal of Zn, Cu, Fe, Ni, Cd, Pb, Mn, and Cr from NaCl-enriched wastewater using two types of lime stones (amended with a mixture of mushroom compost, leaf litter, and topsoil). They determined that different substrates resulted in different heavy metal removal characteristics in CWs. Haynes (2015) conducted experiments in CWs to determine the efficacy of using various industrial wastes, including steel slag, coal fly ash, furnace slag, alum, and water treatment sludge, to treat complicated saline water with phosphates and metals. In addition to increasing P and metal retention, and getting around the problem of limited-duration adsorption induced by the presence of Fe/Al hydrous oxide and $CaCO_3$ adsorption surfaces, all of these industrial wastes showed promise as a potential active filter medium. When deciding on a substrate, it is important to think about anything from the wastewater type to the amount of nitrogen and organic matter present. The following factors should be taken into consideration when choosing a substrate for CWs: low cost, strong support for plant development, local availability, high efficacy at removing a particular type of wastewater, nontoxicity to microorganisms, durability under washing, and consistent hydraulic conductivity. Tourmaline, iron-carbon alloy (Fe-C alloy), and industrial waste such as steel slag, blast furnace slag, and alum water treatment sludge are some of the more uncommon substrates that have been used in CWs in the past. Granular materials like gravel, sand, clay, zeolite, activated carbon biological ceramsite, and, more recently, oil palm shells have become more popular substrates (Haynes 2015; Liu et al. 2014; Delkash et al. 2015). Research into the efficacy of these substrates for treating wastewater in CWs has shown notable results, which can and should inform future research. It is also unknown how factors such as P and metal removal interact with one another in terms of competition, which is another area worthy of investigation. Substrate blockage may occur in CWs with prolonged use, resulting in subpar performance and a shortened CWs' life span. Substrate clogging issues can be solved in several ways, including gravel cleaning, chemical treatment, and the introduction of earthworms (Nivalaet al. 2012). Substrate clogging issues are a given with saline wastewater, just as they are with nonsaline wastewater. When evaluating pollution control methods, it is important to think about more than just how well they remove pollutants.

13.5.4 Factors Affecting Tannery Wastewater Treatment in Constructed Wetlands

Many factors, including the plants, microorganisms, substrates, wastewater kinds, environmental pH, temperature, structure, operation parameters, and humidity, might influence a CW's efficiency in treating either salty or fresh water wastewater. Pilot-scale horizontal subsurface flow CWs' nitrogen and ammonium removal efficiency declined to 58.5% and 37.9%, respectively, below 15°C (Akratos and Tsihrintzis 2007). Consistency with a horizontal subsurface circulation Garca et al. (2005) observed that the average COD removal rate in CWs with a water depth of 0.27 m was 60%, whereas, in CWs with a water depth of 0.5 m, it was 75%. In a study of a mesocosm CW for wastewater treatment, Vera et al. (2016) reported that a 10% increase in NH3-N, TP, and TN removal was achieved by raising the HRT from 3.5 days to 7 days. It has been demonstrated that temperature can influence some physical and biochemical processes that regulate nutrient removal in CWs. Temperature and humidity have a crucial role in the purification of salty wastewater. The evaporation rates of CWs can be altered and hence their physical performance is altered by the combination of humidity and temperature. The effluent of a horizontal subsurface flow CW was found to have higher salt concentrations than the influent, for instance, in arid climates where humidity and warmth promoted evapotranspiration (Freedman et al. 2014). In addition, the diversity and abundance of microorganisms in CWs may have a significant biological impact on a wide range of biochemical processes, including N mineralization, nitrification, and denitrification. Most biochemical processes involving microorganisms have a preferred temperature range, and this can have an effect on the outcomes you observe; for denitrification, for example, the ideal temperature range is 20–40°C (Parket al. 2015; Song et al. 2014; Warneke et al. 2011). Temperature not only affects the efficiency of these mechanisms, but it also has physiological effects on the formation and growth of halophytes in CWs by regulating processes such as photosynthesis and respiration (Leeet al. 2009; Bezbaruah and Zhang 2004; Kadlec and Reddy 2001). Many activities, including biological reactions, in CWs, are pH-dependent, and as a result, their ability to remove contaminants is affected. As their activity can cause a pH drop to below 7.0, pH is critically important for the nitrification process, although denitrification works best at a pH of roughly 8.0 (Lee et al. 2009). In addition, the pH of the solution is the most important element in determining the heavy metal concentrations and adsorption efficiencies. It has been determined via numerous research that apatite's ability to adsorb aqueous Pb, Cd, and Zn is dependent on the pH of the solution. Zinc oxide (ZnO) forms in alkaline environments, while zinc phosphate dihydrate ($Zn_3(PO_4)24H_2O$) forms in acidic environments (Chen et al. 1997). When certain metals are removed using biosorption procedures, the pH varies dramatically and also affects the substrate's ability to adsorb the metals (Fourest and Roux 1992). In conclusion, pH has a significant impact on the efficiency with which salty wastewater is removed from integrated CWs, particularly those containing heavy metals. The structure also has a significant role in determining how well CWs treat wastewater, whether it is salty or fresh. Two primary hydrological classifications of CWs are FWSCW (free-water surface flow CWs) and SSFCW (subsurface flow CWs). The subset of SSFCW known as HFCW (horizontal flow CW) can be further subdivided into the VFCW (vertical flow CW) and S (VFCW). In addition, hybrid CWs-CWs that combine HFCW and VFCW is feasible (Vymazal 2005). An FWSCW is a closed basin with a short depth of water (typically 20–40 cm) through the unit, making it susceptible to climatic changes; yet, it has a strong re-aerating ability due to air diffusion (Vymazal 2014). The bed of an HFCW is made out of gravel or another porous media, and wastewater runs through the bed in a horizontal direction. These CWs have a lower concentration of DO but higher SS and ion removal efficiency. Since the water is added to the VFCW in huge quantities, the oxygen content of the water increases as it percolates through the medium, making the VFCW far more aerobic than an HFCW. Since the higher evaporation rate could lead to further salinization of the wastewater, we do not advise using an FWSCW with saline wastewater. The performance of CWs is also affected by the operation patterns and characteristics, such as hydraulic loading rate, HRT, water table depth, and continuous or intermittent flow. Many activities involving oxidation-reduction reactions are affected by the water table depth, which in turn affects the redox potential of CWs (Garca et al. 2005). The evaporation rates of wastewater in CWs are also affected by the depth of the water table when dealing with saltwater. Depending on whether the flow is steady or sporadic, the re-aeration process will result in a range of redox potential and DO concentrations. HRT is a crucial factor that dictates the duration of all CWs processes and, by extension, their efficacy in waste removal (Kottiet al. 2010). There is an intricate web of interplay between all these elements. High temperatures enhance evapotranspiration and lead to higher salt concentrations; in this instance, the situation could be remedied by, for example, lowering HRT and changing the water table depth. As a result, these considerations must be taken into account thoroughly and holistically during the design and operation of CWs.

13.6 Mechanisms of Chromium Remediation in Constructed Wetlands

Around 8% of all effluent comes from chrome tanning. Wastewater of this type often has chromium concentrations in the range of 3000 to 6000 mg/L of trivalent chromium (Cr^{3+}) (Gao and Su 2001). Cells undergo metamorphosis when chromium and protein are present. When activated sludge enters the biological treatment system, its flocculability is destroyed, which might have harmful or inhibitory consequences on the biochemical treatment that follows. In extreme cases, sludge loss occurs as a result of the entire breakdown of activated sludge (Li et al. 2003). Six chromium (Cr^{6+}) is readily absorbed by the human body despite being a potent carcinogenic and mutagenesis-causing agent. When compared to Cr^{3+}, its toxicity is one hundred times higher (Meng 2000). Physical chemistry and biological methods are used together to remediate wastewater containing chromium. Some examples of physical chemistry processes are alkaline precipitation, adsorption, ion exchange, electrolysis, and membrane technology. Physical chemistry is the first stage of wastewater treatment (Li 2014; de Mara Guillén-Jiménez et al. 2008; Hashem et al. 2019). Plants adapted to wetland conditions play a crucial role in the removal of pollutants from wastewater in constructed wetlands. Plants native to wetlands have the ability to both immediately absorb nutrients and store them as Cr in their roots and shoots (Liang et al. 2017). Root exudates from wetland plants can alter metal toxicity and mobility, root surfaces provide more room for microbial development, and root tissues aid in Cr buildup. Wetland plant roots are the most concentrated area of metals due to the slow translation of metals from roots to buds and the coupling of carboxyl groups with compounds that may impede the migration of dangerous metals to buds (Zhang et al. 2010). Cr ions can bond with the cell wall of plant tissues, according to research by Yadav et al. (2010), which inhibits the development of new shoots from the roots. Deposition on the roots of wetland plants contributes to the buildup of metals like Cr in tannery effluent in substrate medium (Kidd et al. 2009). The organic acids produced by wetland plants, such as malate, citrate, malonate, oxalate, acetate, and fumarate, are essential for metabolic reactions and the formation of complexes with metals in wastewater. Combining Cr ions with chelating substances including glutathione, organic acids, metallothioneins, and phytochelatin, wetland plants can reduce the intracellular toxicity of metals (Sultana et al. 2014). Cells in the plant's root system are symplastic receptors for both Cr forms (Cr^{3+} and Cr^{6+}), and when Cr^{6+} is depleted, Cr^{3+} is produced and accumulates in the root cells. Root membranes are susceptible to injury from Cr^{6+}, which can stunt plant growth. Plants are unable to take in critical nutrients including Fe (iron), Mn (manganese), K (potassium), P (phosphorus), Mg (magnesium), and Ca (calcium) because Cr^{6+} has a similar ionic structure (Sultana et al. 2014). Senescence of leaves is a key method by which wetland plants detoxify or remove Cr from their systems. Plants can undergo a degeneration process similar to senescence if they are subjected to heavy metal toxicity, such as Cr stress. As the leaves age, heavy metals like Cr are liberated from apoproteins and moved to other binding sites in the plant (Ouelhadj et al. 2006). The buildup of reactive oxygen species during the exposure of marsh plants to hazardous levels of Cr appears to induce leaf senescence (ROS). When Cr^{6+} is converted to Cr^{3+} via the Fenton reaction, reactive oxygen species (ROS) are formed (Wakeel et al. 2020). According to Brezinov'a and Vymazal, biofilms formed as wastewater flows through a plant formation play a key role in the removal of toxins in CW systems (2014). Plant australis (*P. australis*) was used to treat residential wastewater by Lesage et al. (2007), and they found that the position of plants in CWs, in addition to Zn and Cr, can have a substantial impact on removal efficiency. According to research by Calheiros et al. (2008a, b), most of the Cr in common reed is concentrated in its roots, while only a small percentage is kept in its aerial parts. Plant metal extraction was found to be critically vital by Zayed and Terry (2003) when a greater amount of Cr was transported to the tissues via the roots. In addition, plant roots can contribute to metal deposition in a substrate medium (Tangahu et al. 2011). Wastewater that contains Cr is often treated using cattails. Since Cr accumulates in different spots in the wetland growing media, several researchers have noticed that plant position in the CWs has little bearing on Cr storage. There was no significant difference between leaf and stem Cr concentrations; however, the accumulation rate of the air portion was less than that of the underwater region (Calheiros et al. 2008a, b).In natural settings, huge plants that were either floating or submerged acquired more Cr than those that were just beginning to emerge (Vymazal and Kr opfelov'a 2008a, b). To immobilize metals in biomass before disposal at landfill sites, CWs produce plant biomass, which can be burned or exchanged. Root extract organic acids have been discovered to increase plant Cr absorption through complex formation with Cr (Srivastava et al. 1998; Barlett and James 1988). While Cr transport is restricted to the body's upper regions, its accumulation and distribution deep into the tissue are determined by the chemical makeup of the tissue itself (James and Barlett 1983). Plant root membranes are sensitive to Cr (VI) because of the compound's strong oxidizing potential. In addition, because of their comparable ionic forms, Cr (VI) prevents the absorption of several vital elements by plants (including Mg, P, Fe, Mn, Ca, and K) (Gardea-Torresdey et al. 2005; Pandey and Sharma 2003). Because of their impact on metal mobilization, toxicity, and bioavailability, microorganisms play a crucial role in CW operation. Microbe-metal interactions are most complicated and intense at the root zone of artificial wetlands. Generally speaking, bacteria can either actively, through a process called bioaccumulation, or passively, through a process called biosorption, concentrate metals (Unz and Shuttleworth, 1996). Metal speciation, biosorption, precipitation via sulfate reduction, and methylation are only some of the heavy metal properties that microorganisms in engineered wetlands can alter (Kosolapov et al. 2004). Metal sorption is controlled by numerous mechanisms and interactions inside microorganisms, including ion exchange, chelation, adsorptions, and entrapment (Kosolapov et al. 2004; Gadd 2004). Bacteria can create amorphous mineral inclusions, which they use to store metals inside the cell (Vainshtein et al. 2002). Because of dissimilatory reduction or microbial metabolic interaction, they can also precipitate hazardous metals, rendering them immobile (Kosolapov et al. 2004). Symbiotic bacteria and fungi are living in plant roots, and they play a protective role by reducing metal toxicity and boosting the efficacy of phytoremediation through the accumulation of metals in plant tissues (Marchand et al. 2010; de Souza et al. 1999;). Bacterial activity in CWs is responsible for the conversion of Cr to its immobile state (Sinicrope et al. 1992; Nelson et al. 2002; Schiffer 1989; Adriano 2001). Mycorrhizae are symbiotic fungus that forms a link between plant roots and soil, enhancing the absorptive surface area of root hairs and degrading potentially dangerous metals in the soil (Meharg and Cairney 2000). These symbionts prevent metal absorption by plants by sequestering them in fungal tissue (Khan et al. 2000). Lakatos et al. found that *Phragmites australis* in freshwater wetlands benefit from symbionts (i.e., periphyton) that increase the reeds' capacity to acquire and retain metals (1999) (Fig.13.2).

13.7 Prospects and Challenges

Remediation using man-made wetland systems is gentle on the environment. Although many studies have been conducted in the field of constructed wetlands, more research is needed to meet the need for further research on sustainable media and plant species for the constructed wetland treatment that has a long-term performance, higher retention of organic substitutes, and removal of heavy metals and nutrients. It has been established that trivalent chromium and sulphide are both extremely dangerous contaminants. Prioritization should be given to reducing COD and eliminating nitrogen and phosphorus. By precisely locating the baffle walls and the hybrid baffle system, improved performance can be attained. Before and after the constructed wetland treatment, wastewater must undergo pre- and posttreatment, respectively. The return on investment (ROI) of different wastewater treatment methods, such as constructed wetlands, should be calculated. To keep costs down and prevent secondary contamination, it is important to carefully assess the external supply of organic chemicals and/or the use of organic materials as the substrate. To properly treat enormous industrial effluent streams, a sizable tract of land is sometimes required. Plant biomass has several potential commercial uses, such as energy production, construction materials, and paper production. Optimizing land size, medium, plants, engineering design, and automation in a man-made wetland requires more research. The most crucial step in achieving a high removal efficiency of Cr is making the correct choice of plants and growth (bedding) material.

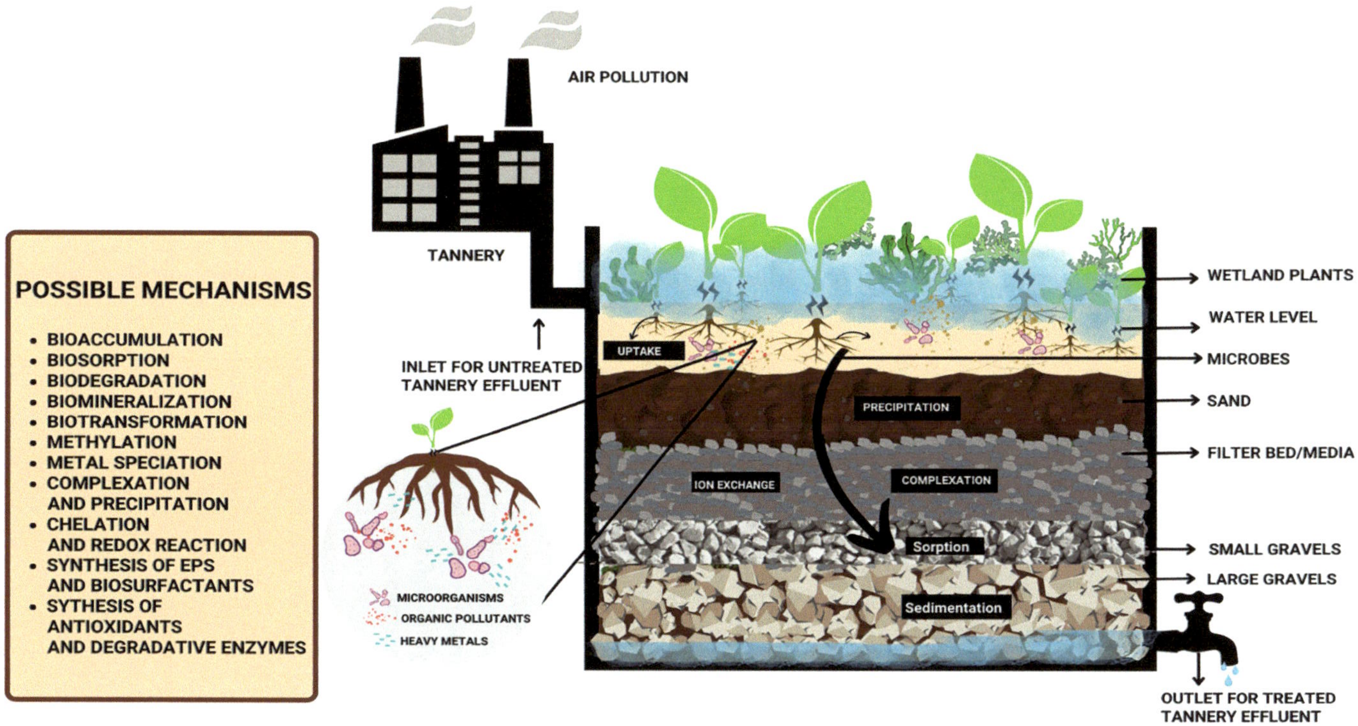

Fig. 13.2 Possible mechanisms of tannery effluent remediation in constructed wetlands

13.8 Conclusions and Recommendations

In the context of industrial wastewater, TWW ranks among the most polluted options. Unfortunately, the manufacture of leather in many developing nations still employs inefficient, chemical- and water-intensive, traditional processes. To make tanning more eco-friendly, it is essential to develop tanning chemicals that can efficiently replace chromium. As more research is needed on sulfide's hazardous mechanism, the right technology for removing H_2S from the air must be put into place. The most eco-friendly options for TWW treatment and management are membrane bioreactors and constructed wetlands, although both have limitations and need further work before they can be fully implemented. When treating TWW, it may be best to use a combination of physical and chemical methods rather than just one or the other. Additionally, research is being conducted on the use of cutting-edge treatment modalities such as membrane filtration and oxidation processes are promising to eliminate the obstinate organic contaminants, but these still need to be optimized for the greatest economic benefit. The viability of the novel anammox method for the anaerobic removal of ammonia from TWW requires further investigation. To develop the most effective treatment plans for future tanneries, a thorough study of the toxicity profiles of TWW may prove useful. In addition to bolstering discharge limitations for TWW, one frequent method for reducing environmental pollution is to place LIs in a predetermined industrial area. Environmental challenges can be overcome by the use of eco-friendly chemicals, water minimization technologies, wastewater treatment/purification and recycling, and other similar techniques, as outlined in the EU's integrated pollution control strategy and greening policy. This implies that the ideal treatment for TWW and associated compounds are not yet accessible. However, further work is needed to identify TWW therapy choices that are more efficient. To provide the best pollution control alternative for future tanneries, policymakers and water sector experts will continue to give priority to the implementation of emerging treatment technologies like AOPs in conjunction with biological treatment processes.

References

Abou-Elela SI, Kamel MM, Fawzy ME (2010) Biological treatment of saline wastewater using a salt-tolerant microorganism. Desalination 250:1–5

Ackerley DF, Gonzalez CF, Keyhan M, Blake IIR, Matin A (2004) Mechanism of chromate reduction by the E. coli protein, NfsA, and the role of different chromate reductases in minimizing oxidative stress during chromate reduction. Environ Microbiol 6(8):851–860

Adriano DC (2001) Trace Elements in Terrestrial Environments: Biogeochemistry, Bioavailability and Risk of Metals, 2nd edn. Springer, New York, p 867

Afaq S, Rana KS (2009) Impact of leather dyes on total protein of fresh water teleost, Cirrhinus mrigala (Ham.). Asian J Exp Sci 23(1):299–302

Aguilar JRP, Cabriales JJP (2008) Vega MM (2008) Identification and characterization of sulfur-oxidizing bacteria in an artificial wetland that treats wastewater from a tannery. Int J Phytoremediation 10:359–370

Akinici IE, Akinci S (2010) Effect of chromium toxicity on germination and early seedling growth in melon (Cucumis melo L.). Afr J Biotechnol 9(29):4589–4594

Akratos CC, Tsihrintzis VA (2007) Effect of temperature, HRT, vegetation and porous media on removal efficiency of pilot-scale horizontal subsurface flow constructed wetlands. Ecol Eng 29: 173–191

Alam MZ, Ahmad S, Malik A (2009) Genotoxic and mutagenic potential of agricultural soil irrigated with tannery effluents at Jajmau (Kanpur), India. Achiev Environ Contam Toxicol 57(3):463–476

Alam MZ, Ahmad S, Malik A, Ahmad M (2010) Mutagenicity and genotoxicity of tannery effluents used for irrigation at Kanpur, India. Ecotoxicol Environ Saf 73(5):1620–1628

Allende KL, Fletcher TD, Sun G (2012) The effect of substrate media on the removal of arsenic, boron and iron from an acidic wastewater in planted column reactors. Chem Eng J 179:1190–1130

Aloui F, Khoufi S, Loukil S, Sayadi S (2009) Performances of an activated sludge process for the treatment of fish processing saline wastewater. Desalination 246:389–396

Alvarez-Bernal D, Contreras-Ramos SM, Trujillo-Tapia N, Olalde-Portugal V, Frias-Hernandez JT, Dendooven L (2006) Effects of tanneries wastewater on chemical and biological soil characteristics. Appl Soil Ecol 33:269–277

Apaydin O, Kurt U, Gonullu MT (2009) An investigation on tannery wastewater by electrocoagulation. Glob Nest J 11(4):546–555

Aravindhan R, Madhan B, Rao R, Nair B, Ramasami T (2004) Bioaccumulation of chromium from tannery wastewater an approach for chrome recovery and reuse. Environ Sci Technol 38(1): 300–306

Aregu MB, Asfaw SL, Khan MM (2021) Developing horizontal subsurface flow constructed wetland using pumice and Chrysopogon zizanioides for tannery wastewater treatment. Environ Syst Res 10:33

Asfaw A, Sime M, Itanna F (2012) Determining the effect of tannery effluent on seeds germination of some vegetable in Ejersa areas of east Shoa. Ethiopia Int J Sci Res 2(12):1–10

Ashraf A, Bibi I, Niazi N, Ok YS, Murtaza G, Shahid M, Kunhikrishnan A, Li D, Mehmood T (2016) Chromium (VI) sorption efficiency of acid-activated banana peel over organo-montmorillonite in aqueous solutions. Int J Phytoremediation 19:605–613

Ashraf S, Afzal M, Naveed M, Shahid M, Ahmad Zahir Z (2018a) Endophytic bacteria enhance remediation of tannery effluent in constructed wetlands vegetated with Leptochloa fusca. Int J Phytoremediation 20:121–128

Ashraf S, Afzal M, Rehman K, Naveed M, Zahir ZA (2018b) Plant-endophyte synergism in constructed wetlands enhances the remediation of tannery effluent. Water Sci Technol 77: 1262–1270

Ashraf S, Afzal M, Rehman K, Tahseen R, Naveed M, Zahir ZA (2018c) Enhanced remediation of tannery effluent in constructed wetlands augmented with endophytic bacteria. Desalin Water Treat 102:93–100

Ashraf S, Naveed M, Afzal M, Ashraf S, Ahmad SR, Rehman K, Zahir ZA, Núñez-Delgado A (2020a) Evaluation of toxicity on ctenopharyngodon idella due to tannery effluent remediated by constructed wetland technology. PRO 8:612

Ashraf S, Naveed M, Afzal M, Seleiman MF, Al-Suhaibani NA, Zahir ZA, Mustafa A, Refay Y, Alhammad BA, Ashraf S (2020b) Unveiling the potential of novel macrophytes for the treatment of tannery effluent in vertical flow pilot constructed wetlands. Water 12:549

Bakare AA, Okunola AA, Adetunji OA, Jenmi HB (2009) Genotoxicity assessment of a pharmaceutical effluent using four bioassays. Genet Mol Biol 32(2):373–381

Barlett RJ, James BR (1988) Mobility and Bioavailability of Chromium in Soils. In: Nriagu JO, Nieboer E (eds) Chromium in Natural and Human Environment. John Wiley and Sons Inc., New York, pp 267–304

Bezbaruah AN, Zhang TC (2004) pH, redox, and oxygen microprofiles in rhizosphere of bulrush (Scirpus validus) in a constructed wetland treating municipal wastewater. Biotechnol Bioeng 88: 60–70

Bharagava RN, Yadav S, Chandra R (2014) Antibiotic and heavy metal resistance properties of bacteria isolated from the aeration lagoons of common effluent treatment plant (CETP) of tannery industries (Unnao, India). Indian J Biotechnol 13(4):514–519

Bharagava RN, Saxena G, Chowdhary P (2017) Constructed wetlands: an emerging phytotechnology for degradation and detoxification of industrial wastewaters. In: Environmental pollutants and their bioremediation approaches. CRC Press, pp 397–426

Bhatnagar MK, Singh R, Gupta S, Bhatnagar P (2013) Study of tannery effluents and its effects on sediments of river Ganga in special reference to heavy metals at Jajmau, Kanpur, India. J Environ Res Dev 8(1):56–59

Bibi I, Niazi NK, Choppala G, Burton ED (2018) Chromium (VI) removal by siderite (FeCO3) in anoxic aqueous solutions: an X-ray absorption spectroscopy investigation. Sci Total Environ 640:1424–1431

Borba FH, Seibert D, Pellenz L, Espinoza-Quiñones FR, Borba CE, Módenes AN, Bergamasco R (2018) Desirability function applied to the optimization of the Photoperoxi-Electrocoagulation process conditions in the treatment of tannery industrial wastewater. J Water Process Eng 23: 207–216

Bosinc M, Buljan J, Daniels RP (2000) Regional program for pollution Control. Tanning Industry US/RAS/92/120, South-East Asia: 1–14

Březinová T, Vymazal J (2014) Competition of Phragmites australis and Phalaris arundinacea in constructed wetlands with horizontal subsurface flow–does it affect BOD5, COD and TSS removal? Ecol Eng 73:53–57

Brooks AS, Rozenwald MN, Geohring LD, Lion LW, Steenhuis TS (2000) Phosphorus removal by wollastonite: a constructed wetland substrate. Ecol Eng 15:121–132

Buljan J, Kral I (2011) Introduction to treatment of tannery effluents. United Nations Industrial Development Organization (UNIDO), Vienna

Calheiros CSC, Rangel AOSS, Castro PML (2007) Constructed wetland systems vegetated with different plants applied to the treatment of tannery wastewater. Water Res 41(8):1790–1798

Calheiros CS, Rangel AO, Castro PM (2008a) Evaluation of different substrates to support the growth of Typha latifolia in constructed wetlands treating tannery wastewater over long-term operation. Bioresour Technol 99:6866–6877

Calheiros CSC, Rangel AOSS, Castro PML (2008b) The effects of tannery wastewater on the development of different plant species and chromium accumulation in Phragmites australis. Arch Enviro Con Tox 55(3):404–414

Calheiros CSC, Rangel AOSS, Castro PML (2009) Treatment of industrial wastewater with two-stage constructed wetlands planted with Typha latifolia and Phragmites australis. Bioresour Technol 100(13):3205–3213

Calheiros CSC, Duque AF, Moura A, Henriques IS, Correia A, Rangel AOSS, Castro PML (2009a) Changes in the bacterial community structure in two-stage constructed wetlands with different plants for industrial wastewater treatment. Bioresour Technol 100(13):3228–3235

Calheiros CSC, Duque AF, Moura A, Henriques IS, Correia A, Rangel AOSS, Castro PML (2009b) Substrate effect on bacterial communities from constructed wetlands planted with Typha latifolia treating industrial wastewater. Ecol Eng 35(5):744–753

Calheiros CSC, Quitério PVB, Silva G, Crispim LFC, Moura SC, Brix H, Castro PML (2012a) Use of constructed wetland systems with Arundo and Sarcocornia for polishing high salinity tannery wastewater. J Environ Manag 95(1):66–71

Calheiros CSC, Silva G, Quitério PVB, Crispim LFC, Moura SC, Brix H, Castro PML (2012b) Toxicity of high salinity tannery wastewater and effects on constructed wetland plants. Int J Phytoremediation 14(7):669–680

Carpenter J, Sharma S, Sharma AK, Verma S (2013) Adsorption of dye by using the solid waste from leather industry as an adsorbent. Int J Eng Sci Invent 2(1):64–69

Cassano A, Molinali R, Romomo M, Drioli E (2001) Treatment of aqueous effluents of the leather industry by membrane processes a review. J Membr Sci 181:111–126

Chandra P, Kulshreshtha K (2004) Chromium accumulation and toxicity in aquatic vascular plants. Bot Rev 70:313–327

Chandra R, Bharagava RN, Yadav S, Mohan D (2009) Accumulation and distribution of toxic metals in wheat (Triticum aestivum L.) and Indian mustard (Brassica campestris L.) irrigated with distillery and tannery effluents. J Hazard Mater 162:1514–1521

Chandra R, Bharagava RN, Kapley A, Purohit HJ (2011) Bacterial diversity, organic pollutants and their metabolites in two aeration lagoons of common effluent treatment plant (CETP) during the degradation and detoxification of tannery wastewater. Bioresour Technol 102(3):2333–2341

Chen, Wright JV, Conca JL, Peurrung LM (1997) Effects of pH on heavy metal sorption on mineral apatite. Environ Sci Technol 31:624–631

Chen G, Huang MH, Chen L, Chen DH (2011a) A batch decolorization and kinetic study of Reactive Black 5 by a bacterial strain Enterobacter sp. GY-1. Int Biodeterior. Biodegradation 65(6):790–796

Chen ZG, Zhang SH, Huang YJ, Su CY, Qi JP, Sun XF, Yu LM, Xie SH, Liu HX (2011b) Explanation of "Wastewater emission standard in tannery and fur manufacturing industry". China Leather Association and Cleaner Production Center of China Light Industry, Beijing

Chen Y, Wen Y, Zhou Q, Vymazal J (2014) Effects of plant biomass on denitrifying genes in subsurface-flow constructed wetlands. Bioresour Technol 157:341–345

Cheng H, Xu W, Liu L, Zhao Q, Chen G (2007) Application of composted sewage sludge (CSS) as a soil amendment for turfgrass growth. Ecol Eng 29:96–104

Chidambaram AP, Sundaramoorthy A, Murugan K, Baskaran SGL (2009) Chromium induced cytotoxicity in black gram (Vigna mungo L). Iran J Environ Health Sci Eng 6(1):17–22

Choo TP, Lee CK, Low KS, Hishamuddin O (2006) Accumulation of chromium (VI) from aqueous solutions using water lilies (Nymphaea spontanea). Chemosphere 62:961–967

Cooman K, Gajardo M, Nieto J, Bornhardt C, Vidal G (2003) Tannery wastewater characterization and toxicity effects on Daphnia spp. Environ Toxicol 18(1):45–51

Costa CR, Olivi P (2009) Effect of chloride concentration on the electrochemical treatment of a synthetic tannery wastewater. Electrochim Acta 54:2046–2052

CSIRO (2001) Salinity reduction in tannery effluents in India & Australia. Project proposal to ACIAR by CSIRO Textile and Fibre Technology, Leather Research Centre

DasSarma S, DasSarma P (2001) Halophiles e LS. John Wiley & Sons, Ltd., Hoboken, NJ

Dargo H, Ayalew A (2014) Tannery waste water treatment: a review. Int J Emerg Trends Sci Technol 01(09):1488–1494

DBT (2019) Manual on constructed wetland as an alternative technology for sewage management in India. Department of Bio Technology (DBT), New Delhi

de Souza MP, Huang CPA, Chee N, Terry N (1999) Rhizosphere bacteria enhance the accumulation of selenium and mercury in wetland plants. Planta 209:259–263

Deinlein U, Stephan AB, Horie T, Luo W, Xu GH, Schroeder JI (2014) Plant salt-tolerance mechanisms. Trends Plant Sci 19:371–379

Delkash M, Bakhshayesh BE, Kazemian H (2015) Using zeolitic adsorbents to cleanup special wastewater streams: a review. Microporous Mesoporous Mater 214:224–241

Dixit S, Yadav A, Dwivedi PD, Das M (2015) Toxic hazards of leather industry and technologies to combat threat: a review. J Clean Prod 87:39–49

Dotro G, Larsen D, Palazolo P (2011a) Preliminary evaluation of biological and physical–chemical chromium removal mechanisms in gravel media used in constructed wetlands. Water Air Soil Pollut 215:507–515

Dotro G, Larsen D, Palazolo P (2011b) Treatment of chromium-bearing wastewaters with constructed wetlands. Water Environ J 25:241–249

Dotro G, Castro S, Tujchneider O, Piovano N, Paris M, Faggi A, Palazolo P, Larsen D, Fitch M (2012) Performance of pilot-scale constructed wetlands for secondary treatment of chromium-bearing tannery wastewaters. J Hazard Mater 239-240:142–151

Duarte AALS, Canais-Seco T, Peres JA, Bentes I, Pinto J (2010) Sustainability indicators of subsurface flow constructed wetlands in Portuguese small communities. WSEAS Trans Environ Dev 9(6):625–634
Dunbabin JS, Bowmer KH (1992) Potential use of constructed wetlands for treatment of industrial wastewaters containing metals. Sci Total Environ 111:151–168
Durai G, Rajasimman M (2011) Biological treatment of tannery waste water, a review. J Environ Sci Technol 4(1):1–17
Fourest E, Roux J-C (1992) Heavy metal biosorption by fungal mycelial by-products: mechanisms and influence of pH. Appl Microbiol Biotechnol 37:399–403
Freedman A, Gross A, Shelef O, Rachmilevitch S, Arnon S (2014) Salt uptake and evapotranspiration under arid conditions in horizontal subsurface flow constructed wetland planted with halophytes. Ecol Eng 70:282–286
Fritioff Å, Kautsky L, Greger M (2005) Influence of temperature and salinity on heavy metal uptake by submersed plants. Environ Pollut 133:265–274
Gadd GM (2004) Microbial influence on metal mobility and application for bioremediation. Geoderma 122:109–119
Gao ZB, Su CY (2001) Wastewater treatment of tannery industry. Chemical Industry Press, Beijing
Gao F, Yang ZH, Li C, Jin WH (2015) Saline domestic sewage treatment in constructed wetlands: study of plant selection and treatment characteristics. Desalin Water Treat 53:593–602
García J, Aguirre P, Barragán J, Mujeriego R, Matamoros V, Bayona JM (2005) Effect of key design parameters on the efficiency of horizontal subsurface flow constructed wetlands. Ecol Eng 25:405–418
García-Valero A, Martínez-Martínez S, Faz Á, Terrero MA, Muñoz MÁ, Gómez-López MD, Acosta JA (2020) Treatment of wastewater from the tannery industry in a constructed wetland planted with Phragmites australis. Agronomy 10:176
Gardea-Torresdey JL, de la Rosa G, Peralta-Videa JR, Montes M, CruzJimenez G, Cano-Aguilera I (2005) Differential uptake and transport of trivalent and hexavalent chromium by tumbleweed (Salsola kali). Arch Environ Contam Toxicol 48:225–232
Goutam SP, Saxena G, Singh V, Yadav AK, Bharagava RN, Thapa KB (2018) Green synthesis of TiO2 nanoparticles using leaf extract of Jatropha curcas L. for photocatalytic degradation of tannery wastewater. Chem Eng J 336:386–396
Goutam SP, Saxena G (2021) Biogenic nanoparticles for removal of heavy metals and organic pollutants from water and wastewater: advances, challenges, and future prospects. Bioremediation Environ Sustain:623–636
Guillén-Jiménez FDM, Morales-Barrera L, Morales-Jimrera J, Cristiani-Urbina E (2008) Modulation of tolerance to Cr(VI) and Cr(VI) reduction by sulfate ion in a Candida yeast strain isolated from tannery wastewater. J Ind Microbiol Biotechnol 35(11):1277–1287
Gupta B, Huang BR (2014) Mechanism of salinity tolerance in plants: physiological, biochemical, and molecular characterization. Int J Genomics 2014:701596
Gupta K, Gaumat S, Mishra K (2011) Chromium accumulation in submerged aquatic plants treated with tannery effluent at Kanpur, India. J Environ Biol 32:591–597
Gupta K, Gaumat S, Mishra K (2012) Studies on phyto-genotoxic assessment of tannery effluent and chromium on Allium cepa. J Environ Biol 33(3):557–563
Hadad HR, Maine MA, Bonetto CA (2006) Macrophyte growth in a pilot-scale constructed wetland for industrial wastewater treatment. Chemosphere 63:1744–1753
Haddad M, Mizyed N (2011) Evaluation of various hydroponic techniques as decentralised wastewater treatment and reuse systems. Int J Environ Stud 68(4):461–476
Hasan SH, Talat M, Rai S (2007) Sorption of zinc and cadmium from aqueous solutions by water hyacinth (Eichhornia crassipes). Bioresour Technol 98:918–928
Hashem MA, Momen MA, Hasan M (2019) Chromium removal from tannery wastewater using Syzygium cumini bark adsorbent. Int J Environ Sci Technol 16(3):1395–1404

Haynes RJ (2015) Use of industrial wastes as media in constructed wetlands and filter beds-prospects for removal of phosphate and metals from wastewater streams. Crit Rev Env Sci Technol 45:1041–1103
Hill CM, Duxbury J, Geohring L, Peck T (2000) Designing constructed wetlands to remove phosphorus from barnyard runoff: a comparison of four alternative substrates. J Environ Sci Heal A 35:1357–1375
Hu Y, Schmidhalter U (2005) Drought and salinity: a comparison of their effects on mineral nutrition of plants. J Plant Nutr Soil Sci 168:541–549
Hussain F, Malik SA, Athar M, Bashir N, Younis U, Mahmood-ul-Hassan MS (2010) Effect of tannery effluents on seed germination and growth of two sunflower cultivars. Afri J Biotechnol 9(32):5113–5120
Islam BI, Musa AE, Ibrahim EH, Sharafa SAA, Elfaki BM (2014) Evaluation and characterization of tannery wastewater. J For Prod Ind 3:141–150
Jain RK, Kapur M, Labana S, Lal B, Sarma PM, Bhattacharya D, Thakur IS (2005) Microbial diversity: application of microorganisms for the biodegradation of xenobiotics. Curr Sci 89(1): 101–112
James BR, Barlett RJ (1983) Behavior of chromium in soils VII. Adsorption and reduction of hexavalent forms. J Environ Qual 12:177–181
Kadlec RH, Reddy KR (2001) Temperature effects in treatment wetlands. Water Environ Res 73 (5):543–557
Kaasik A, Vohla C, Motlep R, Mander U, Kirsimae K (2008) Hydrated calcareous oil-shale ash as potential filter media for phosphorus removal in constructed wetlands. Water Res 42:1315–1323
Karajić M, Lapanje A, Razinger J, Zrimec A, Vrhovšek D (2010) The effect of the application of halotolerant microorganisms on the efficiency of a pilot-scale constructed wetland for saline wastewater treatment. J Serb Chem Soc 75:129–142
Kargi F (2002) Enhanced biological treatment of saline wastewater by using halophilic bacteria. Biotechnol Lett 24:1569–1572
Katsenovich YP, Hummel-Batista A, Ravinet AJ, Miller JF (2009) Performanceevaluation of constructed wetlands in a tropical region. Ecol Eng 35:1529–1537
Khan AG, Kuek C, Chaudhry TM, Koo CS, Hayes W (2000) Role of plants, mycorrhizae and phytochelators in heavy metal contaminated land remediation. Chemosphere 41:197–207
Khan MS, Ahmed S, Evans AE, Chadwick M (2009) Methodology for performance analysis of textile effluent treatment plants in Bangladesh. Chem Eng Res Bull 13(2):61–66
Kidd P, Barceló J, Bernal MP, Navari-Izzo F, Poschenrieder C, Shilev S, Clemente R, Monterroso C (2009) Trace element behaviour at the root–soil interface: Implications in phytoremediation. Environ Exp Bot 67:243–259
Kongjao S, Damronglerd S, Hunsom M (2008) Simultaneous removal of organic and inorganic pollutants in tannery wastewater using electrocoagulation technique. Korean J Chem Eng 25(4): 703–709
Kosolapov DB, Kuschk P, Vainshtein MB, Vatsourina AV, Wiebner A, Kastner M, Muller RA (2004) Microbial processes of heavy metal removal from carbondeficient effluents in constructed wetland Eng. Life Sci 4:403–411
Kotti IP, Gikas GD, Tsihrintzis VA (2010) Effect of operational and design parameters on removal efficiency of pilot-scale FWS constructed wetlands and comparison with HSF systems. Ecol Eng 36:862–875
Kucuk OS, Sengul F, Kapdan IK (2003) Removal of ammonia from tannery effluents from a red bed constructed wetland. Water Sci Technol 48:179–186
Kucuk O, Sengul F, Kapdan I (2004) Removal of ammonia from tannery effluents in a reed bed constructed wetland. Water Sci Technol 48:179–186
Kumar V, Majumdar C, Roy P (2008) Effects of endocrine disrupting chemicals from leather industry effluents on male reproductive system. J Steroid Biochem Mol Biol 111(3–5):208–216
Kumari A, Das P, Parida AK, Agarwal PK (2015) Proteomics, metabolomics, and ionomics perspectives of salinity tolerance in halophytes. Front Plant Sci 6:537

Lakatos G, Kiss M, Mezzaros I (1999) Heavy metal content of common reed (Phragmites australis/ Cav./Trin. ex Steudel) and its periphyton in Hungarian shallow standing waters. Hydrobiologia 415:47–53
Lanyi JK (1974) Salt-dependent properties of proteins from extremely halophilic bacteria. Bacteriol Rev 38:212–290
Larsen H (1986) Halophilic and halotolerant microorganisms-an overview and historical–perspective. Ferms Microbiol Lett 39:3–7
Lee B-H, Scholz M (2006) Application of the self-organizing map (SOM) to assess the heavy metal removal performance in experimental constructed wetlands. Water Res 40:3367–3374
Lee C-g, Fletcher TD, Sun G (2009) Nitrogen removal in constructed wetland systems. Eng Life Sci 9:11–22
Lesage E, Rousseau D, Meers E, Tack F, De Pauw N (2007) Accumulation of metals in a horizontal subsurface flow constructed wetland treating domestic wastewater in Flanders, Belgium. Sci Total Environ 380:102–115
Lesley Sounderraj SF, Lesley N, Senthilkumar P (2012) Studies on the effect of tannery effluent and chromium accumulation in common crop Tilapia mossambica. Int J Pharm Biol Arch 3(4): 978–985
Li H (2014) Study on treatment of tannery wastewater and chromium sludge for Cr (III) recovery. Doctoral Dissertation of Harbin Institute of Technology, Harbin
Li XX, Yu CZ, Ma XY (2003) Progress of study on treatment of effluent from tannery. China Leather 32(19):26–31
Li G, Wu Z, Cheng S, Liang W, He F, Fu G, Zhong F (2007) Application ofconstructed wetlands on wastewater treatment for aquaculture ponds. Wuhan Univ J Nat Sci 12:1131–1135. in Chinese
Liang Y, Zhu H, Bañuelos G, Yan B, Shutes B, Cheng X, Chen X (2017) Removal of nutrients in saline wastewater using constructed wetlands: plant species, influent loads and salinity levels as influencing factors. Chemosphere 187:52–61
Liu Y, Liu YJ (2008) Biosorption isotherms, kinetics and thermodynamics. Sep Purif Technol 61: 229–242
Liu M, Wu S, Chen L, Dong R (2014) How substrate influences nitrogen transformations in tidal flow constructed wetlands treating high ammonium wastewater? Ecol Eng 73:478–486
Lofrano G, Aydin E, Russo F, Guida M, Belgiorno V, Meric S (2008) Characterization, fluxes and toxicity of leather tanning bath chemicals in a large tanning district area (IT). Water Air Soil Pollut 8:529–542
Lofrano G, Meriç S, Zengin GE, Orhon D (2013) Chemical and biological treatment technologies for leather tannery chemicals and wastewaters: a review. Sci Total Environ 461:265–281
Lopez-Luna J, Gonzalez-Chavez MC, Esparza-Garcia FJ, Rodriguez-Vazquez R (2009) Toxicity assessment of soil amended with tannery sludge, trivalent chromium and hexavalent chromium, using wheat, oat and sorghum plants. J Hazard Mater 163(23):829–883
Lymbery AJ, Doupe RG, Bennett T, Starcevich MR (2006) Efficacy of a subsurface-flow wetland using the estuarine sedge Juncus kraussii to treat effluent from inland saline aquaculture. Aquac Eng 34:1–7
Ma RH (2004) Clearer production and pollution control technology in tannery industry. Chemical Industry Press, Beijing
Machemer SD, Wildeman TR (1992) Adsorption compared with sulfide precipitation as metal removal processes from acid mine drainage in a constructed wetland. J Contam Hydrol 9:115–131
Maine MA, Sune N, Hadad H, Sánchez G, Bonetto C (2009) Influence of vegetation on the removal of heavy metals and nutrients in a constructed wetland. J Environ Manag 90:355–363
Marchand L, Mench M, Jacob DL, Otte ML (2010) Metal and metalloid removal in constructed wetlands, with emphasis on the importance of plants and standardized measurements: a review. Environ Pollut 158:3447–3461
Margesin R, Schinner F (2001) Potential of halotolerant and halophilic microorganisms for biotechnology. Extremophiles 5:73–83

Matsumoto ST, Mnlovani SM, Malaguttii MIA, Dias AL, Fonseca IC, Morales MAM (2006) Genotoxicity and mutagenicity of water contaminated with tannery effluent, as evaluated by the micronucleus test and comet assay using the fish Oreochromis niloticus and chromosome aberrations in onion root tips. Genet Mol Biol 29(1):148–158

Meharg AA, Cairney JW (2000) Co-evolution of mycorrhizal symbionts and their hosts to metal-contaminated environments. Adv Ecol Res 30:69–112

Mei B, Puryear JD, Newton RJ (2002) Assessment of Cr tolerance and accumulation in selected plant species. Plant Soil 247:223–231

Meng ZQ (2000) Environmental toxicology. China Environmental Science Press, Beijing

Metcalf and Eddy (1979) Wastewater engineering: treatment, disposal, reuse, 2nd edn. McGraw-Hill, New York

Mishra VK, Tripati BD (2009) Accumulation of chromium and zinc from aqueous solutions using water hyacinth (Eichhornia crassipes). J Hazard Mater 164:1059–1063

Mitchell LK, Karathanasis AD (1995) Treatment of metal-chloride-enriched waste-water by simulated constructed wetlands. Environ Geochem Health 17:119–126

Mohanta MK, Salam MA, Saha AK, Hasan A, Roy AK (2010) Effects of tannery effluents on survival and histopathological changes in different organs of Channa punctatus. Asian J Exp Biol Sci 1(2):294–302

Mondal A, Banerjee PK, Bhattacharjee C, Saha PD (2012) Treatment of chromium present in tannery wastewater using chemical & biological techniques. Elixir Pollut 49:9832–9835

Money CA (2008) Salinity reduction in tannery effluents in India and Australia. Final report on project AS1/2001/005. ACIAR, Canberra, ACT

Mwinyihija M (2010) Main pollutants and environmental impacts of the tanning industry. In: Ecotoxicological diagnosis in the Tanning Industry. Springer, New York, NY

Nelson EA, Specht WL, Bowers JA, Gladden JB (2002) Constructed wetlands for removal of heavy metals from NPDES outfall effluent. In: Proceedings of the 2002 National Conference on Environmental Science and Technology. Battelle Press. http://sti.srs.gov/fulltext/ms2002600/ms2002600.html.

Nivala J, Knowles P, Dotro G, García J, Wallace S (2012) Clogging in subsurface-flow treatment wetlands measurement, modeling and management. Water Res 46:1625–1640

Oncescu T, Oancea P, Enache M, Popescu G, Dumitru L, Kamekura M (2007) Halophilic bacteria are able to decontaminate dichlorvos, a pesticide, from saline environments. Central Eur J Biol 2:563–573

Ong SA, Uchiyama K, Inadama D, Yamagiwa K (2009) Simultaneous removal of color, organic compounds and nutrients in azo dye-containing wastewater using up-flow constructed wetland. J Hazard Mater 165:696–703

Oral R, Meric S, De Nicola E, Petruzzelli D, Della Rocca C, Pagano G (2007) Multi-species toxicity evaluation of a chromium-based leather tannery wastewater. Desalination 211(1–3):48–57

Ouelhadj A, Kuschk P, Humbeck K (2006) Heavy metal stress and leaf senescence induce the barley gene HvC2d1 encoding a calcium-dependent novel C2 domain-like protein. New Phytol 170:261–273

Outridge PM, Noller BN (1991) Accumulation of toxic trace elements by freshwater vascular plants. Rev Environ Contam Toxicol 121:1–63

Pandey N, Sharma CP (2003) Chromium interference in iron nutrition and water relations of cabbage. Environ Exp Bot 49:195–200

Papaevangelou VA, Gikas GD, Tsihrintzis VA (2017) Chromium removal from wastewater using HSF and VF pilot-scale constructed wetlands: overall performance, and fate and distribution of this element within the wetland environment. Chemosphere 168:716–730

Park JH, Kim SH, Delaune RD, Cho JS, Heo JS, Ok YS, Seo DC (2015) Enhancement of nitrate removal in constructed wetlands utilizing a combined autotrophic and heterotrophic denitrification technology for treating hydroponic wastewater containing high nitrate and low organic carbon concentrations. Agric Water Manag 162:1–14

Patil DB, Jadhav SH, Bhamburdekar SB (2012) Interactive effect of boron and NaCl on gemination performance in soybean. Int J Appl Biol Pharm Technol 3:366–368

Pradhan D, Sukla LB, Sawyer M, Rahman PK (2017) Recent bioreduction of hexavalent chromium in wastewater treatment: a review. J Ind Eng Chem 55:120

Praveena M, Sandeep V, Kavitha N, Jayantha Rao K (2013) Impact of tannery effluent, chromium on hematological parameters in a fresh water fish, Labeo Rohita (Hamilton). Res J Animal Vet Fish Sci 1(6):1–5

Qiu J, Wang J, Zhang Y (2009) Applicability of constructed wetlands to the treatment of wastewater with high salinity and its research progress. Indus Water Treatment 29:1–3. in Chinese

Rabbani A, Zainith S, Deb VK, Das P, Bharti P, Rawat DS et al (2021) Microbial technologies for environmental remediation: potential issues, challenges, and future prospects. In: Microbe mediated remediation of environmental contaminants, pp 271–286

Rai UN, Dwivedi S, Tripathi RD, Shukla OP, Singh NK (2005) Algal biomass: an economical method for removal of chromium from tannery effluent. Bull Environ Contam Toxicol 75(2): 297–303

Rai UN, Tripathi RD, Singh NK, Upadhyay AK, Dwivedi S, Shukla MK, Mallick S, Singh SN, Nautiyal CS (2013) Constructed wetland as an ecotechnological tool for pollution treatment for conservation of Ganga river. Bioresour Technol 148:535–541

Raj A, Kumar S, Haq I, Kumar M (2014) Detection of tannery effluents induced DNA damage in mung bean by use of random amplified polymorphic DNA markers. ISRN Biotechnol 2014: 727623

Rameshraja D, Suresh S (2011) Treatment of tannery wastewater by various oxidation and combined processes. Int J Environ Resour 5(2):349–360

Ramteke PW, Awasthi S, Srinath T, Joseph B (2010) Efficiency assessment of common effluent treatment plant (CETP) treating tannery effluents. Environ Monit Assess 169(1–4):125–131

Robinson MF, Very AA, Sanders D, Mansfield TA (1997) How can stomata contribute to salt tolerance? Ann Bot 80:387–393

Russell NJ (1989) Adaptive modifications in membranes of halotolerant and halophilic microorganisms. J Bioenerg Biomembr 21:93–113

Saeed T, Afrin R, Muyeed AA, Sun G (2012) Treatment of tannery wastewater in a pilot-scale hybrid constructed wetland system in Bangladesh. Chemosphere 88:1065–1073

Sahasranaman A, Jackson M (2005) Salinity reduction tannery effluents in India and Australia: project review. ACIAR, Canberra, ACT

Sahu RK, Katiyar S, Tiwari J, Kisku GC (2007) Assessment of drain water receiving effluent from tanneries and its impact on soil and plants with particular emphasis on bioaccumulation of heavy metals. J Environ Biol 28(3):685–690

Satyakala G, Jamil K (1992) Chromium induced biochemical changes in Eichhornia crassipes (Mart) Solms. and Pistia stratiotes L. Bull Environ Contam Toxicol 48:921–928

Saxena G, Bharagava RN (2015) Persistent organic pollutants and bacterial communities present during the treatment of tannery wastewater. In: Chandra R (ed) Environmental waste management. CRC Press, Taylor and Francis Group, Boca Raton, FL, pp 217–247. https://doi.org/10.1201/b19243-10

Saxena G, Chandra R, Bharagava RN (2017) Environmental pollution, toxicity profile and treatment approaches for tannery wastewater and its chemical pollutants. Rev Environ Contam Toxicol 240:31–69

Saxena G (2020) Bacterial degradation of persistent organic pollutants from tannery wastewater after secondary treatment process. PhD Thesis. Babasaheb Bhimrao Ambedkar University, India

Saxena G, Purchase D, Bharagava RN (2020a) Environmental hazards and toxicity profile of organic and inorganic pollutants of tannery wastewater and bioremediation approaches. In: Bioremediation of industrial waste for environmental safety: volume I: industrial waste and its management, pp 381–398

Saxena G, Purchase D, Mulla SI, Bharagava RN (2020b) Degradation and detoxification of leather tannery effluent by a newly developed bacterial consortium GS-TE1310 for environmental safety. J Water Process Eng 38:101592
Saxena G, Purchase D, Mulla SI, Saratale GD, Bharagava RN (2020c) Phytoremediation of heavy metalcontaminated sites: eco-environmental concerns, field studies, sustainability issues, and future prospects. Rev Environ Contam Toxicol 249:71–131
Schiffer DM (1989) Water quality variability in a central Florida wetland receiving highway runoff. In: Davis FE (ed) Water: laws and management. American Water Research Association, Bethesda, MD, pp 7A-1–7A-11
Schilling K, Ulrike B, Helmut K, Zessner M (2012) Adapting the Austrian Edict on wastewater emissions for tanneries as consequence of foam formation on surface waters. Environ Sci Pol 23:68–73
Schrank SG, Jose HJ, Moreira RFPM, Schroder HF (2004) Comparison of different advanced oxidation process to reduce toxicity and mineralisation of tannery wastewater. Water Sci Technol 50(5):329–334
Schröder P, Navarro-Aviñó J, Azaizeh H, Goldhirsh AG, DiGregorio S, Komives T, Langergraber G, Lenz A, Maestri E, Memon AR, Ranalli A, Sebastiani L, Smrcek S, Vanek T, Vuilleumier S, Wissing F (2007) Using phytoremediation technologies to upgrade waste water treatment in Europe. Environ Sci Pollut R 14(7):490–497. https://doi.org/10.1065/espr2006.12.373
Shahandeh H, Hossner LR (2000) Enhancement of Cr(III) phytoaccumulation. Int J Phytoremediation 2:269–286
Shahid M, Shamshad S, Rafiq M, Khalid S, Bibi I, Niazi NK, Dumat C, Rashid M (2017) Chromium speciation, bioavailability, uptake, toxicity and detoxification in soil-plant system: a review. Chemosphere 178:513–533
Shakir L, Ejaz S, Ashraf M, Aziz QN, Ahmad AA, Iltaf I et al (2012) Ecotoxicological risks associated with tannery effluent wastewater. Environ Toxicol Pharmacol 34(2):180–191
Shanker AK, Cervantes C, Loza-Tavera H, Avudainayagam S (2005) Chromium toxicity in plants. Environ Int 31(5):739–753
Singh M, Muller G, Singh IB (2003) Geographic distribution and base line concentration of heavy metals in sediments of Ganga River, India. J Geochem Explor 80:1–17
Singh S, Sinha S (2005) Accumulation of metals and its effects in Brassica juncea (L.) Czern. (cv. Rohini) grown on various amendments of tannery waste. Ecotoxicol Environ Saf 62:118–127
Sinha S, Singh S, Mallick S (2008) Comparative growth response of two varieties of Vigna radiata L. (var. PDM 54 and var. NM 1) grown on different tannery sludge applications: effects of treated wastewater and ground water used for irrigation. Environ Geochem Health 30(22): 407–422
Sinicrope TL, Langis R, Gersberg RM, Busanardo MJ, Zedler JB (1992) Metal removal by wetland mesocosms subjected to different hydroperiods. Ecol Eng 1:309–322
Siqueira IR, Vanzella C, Bianchetti P, Siqueira RMA, Stulp S (2011) Anxiety-like behavior in mice exposed to tannery wastewater: the effect of photo-electro-oxidation treatment. Neurotoxicol Teratol 33(4):481–484
Song K, Hernandez ME, Batson JA, Mitsch WJ (2014) Long-term denitrification rates in created riverine wetlands and their relationship with environmental factors. Ecol Eng 72:40–46
Song Z, Williams CJ, Edyvean GJ (2000) Sedimentation of tannery wastewater. Water Res 34(7): 2171–2176
Srivastava S, Prakash S, Srivastava MM (1998) Fate of trivalent chromium in presence of organic acids: a hydroponic study on the tomato plant. Chem Speciat Bioavailab 10:147–150
Streit KF, Rodrigues MAS, Zoppas Ferreira J (2014) Electrodialysis treatment of tannery wastewater [eds]. Electrodialysis and water reuse. Springer Press, Berlin, Heidelberg, pp 91–99
Su Y, Han FX, Sridhar BB, Monts DL (2005) Phytotoxicity and phytoaccumulation of trivalent and hexavalent chromium in brake fern. Environ Toxicol Chem 24:2019–2026

Suganthi KV, Mahalaksmi M, Balasubramanian N (2013) Development of hybrid membrane bioreactor for tannery effluent treatment. Desalination 309:231–236

Sultana MY, Akratos CS, Pavlou S, Vayenas DV (2014) Chromium removal in constructed wetlands: a review. Int Biodeterior Biodegrad 96:181–190

Sun H, Wang Z, Gao P, Liu P (2013) Selection of aquatic plants for phytoremediation of heavy metal in electroplate wastewater. Acta Physiol Plant 35:355–364

Szpyrkowicz L, Kelsall GH, Kaoul SN, De Faveri M (2001) Performance of electrochemical reactor for treatment of tannery wastewaters. Chem Eng Sci 56(4):1579–1586

Tangahu BV, Abdullah S, Rozaimah S, Basri H, Idris M, Anuar N, Mukhlisin M (2011) A review on heavy metals (As, Pb, and Hg) uptake by plants through phytoremediation. Int J Chem Eng 2011:1–31

Tare V, Gupta S, Bose P (2003) Case studies on biological treatment of tannery wastewater in India. J Air Waste Manage Assoc 53(8):976–982

Tewari CP, Shukla S, Pandey P (2011) Biodegradation of pentachlorophenol (PCP) by consortium of Flavobacterium sp. in tannery effluent. J Environ Res Dev 7(2A):876–882

Tigini V, Giansanti P, Mangiavillano A, Pannocchia A, Varese GC (2011) Evaluation of toxicity, genotoxicity and environmental risk of simulated textile and tannery wastewaters with a battery of biotests. Ecotoxicol Environ Saf 74(4):866–8673

Tripathi M, Vikram S, Jain RK, Garg SK (2011) Isolation and growth characteristics of chromium (VI) and pentachlorophenol tolerant bacterial isolate from treated tannery effluent for its possible use in simultaneous bioremediation. Indian J Microbiol 51(1):61–69

Trujillo-Tapia N, Mondragon CC, Vasquez-Murrieta MS, Cleemput OV, Dendooven L (2008) Inorganic N dynamics and N2O production from tannery effluents irrigated soil under different water regimes and fertilizer application rates: a laboratory study. Appl Soil Ecol 38(3):279–288

Tunay O, Kabdasli I, Orhon D, Ates E (1995) Characterization and pollution profile of leather tanning industry in Turkey. Water Sci Technol 32:1–9

Unnikannan P, Sundaramoorthy P, Baskaran L, Ganesh KS, Chidambaram ALA (2011) Assessment of chromium phytotoxicity in some aquatic weeds. Bot Res Int 4:13–18

Unz RF, Shuttleworth KL (1996) Microbial mobilization and immobilization of heavy metals. Curr Opin Biotechnol 7:307–310

USDHHS (2001) United States Department of Health and Human Services (USDHHS): public health statement. In: Toxicological profile for pentachlorophenol. Prepared by Syracuse Research Corporation, pp. 1–11

USEPA (1986) Guidelines for the health risk assessment of chemical mixtures (PDF) EPA/630/R-98/002

Vainshtein MB, Suzia N, Kudryashova E, Ariskina E (2002) New magnetsensitive structures in bacterial and archaea cells. Biol Cell 94:29–35

Vazquez MD, Poschenrieder C, Barcelo J (1987) Chromium VI induced structural changes in bush bean plants (Phaseolus vulgaris L.). Ann Bot 59:427–438

Vera I, Verdejo N, Chavez W, Jorquera C, Olave J (2016) Influence of hydraulic retention time and plant species on performance of mesocosm subsurface constructed wetlands during municipal wastewater treatment in super-arid areas. J Environ Sci Health A Tox Hazard Subst Environ Eng 51:105–113

Verma T, Maurya A (2013) Isolation of potential bacteria from tannery effluent capable to simultaneously tolerate hexavalent chromium and pentachlorophenol and its possible use in effluent bioremediation. Int J Eng Sci 2:64–69

Verma T, Ramteke PW, Garg SK (2008) Quality assessment of treated tannery wastewater with special emphasis on pathogenic E. coli detection through serotyping. Environ Monit Assess 145(1–3):243–249

Vymazal J (2005) Horizontal sub-surface flow and hybrid constructed wetlands systems for wastewater treatment. Ecol Eng 25:478–490

Vymazal J (2014) Constructed wetlands for treatment of industrial wastewaters: a review. Ecol Eng 73:724–751

Vymazal, J, Kröpfelová L (2008a) Environmental pollution, horizontal flow constructed wetlands, Wastewater Treatment in Constructed Wetlands with Horizontal Sub-Surface Flow, pp. 203–322
Vymazal J, Kröpfelová L (2008b) Wastewater treatment in constructed wetlands with horizontal sub-surface flow. Springer Science & Business Media, Dordrecht
Wakeel A, Xu M, Gan Y (2020) Chromium-induced reactive oxygen species accumulation by altering the enzymatic antioxidant system and associated cytotoxic, genotoxic, ultrastructural, and photosynthetic changes in plants. Int J Mol Sci 21:728
Wang Z, Dong J, Liu L, Zhu G, Liu C (2013) Screening of phosphate-removing substrates for use in constructed wetlands treating swine wastewater. Ecol Eng 54:57–65
Wang Z, Zhang XX, Lu X, Liu B, Li Y, Long C, Li A (2014) Abundance and diversity of bacterial nitrifiers and denitrifiers and their functional genes in tannery wastewater treatment plants revealed by high-throughput sequencing. PLoS One 9:113603
Warneke S, Schipper LA, Bruesewitz DA, McDonald I, Cameron S (2011) Rates, controls and potential adverse effects of nitrate removal in a denitrification bed. Ecol Eng 37:511–522
World Health Organization (2002) The world health report 2002: reducing risks, promoting healthy life. World Health. Organization
Wu X, Du YG, Qu Y, Du DY (2013a) Ternary cycle treatment of high saline wastewater from pesticide production using a salt-tolerant microorganism. Water Sci Technol 67:1960–1966
Wu Y, Tam NFY, Wong MH (2008) Effects of salinity on treatment of municipal wastewater by constructed mangrove wetland microcosms. Mar Pollut Bull 57:727–734
Wu H, Zhang J, Wei R, Liang S, Li C, Xie H (2013b) Nitrogen transformations and balance in constructed wetlands for slightly polluted river water treatment using different macrophytes. Environ Sci Pollut R 20:443–451
Xu D, Xu J, Wu J, Muhammad A (2006) Studies on the phosphorus sorption capacity of substrates used in constructed wetland systems. Chemosphere 63:344–352
Yadav AK, Kumar N, Sreekrishnan TR, Satya S, Bishnoi NR (2010) Removal of chromium and nickel from aqueous solution in constructed wetland: Mass balance, adsorption-desorption and FTIR study. Chem Eng J 160:122–128
Yang JX, Spanjers H, Jeison D, Van Lier JB (2013) Impact of Na+ on biological wastewater treatment and the potential of anaerobic membrane bioreactors: a review. Crit Rev Environ Sci Technol 43:2722–2746
Zapana JSP, Arán DS, Bocardo EF, Harguinteguy CA (2020) Treatment of tannery wastewater in a pilot scale hybrid constructed wetland system in Arequipa, Peru. Int J Environ Sci Technol 17: 4419–4430
Zayed A, Terry N (2003) Chromium in the environment: factors affecting biological remediation. Plant Soil 249:139–156
Zhang D, Gersberg RM, Keat TS (2009) Constructed wetlands in China. Ecol Eng 35:1367–1378
Zhang CB, Wang J, Liu WL, Zhu SX, Liu D, Chang SX, Chang J, Ge Y (2010) Effects of plant diversity on nutrient retention and enzyme activities in a full-scale constructed wetland. Bioresour Technol 101:1686–1692
Zhao KF, Fan H, Ungar IA (2002) Survey of halophyte species in China. Plant Sci 163:491–498
Zhao KF, Li FZ, Fan SJ, Feng LT (1999) Halophytes in China. Chin Bull Bot 16(3):201–207
Zhi W, Ji GD (2012) Constructed wetlands, 1991-2011 A review of research development, current trends, and future directions. Sci Total Environ 441:19–27
Zhou X, Xiang X (2013) Effect of different plants on azo-dye wastewater bio-decolorization. Int Symp Environ Sci Technol 18:540–546
Zhuang XL, Han Z, Bai ZH, Zhuang GQ, Shim HJ (2010) Progress in decontamination by halophilic microorganisms in saline wastewater and soil. Environ Pollut 158:1119–1126

Role of Plant-Bacteria Association in Constructed Wetlands for the Removal of Iron (Fe) from Contaminated Water

14

Nopi Stiyati Prihatini and Soemarno

Abstract

Wetland ecosystems are characterized by an interaction between oxic surface soils and anoxic subsoil, so this ecosystem is very productive. It is a site for fast biogeochemical cycles due to the interaction between the aerobic soil surface and anaerobic subsoil. When constructed wetland (CW) is built for removed certain pollutants, including iron and other heavy metals, a natural process in the wetland will enhance to meet the purpose. Several species of macrophytes were used to improve the process. Macrophytes are morphologically and physiologically adapted to the wetland environment, which is always waterlogged and oxygen-poor conditions. *Eleocharis dulcis* is a type of macrophyte that adapts well to wetland conditions. It is used in CW to reduce excess iron in the water. *E. dulcis* planted in CW treat acid mine drainage (AMD) and show promising efficiency in removing iron. Plant roots create oxidized conditions on anoxic substrates and stimulate the decomposition of organic compounds in an aerobic environment and the bacterial growth. The microbial group associated with plant in CW also take an important role. These microbes play a role in iron-plaque deposition, including bacteria that oxidize Fe (II) (known as FeOB) and bacteria that reduce Fe (III) (known as Ferb), which are found in abundance in the wetland plants' rhizosphere. The role of plant-bacteria association in constructed wetlands, especially for removing iron (Fe) from contaminated water, has been discussed in this chapter.

N. S. Prihatini (✉)
Department of Environmental Engineering, Lambung Mangkurat University, Banjarbaru, South Kalimantan, Indonesia
e-mail: ns.prihatini@ulm.ac.id

Soemarno
Department of Soil Sciences, Brawijaya University, Malang, East Jawa, Indonesia

S. Kumar et al. (eds.), *Aquatic Macrophytes: Ecology, Functions and Services*,
https://doi.org/10.1007/978-981-99-3822-3_14

Keywords

Plant-bacteria association · Constructed wetlands · Iron

14.1 Introduction

In natural wetlands, plants provide attachment sites for organisms which, as they grow, degrade organic compounds and absorb heavy metals. This makes natural wetlands function as wastewater treatment for various contaminants. It is just that with the rapid increase in population growth, the natural wetlands area continues to decrease. In the end, natural wetlands are no longer able to provide environmental services effectively in accordance with quality standards. This is one of the reasons for developing constructed wetlands, especially for wastewater treatment purposes (Zhang et al. 2010).

Constructed wetland (CW) is an alternative wastewater treatment method that mimics natural processes in wetlands to clean polluted water. CW is a system in which each constituent component works together to build a complicated procedure. Performances of CW are affected by these component differences. In a constructed wetland, removal occurs in three main compartments, namely (1) soil and substrate, (2) hydraulics, and (3) vegetation (Sheoran and Sheoran 2006). The interaction of several processes determines the number of ion contaminants that are reduced from the wetlands. It includes precipitation, coprecipitation, sedimentation, sorption, ion exchange, phytoaccumulation, photodegradation, microbial activity, biodegradation, and plant uptake. These processes need the involvement of plants, media, and microorganisms present in CW (Munawar 2007; Vymazal 2010; Vymazal and Kropfelova 2008; Younger et al. 2002). CW uses certain plants that are tolerant and can absorb contaminants. All processes are highly interdependent, making the overall mechanism to remove heavy metals very complex in wetlands (Sheoran and Sheoran 2006).

Iron (Fe) is widely distributed in nature, environmentally friendly, nontoxic, and closely related to pollutant removal (Cheng et al. 2022). However, iron can be a tricky problem over the standard, as in acid mine drainage (Prihatini et al. 2015, 2016; Prihatini and Soemarno 2017, 2021; Yunus and Prihatini 2018). *Eleocharis dulcis* (Burm. f.) Trin. ex Hensch known as Purun tikus are a type of macrophyta that has a very good adaptation to wetland conditions. It is used in CW to reduce excess iron in the water. Purun tikus that are planted in CW treat acid mine drainage (AMD) and show promising efficiency in removing iron. Apart from AMD, well water in several swamp and peat areas also has a high iron content. The study result shows that using different plant species in CW leads to differences in Fe removal (Prihatini et al. 2020, 2022).

14.2 Effect of Plant in Constructed Wetland

14.2.1 Plants Uptake Increases Iron Removal in Constructed Wetland

Selecting vegetation planted in CW is one of the determinants of CW performance. The difference in iron removal in CW without plants and CW with plants shows the effect of the presence of plants on the efficiency of removal by CW (Prihatini and Soemarno 2021). The importance of plants in CW (Wiessner et al. 2006) and the capability of plants to uptake contaminants are one of the mechanisms that determine CW performance (Zhang et al. 2010).

The pollutants metabolism in plants consists of three stages: transformation, compartmentation, and conjugation. Enzymes such as glutathione transferase, cytochrome *P*450, carboxylesterase, *O*-and *N*-malonyl transferase, and *O*-and *N*-glucosyl transferase are involved in these processes. The detoxification final stage includes three stages: release to the cellular vacuole, transport to the extracellular space, and incorporation into cell membranes such as into lignin or other cell membrane components (Stottmeister et al. 2003).

The main route of absorption of heavy metals by wetland plants or macrophytes is through sinking and floating roots. In certain cases where the leaves are entirely or partially immersed, metals are absorbed via leaves and roots. Plants with submerged roots have the potential to extract metals from sediments as well as from water, while plants without roots can only extract metals from water (Sheoran and Sheoran 2006). The number of heavy metals in plant tissues and the speed of plant growth determine the metal removal rate. In herbaceous plants or macrophytes, the metal absorption per unit area of wetlands is usually greater. When leaves take up heavy metals, the aqueous phase moves passively through the cuticle or cracks the cell wall from the stomata to the plasma membrane (Sheoran and Sheoran 2006). When determining the location of absorption of minerals in plants, a passive cation exchange process that allows ions to penetrate the plant was found (Sheoran and Sheoran 2006). The cation exchange site on *Vallisnera spiralis* L. occurs in the cell wall (Sheoran and Sheoran 2006). This location was confirmed by electron microscopic studies of leaf cells of *Potamogeton pectinatus* L. (Sharpe and Denny 1976; Sheoran and Sheoran 2006). These locations on the cell wall were identified and called phytochelatins (Sheoran and Sheoran 2006). Phytochelatins are complexes of heavy metal with peptides consisting of different amino acids (r-glutamic acid-cysteine) n-glycine $n = 3$–7. It participates in the detoxification and heavy metals homeostatic equilibrium in plant cells. Excessive heavy metals bind to the cell wall through metathiolates formation through the mercaptide complex (Sheoran and Sheoran 2006). Although plants have the ability to detoxify foreign substances as described above, their role in directly degrading organic chemicals in sewage treatment plants is small compared to microorganisms (Stottmeister et al. 2003).

14.2.2 Plants Increase Constructed Wetland Performance

Effects of plants that contribute to treatment processes on subsurface flow CW include (Hoffmann et al. 2011) the following: (1) coarse sand substrates maintain their hydraulic conductivity by the root system; (2) plants facilitate the growth of colonies of bacteria and other microorganisms that form biofilms, adhering to root surfaces and substrate particles, and (3) oxygen is transported by plants to the root area so that the roots can survive in a lack of oxygen. Some of the oxygen is available for microbial processes.

The main function of plants in CW in terms of treating wastewater is the physical effect. Macrophyte plants stabilize the media surface, thus providing decent conditions for filtration physically and preventing clogging in the vertical flow system. Plants also protect CW surfaces against the effects of freezing during winter and offer a large surface area for microbial growth. This fact contradicts the previous view that macrophyte growth does not improve the substrate hydraulic conductivity of the subsurface CW soils (Brix 1997). So that the contribution of macrophyte plants to CW is that the root system maintains and increases hydraulic conductivity (Brix 1997; Hoffmann et al. 2011).

All macrophyte plant parts submerged in water provide a wide surface for the microorganism attachment and form biofilms (Brix 1997; Hua 2003). This biofilm is a layer rich in protozoa, bacteria, and photosynthetic algae. At the same time, microorganisms grow in the wetland soil's roots and rhizomes, which become a substrate for their growth. So, biofilms exist in macrophyte plant tissues above ground and underground. Biofilms also form on the surface of dead macrophytes. These biofilms are responsible for most processes involving microbes in the wetlands (Brix 1997).

Macrophytes are morphologically and physiologically adapted to the wetland environment, which is always waterlogged and oxygen-poor conditions. Adaptations that arise can be in the form of large internal air spaces in plant body parts for oxygen transportation from roots and rhizomes. Oxygen internally moves to the lower organs of the plant to provide oxygen for breathing purposes and also supplies the root area by absorbing oxygen through the roots. This affects the oxygen content in the wetland rhizosphere. Plant roots create oxidized conditions on anoxic substrates and stimulate the decomposition of the organic compounds aerobically and the nitrifying bacteria growth. Because of this, aquatic plants are mentioned to play a significant role in treating wastewater in CW with the subsurface flow (Brix 2020).

Complex processes cause iron (Fe) removal in CW, including the interaction of physicochemical and biological processes. These processes involve plants, microorganisms, and media present in CW (Younger et al. 2002; Munawar 2007; Vymazal and Kropfelova 2008; Vymazal 2010; Prihatini and Soemarno 2021). Generally, the most important water purification function of plants in CW is the physical effect induced by the presence of these plants. Plants offer a wide surface area for microorganisms to attach and grow (Hua 2003). Plant and microorganisms' complex interaction is the primary process for metal removal from wastewater (Kosolapov et al. 2004; Samsó and García 2013).

14.3 Bacteria and Wetland Plants

14.3.1 Bacteria Determine the Successful Fe Removal in Constructed Wetland

Microbes play a crucial part in the iron removal process (Stottmeister et al. 2003; Wiessner et al. 2006; Williamson et al. 2006), especially sulfate-reducing bacteria (Munawar 2007). The bacterial community that grows in CW plays a significant role in removing wastewater pollutants, and a stable society is an important factor affecting the performance of these bacteria. Samsó and García (2013) modeled the bacterial community in the CW using BIO_PORE, which showed that the CW was dominated by aerobic bacteria until the 80th day of operation. Then it will be dominated by anaerobic bacteria. Bacterial stability was achieved in the range of 400–700 days. After stability was achieved, sulfate-reducing bacteria were detected as a group of bacteria with the highest biomass (46%). The bacterial community distribution occurs after achieving bacterial stability, determined by the dissolved oxygen concentration and toxicity of H_2S. Once a stable state is achieved, inert solids that accumulate progressively push the active bacterial zone location toward the outlet section (Samsó and García 2013).

Several studies show that bacteria can effectively remove Fe in wastewater (Table 14.1). Sulfate-reducing bacteria (SRB) is a promising bacteria that can be used to remediate acid mine drainage (AMD). The SRB ability can be enhanced by adding Fe^0 (zero-valent iron). SRB with Fe^0 can remove 86% of Fe^{2+} in AMD (Bai et al. 2013) AMD at pH 2.8 and containing metal with high concentration (Fe 463 mg/L, Zn 118 mg/L, Mn 79 mg/L, Cd 58 mg/L, and Cu 76 mg/L) treat with sulfate-reducing bacteria (SRB) that immobilized formed into granules. The bioreactor performance shows pH effluent range from 7.8 to 8.3. The removal efficiency exceeds 9.9%, not including Mn (42.1–99.3%) (Zhang and Wang 2016). Another study, the Fe-Mn oxidizing bacterial consortium was assembled from the AMD site and then adjusted to remove Fe and Mn with high concentrations simultaneously. The Fe removal efficiency reaches 99.80% in 12 days of treatment. The dominant genera found were *Flavobacterium, Brevundimonas, Stenotrophomonas,* and *Thermomonas,* which might play an important role in Fe removal (Hou et al. 2020). Another method using oxidized bacteria in limestone filters can treat groundwater with a Fe concentration of 2.454 mg/L and remove 81.72% Fe (Aziz et al. 2020). In a very saline environment, halophilic sulfate-reducing bacteria (SRBs) can remove 85.30% Fe from liquid synthetic wastewater that contains 350 mg/L Fe (Torbaghan and Khalili Torghabeh 2019).

Table 14.1 Fe removal efficiency of some bacteria

Bacteria	Time	Fe removal efficiency	Type of water being treated	References
Sulfate-reducing bacteria (SRB) with Fe^0	In another	86.20%	Acid mine drainage with Fe concentration 545 mg/L	Bai et al. (2013)
Immobilized sulfa to remove Fe and Mn with high concentration simultaneously uses	7 days	99.90%	Acid mine drainage with Fe concentration 463 mg/L	Zhang and Wang (2016)
Fe-Mn oxidicing bacteria, including *Flavobacterium, Brevundimonas, Stenotrophomonas,* and *Thermomonas*	12 days	99.80%	Acid mine drainage with Fe concentration 100–1300 mg/L	Hou et al. (2020)
Halophilic sulfate-reducing bacteria (SRBs)	24 h	85.30%	Liquid wastewater with Fe concentration 350 mg/L	Torbaghan and Khalili Torghabeh (2019)
Oxidized bacteria in limestone filter	1.5 h	81.72%	Groundwater with Fe concentration 2.454 mg/L	Aziz et al. (2020)

14.3.2 Iron Plaques in the Wetland Plant's Root as a Significant Reserve of Fe (III)

Wetland ecosystems are sites for rapid biogeochemical cycles because of the interaction between the aerobic soil surface and anaerobic subsoil, so this ecosystem is very productive (Weiss et al. 2003; Gutknecht et al. 2006; Hartman et al. 2008; Burgin et al. 2011). Nutrient cycles and methane emissions from wetlands are closely related because the proximity conditions of the oxic-anoxic (Kögel-Knabner et al. 2010) are very distinctive for iron and methane (Laanbroek 2010). The oxidic-anoxic tangent field is expanded by wetland vegetation capable of releasing oxygen from its root system through *radial oxygen loss* (ROL) (Armstrong 1971).

Radial oxygen loss (ROL) is the process of releasing oxygen by the roots of wetland plants which increases the area of the oxic-anoxic interface (Armstrong 1964; Bodelier et al. 2006). In addition, oxygen is also released into the anoxic soil layer through other biological or physical processes, which shows the clear redox gradient nature of wetlands sediments and soil (Coci et al. 2005; Doyle and Otte 1997; Mermillod-Blondin and Lemoine 2010).

Oxygen (O_2) and Fe (II) react to form red-brownish deposits on the surface of the roots, which are usually called iron plaques (Fig. 14.1a). Plaque-iron has widely seen in the roots of aquatic plants and wetland plant species. It can constitute a massive reserve of Fe (III) and can affect metal mobility and nutrient availability, such as calcium and phosphates (Hansel et al. 2001; Yin et al. 2009).

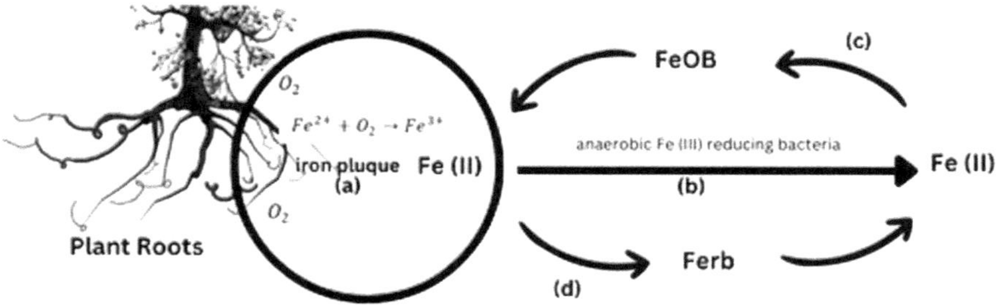

Fig. 14.1 Plant-bacteria association

The roots of *Eleocharis dulcis* (Prihatini and Soemarno 2021) and Bulbous rush (*Juncus bulbosus*) (Küsel et al. 2003) are also covered with iron plaques. In the rhizosphere of these roots, rapid microbial-mediated iron cycling occurs under changing redox conditions. The most likely numbers of aerobic and anaerobic bacteria at pH three that have the ability to consume root exudates were comparable in *Juncus* roots and rhizosphere sediments. Still, the number of aerobic bacteria was considerably higher than that of anaerobic bacteria. At pH 3, the additional organic exudate was mainly aerobically oxidized to CO_2 and was not fermented. However, root exudates, at pH 4.5, were rapidly used even under conditions without oxygen. At pH 4.9, root-associated sulfate reduction was observed but not at pH 3 to 4.5. At all root cultures, pH increased in aerobic and anaerobic conditions. Due to organic root exudates microbial turnover, root surface pH and CO_2 can increase, which favors the colonization of rushes in acidic habitats (Küsel et al. 2003).

Plague-iron development was influenced by the availability of Fe (II), redox potential, ROL, pH, and soil texture (Mendelssohn et al. 1995). Several types of bacteria are also found in iron plaque (Trolldenier 1988), but their role in the development of iron plaque is very diverse. The availability of bacteria oxidizing Fe (II) (FeOB) in the wetland plants' roots (Emerson et al. 1999) indicates that these bacteria are directly linked with the creation of iron plaque. FeOB bacteria can mediate 45–90% of the Fe (II) oxidation process (Emerson and Revsbech 1994; Sobolev and Roden 2001; Neubauer et al. 2002). This suggests that FeOB bacteria play a major role in forming iron plaques in the roots of wetland plants.

14.3.3 Bacteria Contribute a Major Role in the Iron Plaque Formation

In biogeochemical redox reactions Fe play an important role as one of the essential compounds, which interacts with other elemental cycles and also impacts the growth and activity of microorganisms and plants (Fortin and Langley 2005). In a wetland environment, iron oxide is distributed widely and can rapidly be reduced by anaerobic Fe (III)-reducing bacteria (Fig. 14.1b) under anoxic conditions from iron (Fe^{3+}) to ferrous (Fe^{2+}) (Lovley 1997; Burgin et al. 2011). Further, ferrous may be oxidized chemically; although, in wetland water, it is seen that the activity microbes account for the majority of iron oxide production in pH-neutral conditions, even though chemical oxidation in these situations can be faster. This contradiction is triggered by the results of the number of studies that report an association between FeOB with iron oxide in samples from the circumneutral environment (Emerson and Moyer 1997; Weiss et al. 2003; Chan et al. 2009). Mediation by iron oxidation bacteria also occurs, and it does perform a significant role in the formation of iron oxide (Neubauer et al. 2002).

Seasonal changes in O_2 concentration and availability of Fe^{2+} are associated with root growth and root mortality, which controls iron oxide deposition (Sundby et al. 2003). The study by Neubauer et al. (2007) reported that the significant temporal variations in the community of FeOB in microcosmic studies with plant biomass and plant activity perform an important role in the oxidation of Fe (II) in the rhizosphere zone. The composition of FeOB bacterial community changes depends on the hydrology of tidal marshlands and season, although this is linked with oxygen availability (Wang et al. 2011).

Apart from chemical oxidation, there are other biotic processes, such as the oxidation of methane, ammonium, and sulfides which consume oxygen and all the products derived through anaerobic microbial processes. Thermodynamically, Fe (II) oxidation produces high energy than other chemo-lithotrophic reactions during limited oxygen (Thauer et al. 1977), suggesting that sub-oxidic situation may be a niche inhabited by FeOB bacteria in soil (sediment) wetlands. However, there needs to be evidence of the spatial distribution of suitable niches for FeOB bacteria and the competition between other oxygen-consuming organisms and iron-oxidizing bacteria.

In the rhizosphere zone, the availability of O_2 varies, both spatially along with the roots and temporally (daily and seasonal fluctuations); this is the mosaic of anaerobic and aerobic microhabitats. Studies on FeOB bacteria (Emerson et al. 1999) and Ferb bacteria in wetland plant roots (King and Garey 1999) recorded that the mixture of Fe (II) oxidation and Fe (III) reduction processes in the wetland plants' rhizosphere zone stimulates the iron cycle locally (Fig. 14.1c and d). A cycle involving the crystalline form of Fe (III) regenerates via the actions of the FeOB bacteria has significant biogeochemical implications, for example, the repression of methane gas production (Roden and Wetzel 1996; van der Nat and Middelburg 1998).

Bacteria that colonize the rhizosphere zone and roots of root plants and have the effect of improving (triggering) plant growth with certain mechanisms, usually called plant growth-inducing rhizobacteria (PGPR). PGPR has been widely researched and applied to increase plant growth, seed germination, and yields (Herman et al. 2008; Minorsky 2008). A PGPR isolated from the Gramineae root, *Pseudomonas fluorescens* B16, can be colonized in the various types of plants roots to improve plant height growth, number of fruit, number of flowers, and overall fruit weight of plants (Minorsky 2008). PGPR can produce antibacterial compounds that are proven effective against many pathogens (Dey et al. 2004; Minorsky 2008). PGPR is also found in rice and other cereal crops (Yanni et al. 1997; Biswas et al. 2000a, b). Further to the increasing growth of plants, PGPR bacteria directly help to plant by increasing nutrient uptake, phytohormone synthesis, nutrient dissolution, and producing phytosiderophores that chelate iron so that they are available to plant roots (Lalande et al. 1989; Liu et al. 1992; Glick 1995).

The plaque-iron is formed in wetland plants' roots and is a significant Fe (III) reserve in several environments. Forming iron plaque in wetlands with a pH around neutral is assumed to be an abiotic process. Microbial communities are also associated with iron plaque and help during the plaque-iron deposition. In the rhizosphere of wetland plants, the FeOB (oxidizing bacteria) and the Ferb (reducing bacteria) are abundant. A study by Weiss et al. (2003) surveyed 13 wetlands and aquatic habitats of the Mid-Atlantic region and found that 92% of plant specimens, consisting of 25 plant species, contained FeOB in the rhizosphere. The soil had a greater number of bacteria (1.4×10^9 cells/g soil) than roots in bulk (8.6×10^7 cells/g roots; $p < 0.05$). There was no statistically significant difference ($p < 0.05$) between the absolute abundances of aerobic and lithotropic FeOB bacteria in the soil (3.7×10^6/g soil) and in the rhizosphere (5.9×10^5/g roots). In the rhizosphere, it turns out that Ferb bacteria account for 12% on average of all bacterial cells, whereas in the soil, it accounts for <1% of total bacteria. The FeOB bacteria are found abundantly in wetland ecosystems, while Ferb bacteria dominate the rhizosphere bacterial communities of wetland plants. Further, the study concludes that FeOB bacteria are very important in the Fe (II) oxidation process in the rhizosphere, and the amalgamation of FeOB and Ferb plays a vital role in determining iron dynamics in the rhizosphere of wetland plants (Weiss et al. 2003).

Pseudomonas jaduguda JGR2 (LMG25820), *Paenibacillus cookies* JGR8 (MTCC12002), and *Bacillus megaterium* JGR9 (MTCC12001) are FeOBs. Siderophore producers influence iron accumulation in plant roots. Also, *P. pseudoalcaligenes* JGR2 increases the iron content in shoots and overcomes the root-shoot barrier, which allows cattails to exclude metals from their shoots. Of the plant growth promotion (PGP) mechanisms tested, the ability to dissolve phosphate appeared to be the most important for increasing plant biomass (Ghosh et al. 2014). *Desulfosporosinus youngiae* sp. nov. was found in constructed wetland sediment (Lee et al. 2009).

The activity of plant roots has the potential to increase metal solubility. It can change speciation, including acidification/alkalization, modification of redox potentials, exudation of metal chelators and organic ligands (low molecular weight organic acids and phytosiderophores) (Fitz and Wenzel 2002; Wenzel et al. 2003; Jones et al. 2004). Microorganisms can increase metal solubility and change metal speciation through organic matter decomposition, organic ligands production, and metabolites exudation (e.g., organic acids) and siderophore. This substance can form complexes with metal cations (Gadd 2004). In Graminae, phytosiderophores (organic acids) are synthesized by plant roots and released into the rhizosphere, forming complexes with iron. This Fe-phytosiderophore complex is transported across the root cell membrane.

Monocot Gramineae plants have Strategy II (Fe) to absorb iron from the growth medium, where the roots absorb Fe as Fe^{3+}-phytosiderophore. Despite having Strategy II, the rice plant (*Oryza sativa*) also has OsIRT1, which functions as a

Fe^{2+} transporter (Bashir et al. 2013; Lee and An 2009; Meng et al. 2005). Ishimaru et al. (2006) isolated the OsIRT2 gene from rice plants, which is highly homologous to OsIRT1. PCR analysis showed that OsIRT1 and OsIRT2 were mainly found in rice roots, and these transporters were triggered by conditions of low Fe availability (Ishimaru et al. 2006; Yang et al. 2013). OsIRT1 was found in the epidermis and exodermis of the root tip elongation zone and in the cortical tissue of the Fe-deficient mature root zone. Rice plants are apparently able to absorb iron from the growth medium in the form of Fe^{3+}, which is associated with phytosiderofore and free ferrous cations (Fe^{2+}) (Liu et al. 2013; Nozoye et al. 2011). This shows that rice plants can absorb Fe^{2+} directly in adjunct to take up Fe^{3+}-phytosiderophore; this strategy is very beneficial for plant growth in waterlogged conditions (Ishimaru et al. 2006; Bashir et al. 2013; Liu et al. 2013).

14.4 Conclusion

In the constructed wetland system, the removal of iron (Fe) was conducted by the plant-bacteria association. Plants play a secondary role in direct degradation compared to microorganisms, especially bacteria. The activity of plant roots has the potential to increase metal solubility. Siderophore producers influence iron accumulation in plant roots. Iron plaque in macrophytes root constitutes a massive reserve of Fe (III) and can affect metal mobility and nutrient availability, such as calcium and phosphates. Fe (II)-oxidizing bacteria (FeOB) are very important in the Fe (II) oxidation process in the rhizosphere, and the combination of FeOB and Fe (III) reduction bacteria (Ferb) is essential in determining iron dynamics in the rhizosphere of wetland plants.

References

Armstrong W (1964) Oxygen diffusion from the roots of some British bog plants. Nature 204(4960):801–802. https://doi.org/10.1038/204801b0

Armstrong W (1971) Radial oxygen losses from intact rice roots as affected by distance from the apex, respiration and waterlogging. Physiol Plant 25(2):192–197. https://doi.org/10.1111/j.1399-3054.1971.tb01427.x

Aziz HA, Tajarudin HA, Wei THL, Alazaiza MYD (2020) Iron and manganese removal from groundwater using limestone filter with iron-oxidized bacteria. Int J Environ Sci Technol 17(5): 2667–2680. https://doi.org/10.1007/s13762-020-02681-5

Bai H, Kang Y, Quan H, Han Y, Sun J, Feng Y (2013) Treatment of acid mine drainage by sulfate reducing bacteria with iron in bench scale runs. Bioresour Technol 128:818–822. https://doi.org/10.1016/j.biortech.2012.10.070

Bashir K, Nozoye T, Ishimaru Y, Nakanishi H, Nishizawa NK (2013) Exploiting new tools for iron bio-fortification of rice. Biotechnol Adv 31(8):1624–1633. https://doi.org/10.1016/j.biotechadv.2013.08.012

Biswas JC, Ladha JK, Dazzo FB (2000a) Rhizobia inoculation improves nutrient uptake and growth of lowland rice. Soil Sci Soc Am J 64(5):1644–1650. https://doi.org/10.2136/sssaj2000.6451644x

Biswas JC, Ladha JK, Dazzo FB, Yanni YG, Rolfe BG (2000b) Rhizobial inoculation influences seedling vigor and yield of rice. Agron J 92(5):880–886. https://doi.org/10.2134/agronj2000.925880x

Bodelier PLE, Frenzel P, Drake HL, Hurek T, Küsel K, Lovell C, Megonigal P, Reinhold-Hurek B, Sorrell B (2006) Ecological aspects of microbes and microbial communities inhabiting the rhizosphere of wetland plants. Wetlands Nat Resour Manag:205–238. https://doi.org/10.1007/978-3-540-33187-2_10

Brix H (1997) Do macrophytes play a role in constructed treatment wetlands? Water Sci Technol 35(5):11–17. https://doi.org/10.1016/S0273-1223(97)00047-4

Brix H (2020) Macrophyte-mediated oxygen transfer in wetlands: transport mechanisms and rates. In: Moshiri GA (ed) Constructed wetlands for water quality improvement. CRC Press, Boca Raton, FL, pp 391–398. https://doi.org/10.1201/9781003069997-48

Burgin AJ, Yang WH, Hamilton SK, Silver WL (2011) Beyond carbon and nitrogen: How the microbial energy economy couples elemental cycles in diverse ecosystems. Front Ecol Environ 9(1). https://doi.org/10.1890/090227

Chan CS, Fakra SC, Edwards DC, Emerson D, Banfield JF (2009) Iron oxyhydroxide mineralization on microbial extracellular polysaccharides. Geochim Cosmochim Acta 73(13):3807–3818. https://doi.org/10.1016/j.gca.2009.02.036

Cheng M, Li X, Jiao X, Zhao Z, Zhang Y (2022) A critical review on iron-enhanced constructed wetland system: mechanisms and application scope. Water Air Soil Pollut 233(12):524. https://doi.org/10.1007/s11270-022-05985-z

Coci M, Riechmann D, Bodelier PLE, Stefani S, Zwart G, Laanbroek HJ (2005) Effect of salinity on temporal and spatial dynamics of ammonia-oxidising bacteria from intertidal freshwater sediment. FEMS Microbiol Ecol 53(3). https://doi.org/10.1016/j.femsec.2005.01.016

Dey R, Pal KK, Bhatt DM, Chauhan SM (2004) Growth promotion and yield enhancement of peanut (Arachis hypogaea L.) by application of plant growth-promoting rhizobacteria. Microbiol Res 159(4). https://doi.org/10.1016/j.micres.2004.08.004

Doyle MO, Otte ML (1997) Organism-induced accumulation of iron, zinc and arsenic in wetland soils. Environ Pollut 96(1). https://doi.org/10.1016/S0269-7491(97)00014-6

Emerson D, Moyer C (1997) Isolation and characterization of novel iron-oxidizing bacteria that grow at circumneutral pH. Appl Environ Microbiol 63(12). https://doi.org/10.1128/aem.63.12.4784-4792.1997

Emerson D, Revsbech NP (1994) Investigation of an iron-oxidizing microbial mat community located near Aarhus, Denmark: field studies. Appl Environ Microbiol 60(11). https://doi.org/10.1128/aem.60.11.4022-4031.1994

Emerson D, Weiss JV, Megonigal JP (1999) Iron-oxidizing bacteria are associated with ferric hydroxide precipitates (Fe-plaque) on the roots of wetland plants. Appl Environ Microbiol 65(6):2758–2761. https://doi.org/10.1128/aem.65.6.2758-2761.1999

Fitz WJ, Wenzel WW (2002) Arsenic transformations in the soil-rhizosphere-plant system: Fundamentals and potential application to phytoremediation. J Biotechnol 99(3):259–278. https://doi.org/10.1016/S0168-1656(02)00218-3

Fortin D, Langley S (2005) Formation and occurrence of biogenic iron-rich minerals. Earth Sci Rev 72(1–2):1–9. https://doi.org/10.1016/j.earscirev.2005.03.002

Gadd GM (2004) Microbial influence on metal mobility and application for bioremediation. Geoderma 122(2–4 SPEC. IIS):109–119. https://doi.org/10.1016/j.geoderma.2004.01.002

Ghosh UD, Saha C, Maiti M, Lahiri S, Ghosh S, Seal A, MitraGhosh M (2014) Root associated iron oxidizing bacteria increase phosphate nutrition and influence root to shoot partitioning of iron in tolerant plant Typha angustifolia. Plant and Soil 381(1–2):279–295. https://doi.org/10.1007/s11104-014-2085-x

Glick BR (1995) The enhancement of plant growth by free-living bacteria. Can J Microbiol 41(2): 109–117. https://doi.org/10.1139/m95-015

Gutknecht JLM, Goodman RM, Balser TC (2006) Linking soil process and microbial ecology in freshwater wetland ecosystems. Plant Soil 289(1–2):17–34. https://doi.org/10.1007/s11104-006-9105-4

Hansel CM, Fendorf S, Sutton S, Newville M (2001) Characterization of Fe plaque and associated metals on the roots of mine-waste impacted aquatic plants. Environ Sci Tech 35(19):3863–3868. https://doi.org/10.1021/es0105459

Hartman WH, Richardson CJ, Vilgalys R, Bruland GL (2008) Environmental and anthropogenic controls over bacterial communities in wetland soils. Proc Natl Acad Sci U S A 105(46): 17842–17847. https://doi.org/10.1073/pnas.0808254105

Herman MAB, Nault BA, Smart CD (2008) Effects of plant growth-promoting rhizobacteria on bell pepper production and green peach aphid infestations in New York. Crop Prot 27(6):996–1002. https://doi.org/10.1016/j.cropro.2007.12.004

Hoffmann H, Platzer C, von Münch E, Winker M (2011) Technology review of constructed wetlands. Subsurface flow constructed wetlands for greywater and domestic wastewater treatment. Deutsche Gesellschaft fur Internationale Zusammenarbeit, GIZ

Hou D, Zhang P, Wei D, Zhang J, Yan B, Cao L, Zhou Y, Luo L (2020) Simultaneous removal of iron and manganese from acid mine drainage by acclimated bacteria. J Hazard Mater 396: 122631. https://doi.org/10.1016/j.jhazmat.2020.122631

Hua SC (2003) The use of constructed wetlands for wastewater treatment. Wetlands International, Malaysia

Ishimaru Y, Suzuki M, Tsukamoto T, Suzuki K, Nakazono M, Kobayashi T, Wada Y, Watanabe S, Matsuhashi S, Takahashi M, Nakanishi H, Mori S, Nishizawa NK (2006) Rice plants take up iron as an Fe3+ −phytosiderophore and as Fe2+. Plant J 45(3):335–346. https://doi.org/10.1111/j.1365-313X.2005.02624.x

Jones DL, Hodge A, Kuzyakov Y (2004) Plant and mycorrhizal regulation of rhizodeposition. New Phytol 163(3):459–480. https://doi.org/10.1111/j.1469-8137.2004.01130.x

King GM, Garey MA (1999) Ferric iron reduction by bacteria associated with the roots of freshwater and marine macrophytes. Appl Environ Microbiol 65(10):4393–4398. https://doi.org/10.1128/aem.65.10.4393-4398.1999

Kögel-Knabner I, Amelung W, Cao Z, Fiedler S, Frenzel P, Jahn R, Kalbitz K, Kölbl A, Schloter M (2010) Biogeochemistry of paddy soils. Geoderma 157(1–2):1–4. https://doi.org/10.1016/j.geoderma.2010.03.009

Kosolapov DB, Kuschk P, Vainshtein MB, Vatsourina AV, Wießner A, Kästner M, Müller RA (2004) Microbial processes of heavy metal removal from carbon-deficient effluents in constructed wetlands. Eng Life Sci 4(5):403–411. https://doi.org/10.1002/elsc.200420048

Küsel K, Chabbi A, Trinkwalter T (2003) Microbial processes associated with roots of bulbous rush coated with iron plaques. Microb Ecol 46(3):302–311. https://doi.org/10.1007/s00248-002-1054-8

Laanbroek HJ (2010) Methane emission from natural wetlands: Interplay between emergent macrophytes and soil microbial processes. A mini-review. Ann Bot 105(1):141–153. https://doi.org/10.1093/aob/mcp201

Lalande R, Bissonnette N, Coutlée D, Antoun H (1989) Identification of rhizobacteria from maize and determination of their plant-growth promoting potential. Plant and Soil 115(1):7–11. https://doi.org/10.1007/BF02220688

Lee S, An G (2009) Over-expression of OsIRT1 leads to increased iron and zinc accumulations in rice. Plant Cell Environ 32(4):408–416. https://doi.org/10.1111/j.1365-3040.2009.01935.x

Lee Y-J, Romanek CS, Wiegel J (2009) Desulfosporosinus youngiae sp. nov., a spore-forming, sulfate-reducing bacterium isolated from a constructed wetland treating acid mine drainage. Int J Syst Evol Microbiol 59(11):2743–2746. https://doi.org/10.1099/ijs.0.007336-0

Liu ST, Lee LY, Tai CY, Hung CH, Chang YS, Wolfram JH, Rogers R, Goldstein AH (1992) Cloning of an Erwinia herbicola gene necessary for gluconic acid production and enhanced mineral phosphate solubilization in Escherichia coli HB101: Nucleotide sequence and probable

involvement in biosynthesis of the coenzyme pyrroloquinoline quinone. J Bacteriol 174(18): 5814–5819. https://doi.org/10.1128/jb.174.18.5814-5819.1992
Liu D, Zhang C, Chen X, Yang Y, Wang S, Li Y, Hu H, Ge Y, Cheng W (2013) Effects of pH, Fe, and Cd on the uptake of Fe2+ and Cd2+ by rice. Environ Sci Pollut Res 20(12):8947–8954. https://doi.org/10.1007/s11356-013-1855-y
Lovley DR (1997) Microbial Fe(III) reduction in subsurface environments. FEMS Microbiol Rev 20(3–4):305–313. https://doi.org/10.1016/S0168-6445(97)00013-2
Mendelssohn IA, Kleiss BA, Wakeley JS (1995) Factors controlling the formation of oxidized root channels: a review. Wetlands 15(1):37–46
Meng F, Wei Y, Yang X (2005) Iron content and bioavailability in rice. J Trace Elem Med Biol 18(4):333–338. https://doi.org/10.1016/j.jtemb.2005.02.008
Mermillod-Blondin F, Lemoine DG (2010) Ecosystem engineering by tubificid worms stimulates macrophyte growth in poorly oxygenated wetland sediments. Funct Ecol 24(2):444–453. https://doi.org/10.1111/j.1365-2435.2009.01624.x
Minorsky PV (2008) On the inside. Plant Physiol 146(2):323–324. https://doi.org/10.1104/pp.104.900246
Munawar A (2007) Pemanfaatan sumberdaya biologis Lokal Untuk Pengendalian Pasif Air Asam Tambang: Lahan Basah Buatan. Jurnal Ilmu Tanah Dan Lingkungan 7(1):31–42
Neubauer SC, Emerson D, Megonigal JP (2002) Life at the energetic edge: kinetics of circumneutral iron oxidation by lithotrophic iron-oxidizing bacteria isolated from the wetland-plant rhizosphere. Appl Environ Microbiol 68(8):3988–3995. https://doi.org/10.1128/AEM.68.8.3988-3995.2002
Neubauer SC, Toledo-Durán GE, Emerson D, Megonigal JP (2007) Returning to their roots: iron-oxidizing bacteria enhance short-term plaque formation in the wetland-plant rhizosphere. Geomicrobiol J 24(1):65–73. https://doi.org/10.1080/01490450601134309
Nozoye T, Nagasaka S, Kobayashi T, Takahashi M, Sato Y, Sato Y, Uozumi N, Nakanishi H, Nishizawa NK (2011) Phytosiderophore efflux transporters are crucial for iron acquisition in graminaceous plants. J Biol Chem 286(7):5446–5454. https://doi.org/10.1074/jbc.M110.180026
Prihatini NS, Soemarno S (2017) Iron (Fe) bio-concentration in purun tikus (Eleocharis dulcis) planted on the constructed wetland treating the coal acid mine drainage. Int J Biosci 11(3): 69–75. https://doi.org/10.12692/ijb/11.3.69-75
Prihatini NS, Soemarno (2021) Potential of Purun tikus (Eleocharis dulcis (Burm. F.) Trin. ex Hensch) to restore the Iron (Fe) contaminated acid mine drainage by using constructed wetland. In: Bauddh K, Korstad J, Sharma P (eds) Phytorestoration of abandoned mining and oil drilling sites. Elsevier, Amsterdam. https://doi.org/10.1016/B978-0-12-821200-4.00015-7
Prihatini NS, Priatmadi BJ, Masrevaniah A, Soemarno S (2015) Performance of the horizontal subsurface-flow constructed wetland with different operational procedures. Int J Adv Eng Technol 7(6):1620–1629
Prihatini NS, Priatmadi BJ, Masrevaniah A, Soemarno S (2016) Effects of the purun tikus (Eleocharis dulcis (Burm. F.) trin. ex Hensch) planted in the horizontal subsurface flow-constructed wetlands (HSSF-CW) on iron (Fe) concentration of the acid mine drainage. J Appl Environ Biol Sci 6(1):258–264. www.textroad.com
Prihatini NS, Khair RM, Nirtha R., Tanjung RF (2020) Effectiveness of sub surface flow constructed wetland planted with equisetum hyemale and iris pseudacorus to remove iron (Fe) and manganese (Mn) from ground water. In Prosiding Seminar Nasional Lingkungan Lahan Basah (Vol. 5, Issue April)
Prihatini NS, Nirtha I, Khair RM (2022) Study of variation of plant types of vertical subsurface constructed wetland treating groundwater contaminated with iron (Fe). Int J Res Rev 9(1): 635–641. https://doi.org/10.52403/ijrr.20220174
Roden EE, Wetzel RG (1996) Organic carbon oxidation and suppression of methane production by microbial Fe(III) oxide reduction in vegetated and unvegetated freshwater wetland sediments. Limnol Oceanogr 41(8):1733–1748. https://doi.org/10.4319/lo.1996.41.8.1733

Samsó R, García J (2013) Bacteria distribution and dynamics in constructed wetlands based on modelling results. Sci Total Environ 461–462:430–440. https://doi.org/10.1016/j.scitotenv.2013.04.073

Sheoran AS, Sheoran V (2006) Heavy metal removal mechanism of acid mine drainage in wetlands: A critical review. Miner Eng 19(2):105–116. https://doi.org/10.1016/j.mineng.2005.08.006

Sobolev D, Roden EE (2001) Suboxic deposition of ferric iron by bacteria in opposing gradients of Fe(II) and oxygen at circumneutral pH. Appl Environ Microbiol 67(3). https://doi.org/10.1128/AEM.67.3.1328-1334.2001

Stottmeister U, Wießner A, Kuschk P, Kappelmeyer U, Kästner M, Bederski O, Müller RA, Moormann H (2003) Effects of plants and microorganisms in constructed wetlands for wastewater treatment. Biotechnol Adv 22(1–2):93–117. https://doi.org/10.1016/j.biotechadv.2003.08.010

Sundby B, Vale C, Caetano M, Luther GW (2003) Redox chemistry in the root zone of a salt marsh sediment in the Tagus Estuary, Portugal. Aquat Geochem 9(3):257–271. https://doi.org/10.1023/B:AQUA.0000022957.42522.9a

Thauer RK, Jungermann K, Decker K (1977) Energy conservation in chemotrophic anaerobic bacteria. Bacteriol Rev 41(1):100–180. https://doi.org/10.1128/mmbr.41.1.100-180.1977

Torbaghan ME, Khalili Torghabeh GH (2019) Biological removal of iron and sulfate from synthetic wastewater of cotton delinting factory by using halophilic sulfate-reducing bacteria. Heliyon 5(12):e02948. https://doi.org/10.1016/j.heliyon.2019.e02948

Trolldenier G (1988) Visualisation of oxidizing power of rice roots and possible participation of bacteria in iron deposition. Zeitschrift Pflanzenernaehrung Bodenkunde (Germany, F.R.) 151(2):117–121

van der Nat FJW, Middelburg JJ (1998) Seasonal variation in methane oxidation by the rhizosphere of Phragmites australis and Scirpus lacustris. Aquat Bot 61(2):95–110. https://doi.org/10.1016/S0304-3770(98)00072-2

Vymazal J (2010) Constructed Wetlands for Wastewater Treatment. Watermark 2(3):530–549. https://doi.org/10.3390/w2030530

Vymazal J, Kropfelova L (2008) Wastewater treatment in constructed wetlands with horizontal sub-surface flow. Springer Science, Dordrecht

Wang, J., Vollrath, S., Behrends, T., Bodelier, P. L. E., Muyzer, G., Meima-Franke, M., Oudsten, F. den, van Cappellen, P., & Laanbroek, H. J. (2011). Distribution and diversity of Gallionella-like neutrophilic iron oxidizers in a tidal freshwater marsh. Appl Environ Microbiol, 77(7): 2337-2344. doi: https://doi.org/10.1128/AEM.02448-10

Weiss JV, Emerson D, Backer SM, Megonigal JP (2003) Enumeration of Fe(II)-oxidizing and Fe (III)-reducing bacteria in the root zone of wetland plants: Implications for a rhizosphere iron cycle. Biogeochemistry 64(1):77–96

Wenzel WW, Bunkowski M, Puschenreiter M, Horak O (2003) Rhizosphere characteristics of indigenously growing nickel hyperaccumulator and excluder plants on serpentine soil. Environ Pollut 123(1):131–138. https://doi.org/10.1016/S0269-7491(02)00341-X

Wiessner A, Kuschk P, Buddhawong S, Stottmeister U, Mattusch J, Kästner M (2006) Effectiveness of various small-scale constructed wetland designs for the removal of iron and zinc from acid mine drainage under field conditions. Eng Life Sci 6(6):584–592. https://doi.org/10.1002/elsc.200620161

Williamson MA, Kirby CS, Rimstidt JD (2006) Iron Dynamics in Acid Mine Drainage. In: Barnhisel RI (ed) The 7th International Conference on Acid Rock Drainage (ICARD). American Society of Mining and Reclamation (ASMR), St. Louis, MO, pp 2411–2423

Yang A, Li Y, Xu Y, Zhang W-H (2013) A receptor-like protein RMC is involved in regulation of iron acquisition in rice. J Exp Bot 64(16):5009–5020. https://doi.org/10.1093/jxb/ert290

Yanni YG, Rizk RY, Corich V, Squartini A, Ninke K, Philip-Hollingsworth S, Orgambide G, de Bruijn F, Stoltzfus J, Buckley D, Schmidt TM, Mateos PF, Ladha JK, Dazzo FB (1997) Natural endophytic association between Rhizobium leguminosarum bv. trifolii and rice roots and assessment of its potential to promote rice growth. Plant Soil 194(1–2):99–114. https://doi.org/10.1007/978-94-011-5744-5_10

Yin J, Jiang L, Wen Y, Yao Z, Zhou Q (2009) Treatment of polluted landscape lake water and community analysis of ammonia-oxidizing bacteria in constructed wetland. J Environ Sci Health A 44(7):722–731. https://doi.org/10.1080/10934520902847877

Younger PL, Banwart SA, Hedin RS (2002) Passive treatment of polluted mine waters. pp. 311–396. https://doi.org/10.1007/978-94-010-0610-1_5

Yunus R, Prihatini NS (2018) Fe and Mn phytoremediation of acid coal mine drainage using water hyacinth (Eichornia crassipes) and chinese water chestnut (Eleocharis dulcis) on the constructed wetland system. Int J Biosci 12(4):273–282. https://doi.org/10.12692/ijb/12.4.273-282

Zhang M, Wang H (2016) Preparation of immobilized sulfate reducing bacteria (SRB) granules for effective bioremediation of acid mine drainage and bacterial community analysis. Miner Eng 92:63–71. https://doi.org/10.1016/J.MINENG.2016.02.008

Zhang BY, Zheng JS, Sharp RG (2010) Phytoremediation in engineered Wetlands: mechanisms and applications. Procedia Environ Sci 2:1315–1325. https://doi.org/10.1016/J.PROENV.2010.10.142